T0291889

CAMBRIDGE LIBRARY COLLECTION

Books of enduring scholarly value

Mathematical Sciences

From its pre-historic roots in simple counting to the algorithms powering modern desktop computers, from the genius of Archimedes to the genius of Einstein, advances in mathematical understanding and numerical techniques have been directly responsible for creating the modern world as we know it. This series will provide a library of the most influential publications and writers on mathematics in its broadest sense. As such, it will show not only the deep roots from which modern science and technology have grown, but also the astonishing breadth of application of mathematical techniques in the humanities and social sciences, and in everyday life.

Mathematical and Physical Papers

Sir George Stokes (1819-1903) established the science of hydrodynamics with his law of viscosity describing the velocity of a small sphere through a viscous fluid. He published no books, but was a prolific lecturer and writer of papers for the Royal Society, the British Association for the Advancement of Science, the Victoria Institute and other mathematical and scientific institutions. These collected papers (issued between 1880 and 1905) are therefore the only readily available record of the work of an outstanding and influential mathematician, who was Lucasian Professor of Mathematics in Cambridge for over fifty years, Master of Pembroke College, President of the Royal Society (1885-90), Associate Secretary of the Royal Commission on the University of Cambridge and a Member of Parliament for the University.

Cambridge University Press has long been a pioneer in the reissuing of out-of-print titles from its own backlist, producing digital reprints of books that are still sought after by scholars and students but could not be reprinted economically using traditional technology. The Cambridge Library Collection extends this activity to a wider range of books which are still of importance to researchers and professionals, either for the source material they contain, or as landmarks in the history of their academic discipline.

Drawing from the world-renowned collections in the Cambridge University Library, and guided by the advice of experts in each subject area, Cambridge University Press is using state-of-the-art scanning machines in its own Printing House to capture the content of each book selected for inclusion. The files are processed to give a consistently clear, crisp image, and the books finished to the high quality standard for which the Press is recognised around the world. The latest print-on-demand technology ensures that the books will remain available indefinitely, and that orders for single or multiple copies can quickly be supplied.

The Cambridge Library Collection will bring back to life books of enduring scholarly value across a wide range of disciplines in the humanities and social sciences and in science and technology.

Mathematical and Physical Papers

VOLUME 3

GEORGE GABRIEL STOKES

CAMBRIDGE
UNIVERSITY PRESS

CAMBRIDGE UNIVERSITY PRESS

Cambridge New York Melbourne Madrid Cape Town Singapore São Paolo Delhi

Published in the United States of America by Cambridge University Press, New York

www.cambridge.org
Information on this title: www.cambridge.org/9781108002646

© in this compilation Cambridge University Press 2009

This edition first published 1901
This digitally printed version 2009

ISBN 978-1-108-00264-6

MATHEMATICAL

AND

PHYSICAL PAPERS.

𝕷𝖔𝖓𝖉𝖔𝖓: C. J. CLAY AND SONS,
CAMBRIDGE UNIVERSITY PRESS WAREHOUSE,
AVE MARIA LANE.

𝔊𝔩𝔞𝔰𝔤𝔬𝔴: 50, WELLINGTON STREET.

𝔏𝔢𝔦𝔭𝔷𝔦𝔤: F. A. BROCKHAUS.
𝔑𝔢𝔴 𝔜𝔬𝔯𝔨: THE MACMILLAN COMPANY.
𝔅𝔬𝔪𝔟𝔞𝔶: E. SEYMOUR HALE.

MATHEMATICAL

AND

PHYSICAL PAPERS

BY

Sir GEORGE GABRIEL STOKES, Bart.,

M.A., D.C.L., LL.D., F.R.S.,

FELLOW OF PEMBROKE COLLEGE AND LUCASIAN PROFESSOR OF MATHEMATICS
IN THE UNIVERSITY OF CAMBRIDGE.

Reprinted from the Original Journals and Transactions,
with Additional Notes by the Author.

VOL. III.

CAMBRIDGE:
AT THE UNIVERSITY PRESS.
1901

Cambridge:

PRINTED BY J. AND C. F. CLAY,

AT THE UNIVERSITY PRESS.

PREFACE.

WHEN the second volume of my collected papers was published, it was my intention to have entered on some rather elaborate and in part laborious calculations bearing on two of the papers which appear in the present volume. These were however put off from time to time in favour of other matters which claimed my attention; but meanwhile time went on, and I deeply regret to find how long it now is since the second volume appeared. There are other papers which still remain, and I hope, should life and health last, to put these together without delay.

G. G. STOKES.

MALAHIDE, Co. DUBLIN, *September*, 1901.

CONTENTS.

ERRATA IN VOL. II.

    p. 334, equation (10).   Insert " = " before "$\frac{1}{8}\sqrt{-1}$"

    p. 355, equation (66), last term.   For " ·245835 " (log. = $\bar{1}$·390644) read " ·245270 " (log. = $\bar{1}$·389644)

MATHEMATICAL AND PHYSICAL PAPERS.

[From the *Transactions of the Cambridge Philosophical Society*,
Vol. IX. p. [8]].

ON THE EFFECT OF THE INTERNAL FRICTION OF FLUIDS ON THE MOTION OF PENDULUMS.

[Read *December* 9, 1850.]

THE great importance of the results obtained by means of the pendulum has induced philosophers to devote so much attention to the subject, and to perform the experiments with such a scrupulous regard to accuracy in every particular, that pendulum observations may justly be ranked among those most distinguished by modern exactness. It is unnecessary here to enumerate the different methods which have been employed, and the several corrections which must be made, in order to deduce from the actual observations the result which would correspond to the ideal case of a simple pendulum performing indefinitely small oscillations in vacuum. There is only one of these corrections which bears on the subject of the present paper, namely, the correction usually termed the *reduction to a vacuum*. On account of the inconvenience and expense attending experiments in a vacuum apparatus, the observations are usually made in air, and it then becomes necessary to apply a small correction, in order to reduce the observed result to what would have been observed had the pendulum been swung in a vacuum. The most obvious effect of the air consists in a diminution of the moving force, and consequent increase in the time of vibration, arising from the buoyancy of the fluid. The

1

correction for buoyancy is easily calculated from the first principles of hydrostatics, and formed for a considerable time the only correction which it was thought necessary to make for reduction to a vacuum. But in the year 1828 Bessel, in a very important memoir in which he determined by a new method the length of the seconds' pendulum, pointed out from theoretical considerations the necessity of taking account of the inertia of the air as well as of its buoyancy. The numerical calculation of the effect of the inertia forms a problem of hydrodynamics which Bessel did not attack; but he concluded from general principles that a fluid, or at any rate a fluid of small density, has no other effect on the time of very small vibrations of a pendulum than that it diminishes its gravity and increases its moment of inertia. In the case of a body of which the dimensions are small compared with the length of the suspending wire, Bessel represented the increase of inertia by that of a mass equal to k times the mass of the fluid displaced, which must be supposed to be added to the inertia of the body itself. This factor k he determined experimentally for a sphere a little more than two inches in diameter, swung in air and in water. The result for air, obtained in a rather indirect way, was $k = 0·9459$, which value Bessel in a subsequent paper increased to $0·956$. A brass sphere of the above size having been swung in water with two different lengths of wire in succession gave two values of k, differing a little from each other, and equal to only about two-thirds of the value obtained for air.

The attention of the scientific world having been called to the subject by the publication of Bessel's memoir, fresh researches both theoretical and experimental soon appeared. In order to examine the effect of the air by a more direct method than that employed by Bessel, a large vacuum apparatus was erected at the expense of the Board of Longitude, and by means of this apparatus Captain (now Colonel) Sabine determined the effect of the air on the time of vibration of a particular invariable pendulum. The results of the experiments are contained in a memoir read before the Royal Society in March 1829, and printed in the *Philosophical Transactions* for that year. The mean of eight very consistent experiments gave 1 655 as the factor by which for that pendulum the old correction for buoyancy must be multiplied in order to give the whole correction on account of the air. A very remarkable fact was discovered in the course of these experiments. While the effects of air at the atmospheric pressure and under a pressure of about half an atmosphere were found to be as nearly as possible proportional to the densities, it was found that the effect of hydrogen at the atmospheric pressure was much greater, compared with the effect of air, than corresponded with its density.   In fact, it appeared that the ratio of the effects of hydrogen and air on the times of vibration was about 1 to $5\frac{1}{4}$, while the ratio of the densities is only about 1 to 13.   In speaking of this result Colonel Sabine remarks, "The difference of this ratio from that shewn by experiment is greater than can well be ascribed to accidental error in the experiment, particularly as repetition produced results almost identical.   May it not indicate an inherent property in the elastic fluids, analogous to that of viscidity in liquids, of resistance to the motion of bodies passing through them, independently of their density ? a property, in such case, possessed by air and hydrogen gas in very different degrees; since it would appear from the experiments that the ratio of the resistance of hydrogen gas to that of air is more than double the ratio following from their densities. Should the existence of such a distinct property of resistance, varying in the different elastic fluids, be confirmed by experiments now in progress with other gases, an apparatus more suitable than the present to investigate the ratio in which it is possessed by them, could scarcely be devised : and the pendulum, in addition to its many important and useful purposes in general physics, may find an application for its very delicate, but, with due precaution, not more delicate than certain, determinations, in the domain of chemistry."   Colonel Sabine has informed me that the experiments here alluded to were interrupted by a cause which need not now be mentioned, but that as far as they went they confirmed the result of the experiments with hydrogen, and pointed out the existence of a specific action in different gases, quite distinct from mere variations of density.

Our knowledge on the subject of the effect of air on the time of vibration of pendulums has received a most valuable addition from the labours of the late Mr Baily, who erected a vacuum apparatus at his own house, with which he performed many hundreds of careful experiments on a great variety of pendulums. The experiments are described in a paper read before the Royal Society on the 31st of May 1832.   The result for each pendulum is expressed by the value of n, the factor by which the old correction for buoyancy must be multiplied in order to give the whole effect of the air as deduced from observation. Four spheres, not quite $1\frac{1}{2}$ inch in diameter, gave as a mean $n = 1\cdot864$, while three spheres, a little more than 2 inches in diameter, gave only $1\cdot748$. The latter were nearly of the same size as those with which Bessel, by a different method, had obtained $k = 0\cdot946$ or $0\cdot956$, which corresponds to $n = 1\cdot946$ or $1\cdot956$. Among the " Additional Experiments " in the latter part of Baily's paper, is a set in which the pendulums consisted of plain cylindrical rods. With these pendulums it was found that n regularly increased, though according to an unknown law, as the diameter of the rod decreased. While a brass tube $1\frac{1}{2}$ inch in diameter gave n equal to about $2\cdot3$, a thin rod or thick wire only $0\cdot072$ inch in diameter gave for n a value as great as $7\cdot530$.

Mathematicians in the meanwhile were not idle, and several memoirs appeared about this time, of which the object was to determine from hydrodynamics the effect of a fluid on the motion of a pendulum. The first of these came from the pen of the cele- brated Poisson. It was read before the French Academy on the 22nd of August 1831, and is printed in the 11th Volume of the Memoirs. In this paper, Poisson considers the case of a sphere suspended by a fine wire, and oscillating in the air, or in any gas. He employs the ordinary equations of motion of an elastic fluid, simplified by neglecting the terms which involve the square of the velocity; but in the end, in adapting his solution to practice, he neglects, as insensible, the terms by which alone the action of an elastic differs from that of an incompressible fluid, so that the result thus simplified is equally applicable to fluids of both classes. He finds that when insensible quantities are neglected $n = 1\cdot5$, so that the mass which we must suppose added to that of the pendu- lum is equal to half the mass of the fluid displaced. This result does not greatly differ from the results obtained experimentally by Bessel in the case of spheres oscillating in water, but differs ma- terially from the result he had obtained for air. It agrees pretty closely with some experiments which had been performed about fifty years before by Dubuat, who had in fact anticipated Bessel in shewing that the time of vibration of a pendulum vibrating in a fluid would be affected by the inertia of the fluid as well as by its density. Dubuat's labours on this subject had been altogether overlooked by those who were engaged in pendulum experiments;

probably because such persons were not likely to seek in a treatise on hydraulics for information connected with the subject of their researches. Dubuat had, in fact, rather applied the pendulum to hydrodynamics than hydrodynamics to the pendulum.

In the *Philosophical Magazine* for September 1833, p. 185, is a short paper by Professor Challis, on the subject of the resistance to a ball pendulum. After referring to a former paper, in which he had shewn that no sensible error would be committed in a problem of this nature by neglecting the compressibility of the fluid even if it be elastic, Professor Challis, adopting a particular hypothesis respecting the motion, obtains 2 for the value of the factor n for such a pendulum. This mode of solution, which is adopted in several subsequent papers, has given rise to a controversy between Professor Challis and the Astronomer Royal, who maintains the justice of Poisson's result.

In a paper read before the Royal Society of Edinburgh on the 16th of December 1833, and printed in the 13th Volume of the Society's *Transactions,* Green has determined from the common equations of fluid motion the resistance to an ellipsoid performing small oscillations without rotation. The result is expressed by a definite integral ; but when two of the principal axes of the ellipsoid become equal, the integral admits of expression in finite terms, by means of circular or logarithmic functions. When the ellipsoid becomes a sphere, Green's result reduces itself to Poisson's.

In a memoir read before the Royal Academy of Turin on the 18th of January 1835, and printed in the 37th Volume of the memoirs of the Academy, M. Plana has entered at great length into the theory of the resistance of fluids to pendulums. This memoir contains, however, rather a detailed examination of various points connected with the theory, than the determination of the resistance for any new form of pendulum. The author first treats the case of an incompressible fluid, and then shews that the result would be sensibly the same in the case of an elastic fluid. In the case of a ball pendulum, the only one in which a complete solution of the problem is effected, M. Plana's result agrees with Poisson's.

In a paper read before the Cambridge Philosophical Society on the 29th of May 1843, and printed in the 8th Volume of the *Transactions,* p. 105*, I have determined the resistance to a ball

* [*Ante,* Vol. I. p. 179.]

pendulum oscillating within a concentric spherical envelope, and have pointed out the source of an error into which Poisson had fallen, in concluding that such an envelope would have no effect. When the radius of the envelope becomes infinite, the solution agrees with that which Poisson had obtained for the case of an unlimited mass of fluid. I have also investigated the increase of resistance due to the confinement of the fluid by a distant rigid plane. The same paper contains likewise the calculation of the resistance to a long cylinder oscillating in a mass of fluid either unlimited, or confined by a cylindrical envelope, having the same axis as the cylinder in its position of equilibrium. In the case of an unconfined mass of fluid, it appeared that the effect of inertia was the same as if a mass equal to that of the fluid displaced were distributed along the axis of the cylinder, so that $n = 2$ in the case of a pendulum consisting of a long cylindrical rod. This nearly agrees with Baily's result for the long $1\frac{1}{2}$ inch tube ; but, on comparing it with the results obtained with the cylindrical rods, we observe the same sort of discrepancy between theory and observation as was noticed in the case of spheres. The discrepancy is, however, far more striking in the present case, as might naturally have been expected, after what had been observed with spheres, on account of the far smaller diameter of the solids employed.

A few years ago Professor Thomson communicated to me a very beautiful and powerful method which he had applied to the theory of electricity, which depended on the consideration of what he called *electrical images*. The same method, I found, applied, with a certain modification, to some interesting problems relating to ball pendulums. It enabled me to calculate the resistance to a sphere oscillating in presence of a fixed sphere, or within a spherical envelope, or the resistance to a pair of spheres either in contact, or connected by a narrow rod, the direction of oscillation being, in all these cases, that of the line joining the centres of the spheres. The effect of a rigid plane perpendicular to the direction of motion is of course included as a particular case. The method even applies, as Professor Thomson pointed out to me, to the uncouth solid bounded by the exterior segments of two intersecting spheres, provided the exterior angle of intersection be a submultiple of two right angles. A set of corresponding problems, in which the spheres are replaced by long cylinders, may be solved in a similar manner. These results were mentioned at the meeting of the British Association at Oxford in 1847, and are noticed in the volume of reports for that year, but they have not yet been published in detail.

The preceding are all the investigations that have fallen under my notice, of which the object was to calculate from hydrodynamics the resistance to a body of given form oscillating as a pendulum. They all proceed on the ordinary equations of the motion of fluids. They all fail to account for one leading feature of the experimental results, namely, the increase of the factor n with a decrease in the dimensions of the body. They recognize no distinction between the action of different fluids, except what arises from their difference of density.

In a conversation with Dr Robinson about seven or eight years ago on the subject of the application of theory to pendulums, he noticed the discrepancy which existed between the results of theory and experiment relating to a ball pendulum, and expressed to me his conviction that the discrepancy in question arose from the adoption of the ordinary theory of fluid motion, in which the pressure is supposed to be equal in all directions. He also described to me a remarkable experiment of Sir James South's which he had witnessed. This experiment has not been published, but Sir James South has kindly allowed me to mention it. When a pendulum is in motion, one would naturally have supposed that the air near the moving body glided past the surface, or the surface past it, which comes to the same thing if the relative motion only be considered, with a velocity comparable with the absolute velocity of the surface itself. But on attaching a piece of gold leaf to the bottom of a pendulum, so as to stick out in a direction perpendicular to the surface, and then setting the pendulum in motion, Sir James South found that the gold leaf retained its perpendicular position just as if the pendulum had been at rest; and it was not till the gold leaf carried by the pendulum had been removed to some distance from the surface, that it began to lag behind. This experiment shews clearly the existence of a tangential action between the pendulum and the air, and between one layer of air and another. The existence of a similar action in water is clearly exhibited in some experiments of Coulomb's which will be mentioned in the second part of this paper, and indeed might be concluded from several very ordinary phenomena. Moreover Dubuat, in discussing the results of his experiments on the oscillations of spheres in water, notices a slight increase in the effect of the water corresponding to an increase in the time of vibration, and expressly attributes it to the *viscosity* of the fluid.

Having afterwards occupied myself with the theory of the friction of fluids, and arrived at general equations of motion, the same in essential points as those which had been previously obtained in a totally different manner by others, of which, however, I was not at the time aware, I was desirous of applying, if possible, these equations to the calculation of the motion of some kind of pendulum. The difficulty of the problem is of course materially increased by the introduction of internal friction, but as I felt great confidence in the essential parts of the theory, I thought that labour would not be ill-bestowed on the subject. I first tried a long cylinder, because the solution of the problem appeared likely to be simpler than in the case of a sphere. But after having proceeded a good way towards the result, I was stopped by a difficulty relating to the determination of the arbitrary constants, which appeared as the coefficients of certain infinite series by which the integral of a certain differential equation was expressed. Having failed in the case of a cylinder, I tried a sphere, and presently found that the corresponding differential equation admitted of integration in finite terms, so that the solution of the problem could be completely effected. The result, I found, agreed very well with Baily's experiments, when the numerical value of a certain constant was properly assumed; but the subject was laid aside for some time. Having afterwards attacked a definite integral to which Mr Airy had been led in considering the theory of the illumination in the neighbourhood of a caustic, I found that the method which I had employed in the case of this integral* would apply to the problem of the resistance to a cylinder, and it enabled me to get over the difficulty with which I had before been baffled. I immediately completed the numerical calculation, so far as was requisite to compare the formulæ with Baily's experiments on cylindrical rods, and found a remarkably close agreement between theory and observation. These results were mentioned at the meeting of the British Association at Swansea in 1848, and are briefly described in the volume of reports for that year.

The present paper is chiefly devoted to the solution of the problem in the two cases of a sphere and of a long cylinder, and to

* [*Ante*, Vol. ii. p. 328.]

a comparison of the results with the experiments of Baily and others. Expressions are deduced for the effect of a fluid both on the time and on the arc of vibration of a pendulum consisting either of a sphere, or of a cylindrical rod, or of a combination of a sphere and a rod. These expressions contain only one disposable constant, which has a very simple physical meaning, and which I propose to call the *index of friction* of the fluid. This constant we may conceive determined by one observation, giving the effect of the fluid either on the time or on the arc of vibration of any one pendulum of one of the above forms, and then the theory ought to predict the effect both on the time and on the arc of vibration of all such pendulums. The agreement of theory with the experiments of Baily on the time of vibration is remarkably close. Even the rate of decrease of the arc of vibration, which it formed no part of Baily's object to observe, except so far as was necessary for making the small correction for reduction to indefinitely small vibrations, agrees with the result calculated from theory as nearly as could reasonably be expected under the circumstances.

It follows from theory that with a given sphere or cylindrical rod the factor n increases with the time of vibration. This accounts in a good measure for the circumstance that Bessel obtained so large a value of k for air, as is shewn at length in the present paper; though it unquestionably arose in a great degree from the increase of resistance due to the close proximity of a rigid plane to the swinging ball.

I have deduced the value of the index of friction of water from some experiments of Coulomb's on the decrement of the arc of oscillation of disks, oscillating in water in their own plane by the torsion of a wire. When the numerical value thus obtained is substituted in the expression for the time of vibration of a sphere, the result agrees almost exactly with Bessel's experiments with a sphere swung in water.

The present paper contains one or two applications of the theory of internal friction to problems which are of some interest, but which do not relate to pendulums. The resistance to a sphere moving uniformly in a fluid may be obtained as a limiting case of the resistance to a ball pendulum, provided the circumstances be such that the square of the velocity may be neglected. The resistance thus determined proves to be proportional, for a given fluid and a given velocity, not to the surface, but to the radius of the sphere; and therefore the accelerating force of the resistance increases much more rapidly, as the radius of the sphere decreases, than if the resistance varied as the surface, as would follow from the common theory. Accordingly, the resistance to a minute globule of water falling through the air with its terminal velocity depends almost wholly on the internal friction of air. Since the index of friction of air is known from pendulum experiments, we may easily calculate the terminal velocity of a globule of given size, neglecting the part of the resistance which depends upon the square of the velocity. The terminal velocity thus obtained is so small in the case of small globules such as those of which we may conceive a cloud to be composed, that the apparent suspension of the clouds does not seem to present any difficulty. Had the resistance been determined from the common theory, it would have been necessary to suppose the globules much more minute, in order to account in this way for the phenomenon. Since in the case of minute globules falling with their terminal velocity the part of the resistance depending upon the square of the velocity, as determined by the common theory, is quite insignificant compared with the part which depends on the internal friction of the air, it follows that were the pressure equal in all directions in air in the state of motion, the quantity of water which would remain suspended in the state of cloud would be enormously diminished. The pendulum thus, in addition to its other uses, affords us some interesting information relating to the department of meteorology.

The fifth section of the first part of the present paper contains an investigation of the effect of the internal friction of water in causing a series of oscillatory waves to subside. It appears from the result that in the case of the long swells of the ocean the effect of friction is insignificant, while in the case of the ripples raised by the wind on a small pool, the motion subsides very rapidly when the disturbing force ceases to act.

PART I.

ANALYTICAL INVESTIGATION.

SECTION I.

Adaptation of the general equations to the case of the fluid surrounding a body which oscillates as a pendulum. General laws which follow from the form of the equations. Solution of the equations in the case of an oscillating plane.

1. IN a paper "*On the Theories of the Internal Friction of Fluids in Motion, &c.*[*]," which the society did me the honour to publish in the 8th Volume of their *Transactions*, I have arrived at the following equations for calculating the motion of a fluid when the internal friction of the fluid itself is taken into account, and consequently the pressure not supposed equal in all directions:

$$\frac{dp}{dx} = \rho\left(X - \frac{du}{dt} - u\frac{du}{dx} - v\frac{du}{dy} - w\frac{du}{dz}\right) + \mu\left(\frac{d^2u}{dx^2} + \frac{d^2u}{dy^2} + \frac{d^2u}{dz^2}\right)$$
$$+ \frac{\mu}{3}\frac{d}{dx}\left(\frac{du}{dx} + \frac{dv}{dy} + \frac{dw}{dz}\right)\ldots\ldots\ldots\ldots(1),$$

with two more equations which may be written down from symmetry. In these equations u, v, w are the components of the velocity along the rectangular axes of x, y, z; X, Y, Z are the components of the accelerating force; p is the pressure, t the time, ρ the density, and μ a certain constant depending on the nature of the fluid.

The three equations of which (1) is the type are not the general equations of motion which apply to a heterogeneous fluid when internal friction is taken into account, which are those numbered 10 in my former paper, but are applicable to a homogeneous incompressible fluid, or to a homogeneous elastic fluid subject to small variations of density, such as those which accompany sonorous vibrations. It must be understood to be included in the term *homogeneous* that the temperature is uniform throughout the mass, except so far as it may be raised or lowered by sudden condensation or rarefaction in the case of an elastic fluid. The general equations contain the differential coefficients of the quantity μ with respect to x, y, and z; but the equations of the form (1) are in their present shape even more general than is required for the purposes of the present paper.

These equations agree in the main with those which had been previously obtained, on different principles, by Navier, by Poisson, and by M. de Saint-Venant, as I have elsewhere observed*. The differences depend only on the coefficient of the last term, and this term vanishes in the case of an incompressible fluid, to which Navier had confined his investigations.

The equations such as (1) in their present shape are rather complicated, but in applying them to the case of a pendulum they may be a good deal simplified without the neglect of any quantities which it would be important to retain. In the first place the motion is supposed very small, on which account it will be allowable to neglect the terms which involve the square of the velocity. In the second place, the nature of the motion that we have got to deal with is such that the compressibility of the fluid has very little influence on the result, so that we may treat the fluid as incompressible, and consequently omit the last terms in the equations. Lastly, the forces X, Y, Z are in the present case the components of the force of gravity, and if we write

$$p + \Pi + \rho \int (X dx + Y dy + Z dz)$$

for p, we may omit the terms X, Y, Z.

If z' be measured vertically downwards from a horizontal plane drawn in the neighbourhood of the pendulum, and if g be the force of gravity, $\int (X dx + Y dy + Z dz) = g z'$, the arbitrary constant, or arbitrary function of the time if it should be found necessary to suppose it to be such, being included in Π. The part of the whole force acting on the pendulum which depends on the terms $\Pi + g \rho z'$ is simply a force equal to the weight of the fluid displaced, and acting vertically upwards through the centre of gravity of the volume.

* Report on recent researches in Hydrodynamics. Report of the British Association for 1846, p. 16. [*Ante*, Vol. I. p. 182.]

When simplified in the manner just explained, the equations such as (1) become

$$\left.\begin{aligned}\frac{dp}{dx} &= \mu\left(\frac{d^2u}{dx^2}+\frac{d^2u}{dy^2}+\frac{d^2u}{dz^2}\right)-\rho\frac{du}{dt}\\[4pt]\frac{dp}{dy} &= \mu\left(\frac{d^2v}{dx^2}+\frac{d^2v}{dy^2}+\frac{d^2v}{dz^2}\right)-\rho\frac{dv}{dt}\\[4pt]\frac{dp}{dz} &= \mu\left(\frac{d^2w}{dx^2}+\frac{d^2w}{dy^2}+\frac{d^2w}{dz^2}\right)-\rho\frac{dw}{dt}\end{aligned}\right\}\dots\dots\dots(2),$$

which, with the equation of continuity,

$$\frac{du}{dx}+\frac{dv}{dy}+\frac{dw}{dz}=0 \dots\dots\dots\dots\dots\dots(3),$$

are the only equations which have to be satisfied at all points of the fluid, and at all instants of time.

In applying equations (2) to a particular pendulum experiment, we may suppose μ constant; but in order to compare experiments made in summer with experiments made in winter, or experiments made under a high barometer with experiments made under a low, it will be requisite to regard μ as a quantity which may vary with the temperature and pressure of the fluid. As far as the result of a single experiment*, which has been already mentioned, performed with a single elastic fluid, namely air, justifies us in drawing such a general conclusion, we may assert that for a given fluid at a given temperature μ varies as ρ†.

2. For the formation of the equations such as (1), I must refer to my former paper; but it will be possible, in a few words, to enable the reader to form a clear idea of the meaning of the constant μ.

Conceive the fluid to move in planes parallel to the plane of xy, the motion taking place in a direction parallel to the axis of y. The motion will evidently consist of a sort of continuous sliding, and the differential coefficient dv/dz may be taken as a measure of

---

* The first of the experiments described in Col. Sabine's paper, in which the gauge stood as high as 7 inches, leads to the same conclusion; but as the vacuum apparatus had not yet been made stanch it is perhaps hardly safe to trust this experiment in a question of such delicacy.

† [We now know that μ is independent of ρ, until excessive exhaustions are reached, far beyond any that we have here to deal with.]

the rate of sliding. In the theory it is supposed that in general the pressure about a given point is compounded of a normal pressure, corresponding to the density, which being normal is necessarily equal in all directions, and of an oblique pressure or tension, altering from one direction to another, which is expressed by means of linear functions of the nine differential coefficients of the first order of u, v, w with respect to x, y, z, which define the state of relative motion at any point of the fluid. Now in the special case considered above, if we confine our attention to one direction, that of the plane of xy, the total pressure referred to a unit of surface is compounded of a normal pressure corresponding to the density, and a tangential pressure expressed by $\mu\, dv/dz$, which tends to reduce the relative motion.

In the solution of equations (2), μ always appears divided by ρ. Let $\mu = \mu'\rho$. The constant μ' may conveniently be called the *index of friction* of the fluid, whether liquid or gas, to which it relates. As regards its dimensions, it expresses a moving force divided by the product of a surface, a density, and the differential coefficient of a velocity with respect to a line. Hence μ' is the square of a line divided by a time, whence it will be easy to adapt the numerical value of μ' to a new unit of length or of time.

3. Besides the general equations (2) and (3), it will be requisite to consider the equations of condition at the boundaries of the fluid. For the purposes of the present paper there will be no occasion to consider the case of a free surface, but only that of the common surface of the fluid and a solid. Now, if the fluid immediately in contact with a solid could flow past it with a finite velocity, it would follow that the solid was infinitely smoother with respect to its action on the fluid than the fluid with respect to its action on itself. For, conceive the elementary layer of fluid comprised between the surface of the solid and a parallel surface at a distance h, and then regard only so much of this layer as corresponds to an elementary portion dS of the surface of the solid. The impressed forces acting on the fluid element must be in equilibrium with the effective forces reversed. Now conceive h to vanish compared with the linear dimensions of dS, and lastly let dS vanish *. It is evident that the conditions

---

* To be quite precise it would be necessary to say, Conceive h and dS to vanish together, h vanishing compared with the linear dimensions of dS; for so long as dS

of equilibrium will ultimately reduce themselves to this, that the oblique pressure which the fluid element experiences on the side of the solid must be equal and opposite to the pressure which it experiences on the side of the fluid. Now if the fluid could flow past the solid with a finite velocity, it would follow that the tangential pressure called into play by the continuous sliding of the fluid over itself was no more than counteracted by the abrupt sliding of the fluid over the solid. As this appears exceedingly improbable *a priori*, it seems reasonable in the first instance to examine the consequences of supposing that no such abrupt sliding takes place, more especially as the mathematical difficulties of the problem will thus be materially diminished. I shall assume, therefore, as the conditions to be satisfied at the boundaries of the fluid, that the velocity of a fluid particle shall be the same, both in magnitude and direction, as that of the solid particle with which it is in contact. The agreement of the results thus obtained with observation will presently appear to be highly satisfactory. When the fluid, instead of being confined within a rigid envelope, extends indefinitely around the oscillating body, we must introduce into the solution the condition that the motion shall vanish at an infinite distance, which takes the place of the condition to be satisfied at the surface of the envelope.

To complete the determination of the arbitrary functions which would be contained in the integrals of (2) and (3), it would be requisite to put $t = 0$ in the general expressions for u, v, w, obtained by integrating those equations, and equate the results to the initial velocities supposed to be given. But it would be introducing a most needless degree of complexity into the solution to take account of the initial circumstances, nor is it at all necessary to do so for the sake of comparison of theory with experiment. For in a pendulum experiment the pendulum is set swinging and then left to itself, and the first observation is not taken till several oscillations have been completed, during which any irregularities attending the initial motion would have had time to subside. It

remains finite we cannot suppose h to vanish altogether, on account of the curvature of the elementary surface. Such extreme precision in unimportant matters tends, I think, only to perplex the reader, and prevent him from entering so readily into the spirit of an investigation.

will be quite sufficient to regard the motion as already going on, and limit the calculation to the determination of the simultaneous periodic movements of the pendulum and the surrounding fluid. The arc of oscillation will go on slowly decreasing, but it will be so nearly constant for several successive oscillations that it may be regarded as strictly such in calculating the motion of the fluid; and having thus determined the resultant action of the fluid on the solid we may employ the result in calculating the decrement of the arc of oscillation, as well as in calculating the time of oscillation. Thus the assumption of periodic functions of the time in the expressions for u, v, w will take the place of the determination of certain arbitrary functions by means of the initial circumstances.

4.   Imagine a plane drawn perpendicular to the axis of x through the point in the fluid whose co-ordinates are x, y, z. Let the oblique pressure in the direction of this plane be decomposed into three pressures, a normal pressure, which will be in the direction of x, and two tangential pressures in the directions of y, z, respectively. Let P_1 be the normal pressure, and T_3 the tangential pressure in the direction of y, which will be equal to the component in the direction of x of the oblique pressure on a plane drawn perpendicular to the axis of y. Then by the formulæ (7), (8) of my former paper, and (3) of the present,

$$P_1 = p - 2\mu \frac{du}{dx} \dots\dots\dots\dots\dots\dots\dots (4),$$

$$T_3 = -\mu \left(\frac{du}{dy} + \frac{dv}{dx} \right) \dots\dots\dots\dots\dots\dots (5).$$

These formulæ will be required in finding the resultant force of the fluid on the pendulum, after the motion of the fluid has been determined in terms of the quantities by which the motion of the pendulum is expressed.

5.   Before proceeding to the solution of the equations (2) and (3) in particular cases, it will be well to examine the general laws which follow merely from the dimensions of the several terms which appear in the equations.

Consider any number of similar systems, composed of similar solids oscillating in a similar manner in different fluids or in the same fluid. Let a, a', a''... be homologous lines in the different systems; T, T', T''... corresponding times, such for example as the times of oscillation from rest to rest. Let x, y, z be measured from similarly situated origins, and in corresponding directions, and t from corresponding epochs, such for example as the commencements of oscillations when the systems are beginning to move from a given side of the mean position.

The form of equations (2), (3) shews that the equations being satisfied for one system will be satisfied for all the systems provided

$$u \propto v \propto w, \ x \propto y \propto z, \text{ and } p \propto \frac{\mu u}{x} \propto \frac{\rho u x}{t}.$$

The variations $x \propto y \propto z$ merely signify that we must compare similarly situated points in inferring from the circumstance that (2), (3) are satisfied for one system that they will be satisfied for all the systems. If c, c', c''... be the maximum excursions of similarly situated points of the fluids

$$u \propto \frac{c}{T}, \ x \propto a, \ t \propto T,$$

and the sole condition to be satisfied, in addition to that of geometrical similarity, in order that the systems should be dynamically similar, becomes

$$\frac{a^2}{T} \propto \frac{\mu}{\rho} \text{ or } \propto \mu' \dots\dots\dots\dots\dots\dots\dots (6).$$

This condition being satisfied, similar motions will take place in the different systems, and we shall have

$$p \propto \frac{\rho a c}{T^2} \dots\dots\dots\dots\dots\dots\dots (7).$$

It follows from the equations (4), (5); and the other equations which might be written down from symmetry, that the pressures such as P_1, T_3 vary in the same manner as p, whence it appears from (7) that the resultant or resultants of the pressures of the fluids on the solids, acting along similarly situated lines, which vary as $p a^2$, vary as ρa^3 and $c T^{-2}$ conjointly. In other words, these resultants in two similar systems are to one another in a ratio compounded of the ratio of the masses of fluid displaced, and of the ratio of the maximum accelerating effective forces belonging to similarly situated points in the solids.

6. In order that two systems should be similar in which the fluids are confined by envelopes that are sufficiently narrow to

influence the motion of the fluids, it is necessary that the envelopes should be similar and similarly situated with respect to the solids oscillating within them, and that their linear dimensions should be in the same ratio as those of the oscillating bodies. In strictness, it is likewise necessary that the solids should be similarly situated with respect to the axis of rotation. If however two similar solids, such as two spheres, are attached to two fine wires, and made to perform small oscillations in two unlimited masses of fluid, and if we agree to neglect the effect of the suspending wires, and likewise the effect of ·the rotation . of the spheres on the motion of the fluid, which last will in truth be exceedingly small, we may regard the two systems as geometrically similar, and they will be dynamically similar provided the condition (6) be satisfied. When the two fluids are of the same nature, as for example when both spheres oscillate in air, the condition of dynamical similarity reduces itself to this, that the times of oscillation shall be as the squares of the diameters of the spheres.

If, with Bessel, we represent the effect of the inertia of the fluid on the time of oscillation of the sphere by supposing a mass equal to k times that of the fluid displaced added to the mass of the sphere, which increases its inertia without increasing its weight, we must expect to find k dependent on the nature of the fluid, and likewise on the diameter of the sphere. Bessel, in fact, obtained very different values of k for water and for air. Baily's experiments on spheres of different diameters, oscillating once in a second nearly, shew that the value of k increases when the diameter of the sphere decreases. Taking this for the present as the result of experiment, we are led from theory to assert that the value of k increases with the time of oscillation; in fact, k ought to be as much increased as if we had left the time of oscillation unchanged, and diminished the diameter in the ratio in which the square root of the time is increased. It may readily be shewn that the value of k obtained by Bessel's method, ‚by means of a long and short pendulum, is greater than what belongs to the long pendulum, much more, greater than what belongs to the shorter pendulum, which oscillated once in a second nearly. The value of k given by Bessel is in fact considerably larger than that obtained by Baily, by a direct method, from a sphere of nearly the same size as those employed by Bessel, oscillating once in a second nearly.

The discussion of the experiments of Baily and Bessel belongs to Part II. of this paper. They are merely briefly noticed here to shew that some results of considerable importance follow readily from the general equations, even without obtaining any solution of them.

7. Before proceeding to the problems which mainly occupy this paper, it may be well to exhibit the solution of equations (2) and (3) in the extremely simple case of an oscillating plane.

Conceive a physical plane, which is regarded as infinite, to be situated in an unlimited mass of fluid, and to be performing small oscillations in the direction of a fixed line in the plane. Let a fixed plane coinciding with the moving plane be taken for the plane of yz, the axis of y being parallel to the direction of motion, and consider only the portion of fluid which lies on the positive side of the plane of yz. In the present case, we must evidently have $u = 0$, $w = 0$; and p, v will be functions of x and t, which have to be determined. The equation (3) is satisfied identically, and we get from (2), putting $\mu = \mu'\rho$,

$$\frac{dp}{dx} = 0, \quad \frac{dv}{dt} = \mu'\frac{d^2v}{dx^2} \quad \text{......................(8).}$$

The first of these equations gives $p = $ a constant, for it evidently cannot be a function of t, since the effect of the motion vanishes at an infinite distance from the plane; and if we include this constant in Π, we shall have $p = 0$. Let V be the velocity of the plane itself, and suppose

$$V = c \sin nt \quad \text{..........................(9).}$$

Putting in the second of equations (8)

$$v = X_1 \sin nt + X_2 \cos nt \quad \text{...................(10),}$$

we get $\quad nX_1 = \mu'\dfrac{d^2X_2}{dx^2}, \quad nX_2 = -\mu'\dfrac{d^2X_1}{dx^2} = -\dfrac{\mu'^2}{n}\dfrac{d^4X_2}{dx^4} \quad \text{......(11).}$

The last of these equations gives

$$X_2 = \epsilon^{-\sqrt{\frac{n}{2\mu'}}x}\left(A \sin\sqrt{\frac{n}{2\mu'}}x + B\cos\sqrt{\frac{n}{2\mu'}}x\right)$$
$$+ \epsilon^{\sqrt{\frac{n}{2\mu'}}x}\left(C\sin\sqrt{\frac{n}{2\mu'}}x + D\cos\sqrt{\frac{n}{2\mu'}}x\right).$$

2—2

Since X_2 must not become infinite when $x = \infty$, we must have $C = 0$, $D = 0$. Obtaining X_1 from the first of equations (11), and substituting in (10), we get

$$v = \epsilon^{-\sqrt{\frac{n}{2\mu'}}x} \left\{ -A \sin\left(nt - \sqrt{\frac{n}{2\mu'}}x\right) + B \cos\left(nt - \sqrt{\frac{n}{2\mu'}}x\right) \right\}.$$

Now by the equations of conditions assumed in Art. 3, we must have $v = V$ when $x = 0$, whence

$$v = ce^{-\sqrt{\frac{n}{2\mu'}}x} \sin\left(nt - \sqrt{\frac{n}{2\mu'}}x\right) \dots\dots\dots\dots(12).$$

To find the normal and tangential components of the pressure of the fluid on the plane, we must substitute the above value of v in the formulæ (4), (5), and after differentiation put $x = 0$. P_1, T_3 will then be the components of the pressure of the solid on the fluid, and therefore $-P_1$, $-T_3$, those of the pressure of the fluid on the solid. We get

$$P_1 = 0, \quad T_3 = c\rho \sqrt{\frac{n\mu'}{2}} (\sin nt + \cos nt) = \rho \sqrt{\frac{n\mu'}{2}} \left(V + \frac{1}{n}\frac{dV}{dt}\right) \dots(13).$$

The force expressed by the first of these terms tends to diminish the amplitude of the oscillations of the plane. The force expressed by the second has the same effect as increasing the inertia of the plane.

8. The equation (12) shews that a given phase of vibration is propagated from the plane into the fluid with a velocity $\sqrt{(2\mu'n)}$, while the amplitude of oscillation decreases in geometric progression as the distance from the plane increases in arithmetic. If we suppose the time of oscillation from rest to rest to be one second, $n = \pi$; and if we suppose $\sqrt{\mu'} = \cdot116$ inch, which, as will presently be seen, is about its value in the case of air, we get for the velocity of propagation $\cdot2908$ inch per second nearly. If we enquire the distance from the plane at which the amplitude of oscillation is reduced to one half, we have only to put

$$\sqrt{(n/2\mu')}\, x = \log_e 2,$$

which gives, on the same suppositions as before respecting numerical values, $x = \cdot06415$ inch nearly. For water the value of μ' is a good deal smaller than for air, and the corresponding value of x smaller likewise, since it varies *cœteris paribus* as $\sqrt{\mu'}$. Hence if a solid of revolution of large, or even moderately large, dimensions be suspended by a fine wire coinciding with the axis of revolution, and made to oscillate by the torsion of the wire, the effect of the fluid may be calculated with a very close degree of approximation by regarding each element of the surface of the solid as an element of an infinite plane oscillating with the same linear velocity*.

For example, let a circular disk of radius a be suspended horizontally by a fine wire attached to the centre, and made to oscillate. Let r be the radius vector of any element of the disk, measured from its centre, θ the angle through which the disk has turned from its mean position. Then in equation (13), we must put $V = r\,d\theta/dt$, whence

$$T_s = \rho \sqrt{\frac{n\mu'}{2}}\, r \left(\frac{d\theta}{dt} + \frac{1}{n}\frac{d^2\theta}{dt^2}\right).$$

The area of the annulus of the disk comprised between the radii r and $r + dr$ is $4\pi r\,dr$, both faces being taken, and if G be the whole moment of the force of the fluid on the disk, $G = -\, 4\pi \displaystyle\int_0^a r^2 T_s dr$, whence

$$G = -\,\pi\rho a^4 \sqrt{\frac{n\mu'}{2}}\left(\frac{d\theta}{dt} + \frac{1}{n}\frac{d^2\theta}{dt^2}\right).$$

Let $M\gamma^2$ be the moment of inertia of the disk, and let n_1 be what n would become if the fluid were removed, so that $-n_1^2 M\gamma^2\theta$ is the moment of the force of torsion. Then when the fluid is present the equation of motion of the disk becomes

$$\left(M\gamma^2 + \pi\rho a^4 \sqrt{\frac{\mu'}{2n}}\right)\frac{d^2\theta}{dt^2} + \pi\rho a^4 \sqrt{\frac{n\mu'}{2}}\frac{d\theta}{dt} + n_1^2 M\gamma^2\theta = 0\ldots(14),$$

or, putting for shortness

$$\pi\rho a^4 \sqrt{\frac{\mu'}{2n}} = 2\beta M\gamma^2,$$

$$(1 + 2\beta)\frac{d^2\theta}{dt^2} + 2n\beta\frac{d\theta}{dt} + n_1^2\theta = 0,$$

which gives, neglecting β^2,

$$\theta = \theta_0 \epsilon^{-n\beta t} \sin(nt + \alpha) \ldots\ldots\ldots\ldots\ldots (15),$$

where

$$n = n_1(1 - \beta).$$

* [That is, of course, on the supposition that the oscillations are not excessively slow.]

The observation of n and $n_{_1}$, or else the observation of n and of the decrement of the arc of oscillation, would enable us to determine β, and thence μ'. The values of β determined in these two different ways ought to agree.

There would be no difficulty in obtaining a more exact solution, in which the decrement of the arc of oscillation should be taken into account in calculating the motion of the fluid, but I pass on to the problems, the solution of which forms the main object of this paper.

Section II.

Solution of the equations in the case of a sphere oscillating in a mass of fluid either unlimited, or confined by a spherical envelope concentric with the sphere in its position of equilibrium.

9. Suppose the sphere suspended by a fine wire, the length of which is much greater than the radius of the sphere. Neglect for the present the action of the wire on the fluid, and consider only that of the sphere. The motion of the sphere and wire being supposed to take place parallel to a fixed vertical plane, there are two different modes of oscillation possible. We have here nothing to do with the rapid oscillations which depend mainly on the rotatory inertia of the sphere, but only with the principal oscillations, which are those which are observed in pendulum experiments. In these principal oscillations the centre of the sphere describes a small arc of a curve which is very nearly a circle, and which would be rigorously such, if the line joining the centre of gravity of the sphere and the point of attachment of the wire were rigorously in the direction of the wire. In calculating the motion of the fluid, we may regard this arc as a right line. In fact, the error thus introduced would only be a small quantity of the second order, and such quantities are supposed to be neglected in the investigation. Besides its motion of translation, the sphere will have a motion of rotation about a horizontal axis, the angular motion of the sphere being very nearly the same as that of the suspending wire. This motion, which would produce absolutely no effect on the fluid according to the common theory of hydro-

dynamics, will not be without its influence when friction is taken into account; but the effect is so very small in practical cases that it is not worth while to take it into account. For if a be the radius of the sphere, and l the length of the suspending wire, the velocity of a point in the surface of the sphere due to the motion of rotation will be a small quantity of the order a/l compared with the velocity due to the motion of translation. In finding the moment of the pressures of the fluid on the pendulum, forces arising from these velocities, and comparable with them, have to be multiplied by lines which are comparable with a, l, respectively. Hence the moment of the pressures due to the motion of rotation of the sphere will be a small quantity of the order a^2/l^2, compared with the moment due to the motion of translation. Now in practice l is usually at least twenty or thirty times greater than a, and the whole effect to be investigated is very small, so that it would be quite useless to take account of the motion of rotation of the sphere.

The problem, then, reduces itself to this. The centre of a sphere performs small periodic oscillations along a right line, the sphere itself having a motion of translation simply: it is required to determine the motion of the surrounding fluid.

10.   Let the mean position of the centre of the sphere be taken for origin, and the direction of its motion for the axis of x, so that the motion of the fluid is symmetrical with respect to this axis. Let ϖ be the perpendicular let fall from any point on the axis of x, q the velocity in the direction of ϖ, ω the angle between the line ϖ and the plane of xy. Then p, u, and q will be functions of x, ϖ, and t, and we shall have

$$v = q \cos \omega, \quad w = q \sin \omega, \quad y = \varpi \cos \omega, \quad z = \varpi \sin \omega,$$

whence
$$\varpi^2 = y^2 + z^2, \quad \omega = \tan^{-1} \frac{z}{y}.$$

We have now to substitute in equations (2) and (3), and we are at liberty to put $\omega = 0$ after differentiation. We get

$$\frac{d}{dy} = \cos \omega \frac{d}{d\varpi} - \frac{\sin \omega}{\varpi} \frac{d}{d\omega}, \ = \frac{d}{d\varpi} \text{ when } \omega = 0,$$

$$\frac{d^2}{dy^2} = \frac{d^2}{d\varpi^2} \text{ when } \omega = 0,$$

$$\frac{d}{dz} = \sin \omega \frac{d}{d\varpi} + \frac{\cos \omega}{\varpi} \frac{d}{d\omega}, \; = \frac{1}{\varpi} \frac{d}{d\omega} \text{ when } \omega = 0,$$

$$\frac{d^2}{dz^2} = \frac{1}{\varpi} \frac{d}{d\varpi} + \frac{1}{\varpi^2} \frac{d^2}{d\omega^2} \text{ when } \omega = 0,$$

whence we obtain

$$\frac{dp}{dx} = \mu \left(\frac{d^2u}{dx^2} + \frac{d^2u}{d\varpi^2} + \frac{1}{\varpi} \frac{du}{d\varpi} \right) - \rho \frac{du}{dt} \dots\dots\dots\dots(16),$$

$$\frac{dp}{d\varpi} = \mu \left(\frac{d^2q}{dx^2} + \frac{d^2q}{d\varpi^2} + \frac{1}{\varpi} \frac{dq}{d\varpi} - \frac{q}{\varpi^2} \right) - \rho \frac{dq}{dt} \dots\dots\dots (17),$$

$$\frac{du}{dx} + \frac{dq}{d\varpi} + \frac{q}{\varpi} = 0 \dots\dots\dots\dots\dots(18).$$

Eliminating p from (16) and (17), and putting for μ its equivalent $\mu'\rho$, we get

$$\mu' \frac{d}{d\varpi} \left(\frac{d^2}{dx^2} + \frac{d^2}{d\varpi^2} + \frac{1}{\varpi} \frac{d}{d\varpi} \right) u - \mu' \frac{d}{dx} \left(\frac{d^2}{dx^2} + \frac{d^2}{d\varpi^2} \right.$$

$$\left. + \frac{1}{\varpi} \frac{d}{d\varpi} - \frac{1}{\varpi^2} \right) q - \frac{d}{dt} \left(\frac{du}{d\varpi} - \frac{dq}{dx} \right) = 0,$$

or $\quad \left(\frac{d^2}{dx^2} + \frac{d^2}{d\varpi^2} + \frac{1}{\varpi} \frac{d}{d\varpi} - \frac{1}{\varpi^2} - \frac{1}{\mu'} \frac{d}{dt} \right) \left(\frac{du}{d\varpi} - \frac{dq}{dx} \right) = 0 \dots.. (19).$

By virtue of (18), $\varpi (ud\varpi - qdx)$ is an exact differential. Let then

$$\varpi (ud\varpi - qdx) = d\psi \dots\dots\dots\dots(20).$$

Expressing u and q in terms of ψ, we get

$$\frac{du}{d\varpi} - \frac{dq}{dx} = \frac{1}{\varpi} \left(\frac{d^2}{dx^2} + \frac{d^2}{d\varpi^2} - \frac{1}{\varpi} \frac{d}{d\varpi} \right) \psi.$$

Substituting in (19), and operating separately on the factor ϖ^{-1}, we obtain

$$\left(\frac{d^2}{dx^2} + \frac{d^2}{d\varpi^2} - \frac{1}{\varpi} \frac{d}{d\varpi} - \frac{1}{\mu'} \frac{d}{dt} \right) \left(\frac{d^2}{dx^2} + \frac{d^2}{d\varpi^2} - \frac{1}{\varpi} \frac{d}{d\varpi} \right) \psi = 0 \dots(20').$$

Since the operations represented by the two expressions within parentheses are evidently convertible, the integral of this equation is

$$\psi = \psi_1 + \psi_2* \dots\dots\dots\dots\dots(21),$$

* If we denote for shortness the operation

$$\frac{d^2}{dx^2} + \frac{d^2}{d\varpi^2} - \frac{1}{\varpi} \frac{d}{d\varpi}$$

where ψ_1, ψ_2 are the integrals of the equations

$$\left(\frac{d^2}{dx^2} + \frac{d^2}{d\varpi^2} - \frac{1}{\varpi}\frac{d}{d\varpi}\right)\psi_1 = 0 \ldots\ldots\ldots (22),$$

$$\left(\frac{d^2}{dx^2} + \frac{d^2}{d\varpi^2} - \frac{1}{\varpi}\frac{d}{d\varpi} - \frac{1}{\mu'}\frac{d}{dt}\right)\psi_2 = 0 \ldots\ldots\ldots (23).$$

11. By means of the last three equations, the expression for dp obtained from (16) and (17) is greatly simplified. We get, in the first place,

$$\frac{1}{\rho}\frac{dp}{dx} = \left\{\mu'\left(\frac{d^2}{dx^2} + \frac{d^2}{d\varpi^2} + \frac{1}{\varpi}\frac{d}{d\varpi}\right) - \frac{d}{dt}\right\}\frac{1}{\varpi}\frac{d\psi}{d\varpi} \ldots\ldots (24):$$

but by adding together equations (22) and (23), and taking account of (21), we get

$$\frac{d^2\psi}{dx^2} = -\frac{d^2\psi}{d\varpi^2} + \frac{1}{\varpi}\frac{d\psi}{d\varpi} + \frac{1}{\mu'}\frac{d\psi_2}{dt}.$$

On substituting in (24), it will be found that all the terms in the right-hand member of the equation destroy one another, except those which contain $d\psi/dt$ and $d\psi_2/dt$, and the equation is reduced to

$$\frac{dp}{dx} = -\frac{\rho}{\varpi}\frac{d^2\psi_1}{dt\,d\varpi}.$$

by D, our equation becomes

$$D\left(D - \frac{1}{\mu'}\frac{d}{dt}\right)\psi = 0,$$

which gives by the separation of symbols

$$\psi = \left\{D\left(D - \frac{1}{\mu'}\frac{d}{dt}\right)\right\}^{-1}0 = \left(\frac{1}{\mu'}\frac{d}{dt}\right)^{-1}\left\{\left(D - \frac{1}{\mu'}\frac{d}{dt}\right)^{-1} - D^{-1}\right\}0 \ldots\ldots(a),$$

so that $d\psi/dt$ is composed of two parts, which are separately the integrals of (22), (23). Hence we have for the integral of (20') $\psi = \psi_1 + \psi_2 + \Psi$, Ψ being a function of x and ϖ without t which satisfies the equation $D^2\Psi = 0$. For the equations (22), (23) will not be altered if we put $\int\psi_1 dt$, $\int\psi_2 dt$ for ψ_1, ψ_2, the arbitrary functions which would arise from the integration with respect to t being supposed to be included in Ψ. The function Ψ, which taken by itself can only correspond to steady motion, is excluded from the problem under consideration by the condition of periodicity. But we may even, independently of this condition, regard (21) as the complete integral of (20'), provided we suppose included in (21) terms which would be obtained by supposing ψ at first to vary slowly with the time, employing the integrals of (22) and (23), and then making the rate of variation diminish indefinitely. By treating the symbolical expression in the right-hand member of equation (a) as a vanishing fraction, d/dt being supposed to vanish, we obtain in fact $D^{-2}0$; so that under the convention just mentioned the function Ψ may be supposed to be included in $\psi_1 + \psi_2$. The same remarks will apply to the equation in Section III. which answers to (20').

The equation (17) may be reduced in a similar manner, and we get finally

$$dp = \frac{\rho}{\varpi} \left(\frac{d^2\psi_1}{dt\,dx}\, d\varpi - \frac{d^2\psi_1}{dt\,d\varpi}\, dx \right) \dots\dots\dots\dots (25),$$

which is an exact differential by virtue of (22).

12. Passing to polar co-ordinates, let r be the radius vector drawn from the origin, θ the angle which r makes with the axis of x, and let R be the velocity along the radius vector, Θ the velocity perpendicular to the radius vector : then

$$x = r\cos\theta, \;\; \varpi = r\sin\theta, \;\; u = R\cos\theta - \Theta\sin\theta, \;\; q = R\sin\theta + \Theta\cos\theta.$$

Making these substitutions in (20), (22), (23), and (25), we obtain

$$r\sin\theta\,(R r\,d\theta - \Theta\,dr) = d\psi \dots\dots\dots\dots\dots(26),$$

$$\frac{d^2\psi_1}{dr^2} + \frac{\sin\theta}{r^2}\frac{d}{d\theta}\left(\frac{1}{\sin\theta}\frac{d\psi_1}{d\theta}\right) = 0 \dots\dots\dots\dots\dots(27),$$

$$\frac{d^2\psi_2}{dr^2} + \frac{\sin\theta}{r^2}\frac{d}{d\theta}\left(\frac{1}{\sin\theta}\frac{d\psi_2}{a\theta}\right) - \frac{1}{\mu'}\frac{d\psi_2}{dt} = 0 \dots\dots(28),$$

$$dp = \frac{\rho}{r\sin\theta}\left(\frac{d^2\psi_1}{dt\,dr}\,r\,d\theta - \frac{1}{r}\frac{d^2\psi_1}{dt\,d\theta}\,dr\right) \;\dots\dots\dots(29).$$

We must now determine ψ_1 and ψ_2 by means of (27) and (28), combined with the equations of condition. When these functions are known, p will be obtained by integrating the exact differential which forms the right-hand member of (29), and the velocities R, Θ, if required, will be got by differentiation, as indicated by equation (26). Formulæ deduced from (4) and (5) will then make known the pressure of the fluid on the sphere.

13. Let ξ be the abscissa of the centre of the sphere at any instant. The conditions to be satisfied at the surface of the sphere are that when $r = r_1$, the radius vector of the surface, we have

$$R = \cos\theta\,\frac{d\xi}{dt}, \;\;\; \Theta = -\sin\theta\,\frac{d\xi}{dt}.$$

Now r_1 differs from a by a small quantity of the first order, and since this value of r has to be substituted in functions which are already small quantities of that order, it will be sufficient to put $r = a$. Hence, expressing R and Θ in terms of ψ, we get

$$\frac{d\psi}{dr} = a\sin^2\theta\,\frac{d\xi}{dt}, \;\; \frac{d\psi}{d\theta} = a^2\sin\theta\cos\theta\,\frac{d\xi}{dt}, \;\; \text{when } r = a \;\dots\dots(30).$$

When the fluid is unlimited, it will be found that certain arbitrary constants will vanish by the condition that the motion shall not become infinite at an infinite distance in the fluid. When the fluid is confined by an envelope having a radius b, we have the equations of condition

$$\frac{d\psi}{dr} = 0, \quad \frac{d\psi}{d\theta} = 0, \quad \text{when } r = b \quad \dots\dots\dots\dots(31).$$

14. We must now, in accordance with the plan proposed in Section I., introduce the condition that the function ψ shall be composed, so far as the time is concerned, of the circular functions $\sin nt$ and $\cos nt$, that is, that it shall be of the form

$$P \sin nt + Q \cos nt,$$

where P and Q are functions of r and θ only. An artifice, however, which has been extensively employed by M. Cauchy will here be found of great use. Instead of introducing the circular functions $\sin nt$ and $\cos nt$, we may employ the exponentials $\epsilon^{\sqrt{-1}nt}$, and $\epsilon^{-\sqrt{-1}nt}$. Since our equations are linear, and since each of these exponential functions reproduces itself at each differentiation, it follows that if all the terms in any one of our equations be arranged in two groups, containing as a factor $\epsilon^{\sqrt{-1}nt}$ in one case, and $\epsilon^{-\sqrt{-1}nt}$ in the other, the two groups will be quite independent, and the equations will be satisfied by either group separately. Hence it will be sufficient to introduce one of the exponential functions. We shall thus have only half the number of terms to write down, and half the number of arbitrary constants to determine that would have been necessary had we employed circular functions. When we have arrived at our result, it will be sufficient to put each equation under the form $U + \sqrt{-1}\,V = 0$, and throw away the imaginary part, or else throw away the real part and omit $\sqrt{-1}$, since the system of quantities U, and the system of quantities V must separately satisfy the equations of the problem. Assuming then

$$\frac{d\xi}{dt} = c\epsilon^{\sqrt{-1}nt}, \quad \psi = \epsilon^{\sqrt{-1}nt}\,P,$$

we have to determine P as a function of r and θ.

15.   The form of the equations of condition (30) points out $\sin^2\theta$ as a factor of P, and since the operation $\sin\theta\,\dfrac{d}{d\theta}\,\dfrac{1}{\sin\theta}\,\dfrac{d}{d\theta}$ performed on the function $\sin^2\theta$ reproduces the same function with a coefficient -2, it will be possible to satisfy equations (27) and (28) on the supposition that $\sin^2\theta$ is a factor of ψ_1 and ψ_2.* Assume then

$$\psi_1 = \epsilon^{\sqrt{-1}nt}\sin^2\theta\, f_1(r), \quad \psi_2 = \epsilon^{\sqrt{-1}nt}\sin^2\theta\, f_2(r).$$

Putting for convenience

$$n\sqrt{-1} = \mu'm^2 \dots\dots\dots\dots\dots(32),$$

and substituting in (27) and (28), we get

$$f_1''(r) - \frac{2}{r^2}f_1(r) = 0 \ \dots\dots\dots\dots\dots(33),$$

$$f_2''(r) - \frac{2}{r^2}f_2(r) - m^2 f_2(r) = 0 \ \dots\dots\dots(34).$$

The equations of condition (30), (31) become, on putting $f(r)$ for $f_1(r) + f_2(r)$,

$$f'(a) = ac, \quad f(a) = \tfrac{1}{2}a^2c \dots\dots\dots\dots(35),$$

$$f'(b) = 0, \quad f(b) = 0 \ \dots\dots\dots\dots(36).$$

We may obtain p from (29) by putting for ψ_1 its value $\epsilon^{\mu'm^2t}\sin^2\theta f_1(r)$, replacing after differentiation $2f_1(r)$ by its equiva-

---

* When this operation is performed on the function $\sin\theta\,dY_i/d\theta$, the function is reproduced with a coefficient $-i(i+1)$. Y_i here denotes a Laplace's function of the ith order, which contains only one variable angle, namely θ. Now ψ may be expanded in a series of quantities of the general form $\sin\theta\,dY_i/d\theta$. For, since we are only concerned with the differential coefficients of ψ with respect to r and θ, we have a right to suppose ψ to vanish at whatever point of space we please. Let then $\psi = 0$ when $r = a$ and $\theta = 0$. To find the value of ψ at a distance r from the origin, along the axis of x positive, it will be sufficient to put $\theta = 0$, $d\theta = 0$ in (26), and integrate from $r = a$ to r, whence $\psi = 0$. To find the value of ψ at the same distance r along the axis of x negative, it will be sufficient to leave r constant, and integrate $d\psi$ from $\theta = 0$ to $\theta = \pi$. Referring to (26), we see that the integral vanishes, since the total flux across the surface of the sphere whose radius is r must be equal to zero. Hence ψ vanishes when $\theta = 0$ or $= \pi$, and it appears from (26) that when θ is very small or very nearly equal to π, ψ varies ultimately as $\sin^2\theta$ for given values of r and t. Hence $\psi \operatorname{cosec}\theta$, and therefore $\int\psi \operatorname{cosec}\theta\,d\theta$, is finite even when $\sin\theta$ vanishes, and therefore $\int\psi \operatorname{cosec}\theta\,d\theta$ may be expanded in a series of Laplace's functions, and therefore ψ itself in a series of quantities of the form $\sin\theta\,dY_i/d\theta$. It was somewhat in this way that I first obtained the form of the function ψ.

lent $r^2 f_1''(r)$, and then integrating. It is unnecessary to add an arbitrary function of the time, since any such function may be supposed to be included in Π. We get

$$p = -\rho\mu'm^2\epsilon^{\mu'm^2t}\cos\theta f_1'(r) \quad \ldots\ldots\ldots\ldots\ldots(37).$$

16. The integration of the differential equation (33) does not present the least difficulty, and (34) comes under a well-known integrable form. The integrals of these equations are

$$\left.\begin{aligned} f_1(r) &= \frac{A}{r} + Br^2 \\ f_2(r) &= C\epsilon^{-mr}\left(1 + \frac{1}{mr}\right) + D\epsilon^{mr}\left(1 - \frac{1}{mr}\right) \end{aligned}\right\} \quad \ldots\ldots\ldots(38),$$

and we have to determine A, B, C, D by the equations of condition.

The solution of the problem, in the case in which the fluid is confined by a spherical envelope, will of course contain as a particular case that in which the fluid is unlimited, to obtain the results belonging to which it will be sufficient to put $b = \infty$. As, however, the case of an unlimited fluid is at the same time simpler and more interesting than the general case, it will be proper to consider it separately.

Let $+m$ denote that square root of $\mu'^{-1}n\sqrt{-1}$ which has its real part positive; then in equations (38) we must have $D = 0$, since otherwise the velocity would be infinite at an infinite distance. We must also have $B = 0$, since otherwise the velocity would be finite when $r = \infty$, as appears from (26). We get then from the equations of condition (35)

$$A = \tfrac{1}{2}a^3c + \frac{3a^2c}{2m}\left(1 + \frac{1}{ma}\right), \quad C = -\frac{3ac}{2m}\epsilon^{ma},$$

whence

$$\xi = \frac{c}{\mu'm^2}\epsilon^{\mu'm^2t} \quad \ldots\ldots\ldots\ldots\ldots\ldots(39),$$

$$\psi = \tfrac{1}{2}a^2c\epsilon^{\mu'm^2t}\sin^2\theta\left\{\left(1 + \frac{3}{ma} + \frac{3}{m^2a^2}\right)\frac{a}{r}\right.$$
$$\left. - \frac{3}{ma}\left(1 + \frac{1}{mr}\right)\epsilon^{-m(r-a)}\right\} \quad \ldots\ldots\ldots\ldots(40),$$

$$p = \tfrac{1}{2}\rho ac\mu'm^2\left(1 + \frac{3}{ma} + \frac{3}{m^2a^2}\right)\epsilon^{\mu'm^2t}\cos\theta\frac{a^2}{r^2} \quad \ldots\ldots\ldots(41).$$

17. The symbolical equations (40), (41) contain the solution of the problem, the motion of the sphere being defined by the symbolical equation (39). If we wish to exhibit the actual results by means of real quantities alone, we have only to put the right-hand members of equations (39), (40), (41) under the form $U + \sqrt{-1}\,V$, and reject the imaginary part. Putting for shortness

$$\sqrt{\frac{n}{2\mu'}} = \nu \quad \dotfill (42),$$

we have $m = \nu(1 + \sqrt{-1})$, and we obtain

$$\xi = \frac{c}{n} \sin nt \quad \dotfill (43),$$

$$\psi = \tfrac{1}{2} a^2 c \sin^2 \theta \left\{ \left[\left(1 + \frac{3}{2\nu a}\right) \cos nt + \frac{3}{2\nu a}\left(1 + \frac{1}{\nu a}\right) \sin nt \right] \frac{a}{r} \right.$$

$$- \frac{3}{2\nu a} \epsilon^{-\nu(r-a)} \left[\cos (nt - \nu r + \nu a) \right.$$

$$\left. \left. + \left(1 + \frac{1}{\nu r}\right) \sin (nt - \nu r + \nu a) \right] \right\} \quad \dotfill (44),$$

$$p = -\tfrac{1}{2} \rho a c n \left\{ \left(1 + \frac{3}{2\nu a}\right) \sin nt \right.$$

$$\left. - \frac{3}{2\nu a}\left(1 + \frac{1}{\nu a}\right) \cos nt \right\} \cos \theta \cdot \frac{a^2}{r^2}. \quad \dotfill (45).$$

The reader will remark that the ξ, ψ, p of the present article are not the same as the ξ, ψ, p of the preceding. The latter are the imaginary expressions, of which the real parts constitute the former. It did not appear necessary to change the notation.

When $\mu' = 0$, $\nu = \infty$, and ψ reduces itself to

$$\frac{a^3 c}{2r} \sin^2 \theta \cos nt, \quad \text{or} \quad \frac{a^3}{2r} \sin^2 \theta \frac{d\xi}{dt}.$$

In this case we get from (26)

$$R = a^3 \frac{d\xi}{dt} \frac{\cos \theta}{r^3}, \quad \Theta = \tfrac{1}{2} a^3 \frac{d\xi}{dt} \frac{\sin \theta}{r^3},$$

and $R dr + \Theta r d\theta$ is an exact differential $d\phi$ where

$$\phi = -\tfrac{1}{2} a^3 \frac{d\xi}{dt} \frac{\cos \theta}{r^2},$$

which agrees with the result deduced directly from the ordinary equations of hydrodynamics*.

18. Let us now form the expression for the resultant of the pressures of the fluid on the several elements of the surface of the sphere. Let P_r be the normal, and T_θ the tangential, component of the pressure at any point in the direction of a plane drawn perpendicular to the radius vector. The formulæ (4), (5) are general, and therefore we may replace x, y in these formulæ by x', y', where x', y' are measured in any two rectangular directions we please. Let the plane of $x'\,y'$ pass through the axis of x and the radius vector, and let the axis of x' be inclined to that of x at an angle ϑ, which after differentiation is made equal to θ. Then P_1, T_3 will become P_r, T_θ, respectively. We have

$$u' = R\cos(\theta - \vartheta) - \Theta\sin(\theta - \vartheta),$$
$$v' = R\sin(\theta - \vartheta) + \Theta\cos(\theta - \vartheta),$$

and when $\theta = \vartheta$

$$\frac{d}{dx'} = \frac{d}{dr}, \quad \frac{d}{dy'} = \frac{d}{rd\theta},$$
$$\frac{du'}{dx'} = \frac{dR}{dr}, \quad \frac{du'}{dy'} = \frac{dR}{rd\theta} - \frac{\Theta}{r}, \quad \frac{dv'}{dx'} = \frac{d\Theta}{dr};$$

whence

$$P_r = p - 2\mu\frac{dR}{dr}, \quad T_\theta = -\mu\left(\frac{dR}{rd\theta} + \frac{d\Theta}{dr} - \frac{\Theta}{r}\right)\ldots\ldots(46).$$

In these formulæ, suppose r put equal to a after differentiation. Then P_r, T_θ will be the components in the direction of r, θ of the pressure of the sphere on the fluid. The resolved part of these in the direction of x is

$$P_r\cos\theta - T_\theta\sin\theta,$$

which is equal and opposite to the component, in the direction of x, of the pressure of the fluid on the sphere. Let F be the whole force of the fluid on the sphere, which will evidently act along the axis of x. Then, observing that $2\pi a^2\sin\theta d\theta$ is the area of an elementary annulus of the surface of the sphere, we get

$$F = 2\pi a^2\int_0^\pi(-P_r\cos\theta + T_\theta\sin\theta)_a\sin\theta d\theta\ \ldots\ldots(47),$$

* See *Camb. Phil. Trans.* Vol. VIII. p. 119. [*Ante*, Vol. I. p. 41.]

the suffix a denoting that r is supposed to have the value a in the general expressions for P_r and T_θ.

The expression for F may be greatly simplified, without employing the solution of equations (27), (28), by combining these equations in their original state with the equations of condition (30). We have, in the first place, from (26)

$$R = \frac{1}{r^2 \sin \theta} \frac{d\psi}{d\theta}, \quad \Theta = -\frac{1}{r \sin \theta} \frac{d\psi}{dr} \quad \ldots\ldots\ldots\ldots(48).$$

Now the equations (30) make known the values of ψ and $d\psi/dr$, and of their differential coefficients of all orders with respect to θ, when $r = a$. When the expressions for R and Θ are substituted in (46), the result will contain only one term in which the differentiation with respect to r rises to the second order. But we get from (21), (27), (28)

$$\frac{d^2\psi}{dr^2} = -\frac{\sin \theta}{r^2} \frac{d}{d\theta} \left(\frac{1}{\sin \theta} \frac{d\psi}{d\theta} \right) + \frac{1}{\mu'} \frac{d\psi_2}{dt},$$

and the second of equations (30) gives the value for $r = a$ of the first term in the right-hand member of the equation just written. We obtain from (48) and (30)

$$\left(\frac{dR}{dr} \right)_a = 0,$$

$$\left(\frac{dR}{rd\theta} \right)_a = -\frac{\sin \theta}{a} \frac{d\xi}{dt} = \left(\frac{\Theta}{r} \right)_a,$$

$$\left(\frac{d\Theta}{dr} \right)_a = -\frac{1}{\mu' a \sin \theta} \left(\frac{d\psi_2}{dt} \right)_a.$$

Substituting in (47), and writing $\mu'\rho$ for μ, we get

$$F = 2\pi a \int_0^\pi \left\{ -ap_a \cos \theta + \rho \left(\frac{d\psi_2}{dt} \right)_a \right\} \sin \theta d\theta.$$

With respect to the first term in this expression, we get by integration by parts

$$\int p \cos \theta \sin \theta \, d\theta = \tfrac{1}{2} \sin^2 \theta \, . \, p - \tfrac{1}{2} \int \sin^2 \theta \frac{dp}{d\theta} \, d\theta.$$

The first term vanishes at the limits. Substituting in the second term for $dp/d\theta$ the expression got from (29), and putting $r = a$, we get

$$\int_0^\pi p_a \cos\theta \sin\theta \, d\theta = -\tfrac{1}{2}\rho \frac{d}{dt} \int_0^\pi \left(\frac{d\psi_1}{dr}\right)_a \sin\theta \, d\theta.$$

Substituting in the expression for F, we get

$$F = \pi\rho a \frac{d}{dt} \int_0^\pi \left\{ a\left(\frac{d\psi_1}{dr}\right)_a + 2\,(\psi_2)_a \right\} \sin\theta \, d\theta \ldots\ldots(49).$$

19. The above expression for F, being derived from the general equations (27), (28), combined with the equations of condition (30), holds good, not merely when the fluid is confined by a spherical envelope, but whenever the motion is symmetrical about an axis, and that, whether the motion of the sphere be or be not expressed by a single circular function of the time. It might be employed, for instance, in the case of a sphere oscillating in a direction perpendicular to a fixed rigid plane.

When the fluid is either unconfined, or confined by a spherical envelope concentric with the sphere in its position of equilibrium, the functions ψ_1, ψ_2 consist, as we have seen, of $\sin^2\theta$ multiplied by two factors independent of θ. If we continue to employ the symbolical expressions, which will be more convenient to work with than the real expressions which might be derived from them, we shall have

$$\epsilon^{\sqrt{-1}nt} f_1(r), \quad \epsilon^{\sqrt{-1}nt} f_2(r),$$

for these factors respectively. Substituting in (49), and performing the integration with respect to θ, we get

$$F = \tfrac{4}{3}\pi\rho an \sqrt{-1}\,\{af_1'(a) + 2f_2(a)\}\,\epsilon^{\sqrt{-1}nt} \ldots\ldots\ldots(50).$$

20. Consider for the present only the case in which the fluid is unlimited. The arbitrary constants which appear in equations (38) were determined for this case in Art. 16. Substituting in (50) we get

$$F = -\tfrac{2}{3}\pi\rho a^3 cn \sqrt{-1}\left(1 + \frac{9}{ma} + \frac{9}{m^2 a^2}\right)\epsilon^{\sqrt{-1}nt}.$$

Putting for m its value $\nu(1 + \sqrt{-1})$, and denoting by M' the mass of the fluid displaced by the sphere, which is equal to $\frac{4}{3}\pi\rho a^3$, we get

$$F = -M'cn\left\{\left(\frac{1}{2} + \frac{9}{4\nu a}\right)\sqrt{-1} + \frac{9}{4\nu a}\left(1 + \frac{1}{\nu a}\right)\right\}\epsilon^{\sqrt{-1}nt};$$

whence

$$F = -\left(\frac{1}{2} + \frac{9}{4\nu a}\right)M'\frac{d^2\xi}{dt^2} - \frac{9}{4\nu a}\left(1 + \frac{1}{\nu a}\right)M'n\frac{d\xi}{dt} \ldots\ldots\ldots(51).$$

Since $\sqrt{-1}$ has been eliminated, this equation will remain unchanged when we pass from the symbolical to the real values of F and ξ.

Let τ be the time of oscillation from rest to rest, so that $n\tau = \pi$, and put for shortness k, k' for the coefficients of M' in (51); then

$$\nu = \sqrt{\frac{\pi}{2\mu'\tau}}, \quad k = \frac{1}{2} + \frac{9}{4\nu a}, \quad k' = \frac{9}{4\nu a}\left(1 + \frac{1}{\nu a}\right)\ldots\ldots(52).$$

The first term in the expression for the force F has the same effect as increasing the inertia of the sphere. To take account of this term, it will be sufficient to conceive a mass kM' collected at the centre of the sphere, adding to its inertia without adding to its weight. The main effect of the second term is to produce a diminution in the arc of oscillation: its effect on the time of oscillation would usually be quite insensible, and must in fact be neglected for consistency's sake, because the motion of the fluid was determined by supposing the motion of the sphere permanent, which is only allowable when we neglect the square of the rate of decrease of the arc of oscillation.

If we form the equation of motion of the sphere, introducing the force F, and then proceed to integrate the equation, we shall obtain in the integral an exponential $\epsilon^{-\delta t}$ multiplying the circular function, δ being half the coefficient of $d\xi/dt$ divided by that of $d^2\xi/dt^2$. Let M be the mass of the sphere, $M\gamma^2$ its moment of inertia about the axis of suspension, then

$$nk'M'(l+a)^2 = 2\delta\{M\gamma^2 + kM'(l+a)^2\}.$$

In considering the diminution of the arc of oscillation, we may put $l + a$ for γ. During i oscillations, let the arc of oscillation be diminished in the ratio of A_0 to A_i, then

$$\log_\epsilon\frac{A_0}{A_i} = i\tau\delta = \frac{\pi i}{2}\frac{k'M'}{M + kM'}\ldots\ldots\ldots\ldots\ldots(53).$$

For a given fluid and a given time of oscillation, both k and k' increase as a decreases. Hence it follows from theory, that the smaller be the sphere, its density being supposed given, the more the time of oscillation is affected, and the more rapidly the arc of oscillation diminishes, the alteration in the rate of diminution of the arc due to an alteration in the radius of the sphere being more conspicuous than the alteration in the time of oscillation.

21. Let us now suppose the fluid confined in a spherical envelope. In this case, we have to determine the four arbitrary constants which appear in (38) by the four equations (35) and (36). We get, in the first place,

$$\frac{A}{a} + Ba^2 + C\epsilon^{-ma}\left(1 + \frac{1}{ma}\right) + D\epsilon^{ma}\left(1 - \frac{1}{ma}\right) = \tfrac{1}{2}a^2c \dots\dots\dots(54),$$

$$-\frac{A}{a} + 2Ba^2 - C\epsilon^{-ma}\left(ma + 1 + \frac{1}{ma}\right) + D\epsilon^{ma}\left(ma - 1 + \frac{1}{ma}\right) = a^2c \ (55),$$

$$\frac{A}{b} + Bb^2 + C\epsilon^{-mb}\left(1 + \frac{1}{mb}\right) + D\epsilon^{mb}\left(1 - \frac{1}{mb}\right) = 0 \dots\dots\dots\dots (56),$$

$$-\frac{A}{b} + 2Bb^2 - C\epsilon^{-mb}\left(mb + 1 + \frac{1}{mb}\right) + D\epsilon^{mb}\left(mb - 1 + \frac{1}{mb}\right) = 0 \ (57).$$

Putting a^2cK for $af_1'(a) + 2f_2(a)$, which is the quantity that we want to find, we get from (38) and (54)

$$K = 1 - \frac{3A}{a^3c} \dots\dots\dots\dots\dots\dots\dots(58).$$

Eliminating in succession B from (54) and (55), from (56) and (57), and from (54) and (56), we shall obtain for the determination of A, C, D three equations which remain unchanged when a and b are interchanged, and the signs of A, C, and D changed. Hence $-A, -C, -D$ are the same functions of b and a that A, C, D are of a and b. It will also assist in the further elimination to observe that C and D are interchanged when the sign of m is changed. The result of the elimination is

$$K = 1 - \frac{3b}{2m^2a^2} \cdot \frac{\eta(a, b) - \eta(b, a)}{12mab + \zeta(a, b) + \zeta(b, a)} \dots\dots\dots(59),$$

the functions ζ, η being defined by the equations

$$\left. \begin{aligned} \eta(a, b) &= (m^2a^2 + 3ma + 3)(m^2b^2 - 3mb + 3)\,\epsilon^{m(b-a)} \\ \zeta(a, b) &= \{b(m^2b^2 - 3mb + 3) - a(m^2a^2 + 3ma + 3)\}\,\epsilon^{m(b-a)} \end{aligned} \right\} \dots\dots(60).$$

It turns out that K is a complicated function of m and ab^{-1}, and the algebraical expressions for the quantities which answer to k and k' in Art. 20 would be more complicated still, because $\nu(1 + \sqrt{-1})$ would have to be substituted for m in (60) and (59), and then K reduced to the form $-k + \sqrt{-1}k'$. To obtain numerical results from these formulæ, it would be best to substitute the numerical values of a, b, and ν in (60) and (59), and perform the reduction of K in figures.

22. If the distance of the envelope from the surface of the sphere be at all considerable, the exponential $\epsilon^{\nu(b-a)}$, which arises from $\epsilon^{m(b-a)}$, will have so large a numerical value that we may neglect the terms in the numerator and denominator of the fraction in the expression for K which contain $\epsilon^{-\nu(b-a)}$, as well as the term in the denominator which is free from exponentials, in comparison with the terms which contain $\epsilon^{\nu(b-a)}$. Thus, if $b-a$ be two inches, τ one second, and $\sqrt{\mu'} = \cdot 116$, we have $\epsilon^{\nu(b-a)} = 2424000000$, nearly; and if $b-a$ be only an inch or half an inch, we have still the square or fourth root of the above quantity, that is, about 49234 or 222, for the value of that exponential. Hence, in practical cases, the above simplification may be made, which will cause the exponentials to disappear from the expression for K. We thus get

$$K = 1 - \frac{3b}{2m^2a^2}\frac{(m^2a^2 + 3ma + 3)(m^2b^2 - 3mb + 3)}{b(m^2b^2 - 3mb + 3) - a(m^2a^2 + 3ma + 3)} \dots (61).$$

If we assume

$$3\nu a + 3 + (2\nu^2a^2 + 3\nu a)\sqrt{-1} = A'(\cos\alpha + \sqrt{-1}\sin\alpha),$$

$$-3\nu b + 3 + (2\nu^2b^2 - 3\nu b)\sqrt{-1} = B'(\cos\beta + \sqrt{-1}\sin\beta),$$

$$bB'\cos\beta - aA'\cos\alpha = C'\cos\gamma,$$

$$bB'\sin\beta - aA'\sin\alpha = C'\sin\gamma,$$

we get from (61)

$$K = 1 + \frac{3b\sqrt{-1}}{4\nu^2a^2}\cdot\frac{A'B'}{C'}\{\cos(\alpha+\beta-\gamma) + \sqrt{-1}\sin(\alpha+\beta-\gamma)\},$$

whence

$$k = \frac{3bA'B'}{4\nu^2a^2C'}\sin(\alpha+\beta-\gamma) - 1$$

$$k' = \frac{3bA'B'}{4\nu^2a^2C'}\cos(\alpha+\beta-\gamma);$$

$$\left.\vphantom{\begin{array}{c}1\\1\\1\end{array}}\right\}\dots\dots(62);$$

and, as before, kM' is the imaginary mass which we must conceive to be collected at the centre of the sphere, in order to allow for the inertia of the fluid, and $-k'M'n\, d\xi/dt$ the term in F on which depends the diminution in the arc of oscillation.

23. If we suppose $\mu' = 0$, and therefore $m = \infty$, we get from (61)

$$K = -\frac{b^3 + 2a^3}{2(b^3 - a^3)} \quad\dots\dots\dots\dots\dots\dots (63),$$

and, in this case, k is the same as with K sign changed, and $k' = 0$, which agrees with the result obtained directly from the ordinary equations of hydrodynamics*. If, on the other hand, we make $b = \infty$, we arrive at the results already obtained in Art. 20. In both these cases it becomes rigorously exact to neglect in the expression for $K - 1$ given by (59) all the terms which are not multiplied by $\epsilon^{\nu(b-a)}$.

If the effect of the envelope be but small, which will generally be the case, it will be convenient to calculate k and k' from the formulæ (52), which apply to the case in which $b = \infty$, and then add corrections Δk, $\Delta k'$ due to the envelope. We get from (61)

$$\Delta k - \sqrt{-1}\Delta k' = \frac{3}{2m^2a}\frac{(m^2a^2 + 3ma + 3)^2}{b(m^2b^2 - 3mb + 3) - a(m^2a^2 + 3ma + 3)}\dots(64),$$

which may be treated, if required, as the equation (61) was treated in the preceding article. If, however, we suppose m large, and are content to retain only the most important term in (64), we get simply

$$\Delta k = \frac{3a^3}{2(b^3 - a^3)}, \quad \Delta k' = 0 \dots\dots\dots\dots\dots (65),$$

so that the correction for the envelope may be calculated as if the fluid were destitute of friction.

* See *Camb. Phil. Trans.* Vol. VIII. p. 120. [*Ante*, Vol. I. p. 41.]

Solution of the equations in the case of an infinite cylinder oscillating
in an unlimited mass of fluid, in a direction perpendicular to
its axis.

24.   Suppose a long cylindrical rod suspended at a point in its
axis, and made to oscillate as a pendulum in an unlimited mass of
fluid.   The resistance experienced by any element of the cylinder
comprised between two parallel planes drawn perpendicular to the
axis will manifestly be very nearly the same as if the element
belonged to an infinite cylinder oscillating with the same linear
velocity.   For an element situated very near either extremity of
the rod, the resistance thus determined would, no doubt, be sensibly
erroneous ; but as the diameter of the rod is supposed to be but
small in comparison with its length, it will be easily seen that the
error thus introduced must be extremely small.

Imagine then an infinite cylinder to oscillate in a fluid, in a
direction perpendicular to its axis, so that the motion takes place
in two dimensions, and let it be required to determine the motion
of the fluid.   The mode of solution of this problem will require
no explanation, being identical in principle with that which has
been already adopted in the case of a sphere.   In the present
instance the problem will be found somewhat easier, up to the
formation of the equations analogous to (33) and (34), after which
it will become much more difficult.

25.   Let a plane drawn perpendicular to the axis of the
cylinder be taken for the plane of xy, the origin being situated in
the mean position of the axis of the cylinder, and the axis of x
being measured in the direction of the cylinder's motion.   The
general equations (2), (3) become in this case

$$\left.\begin{aligned}
\frac{dp}{dx} &= \mu\left(\frac{d^2u}{dx^2} + \frac{d^2u}{dy^2}\right) - \rho\frac{du}{dt} \\
\frac{dp}{dy} &= \mu\left(\frac{d^2v}{dx^2} + \frac{d^2v}{dy^2}\right) - \rho\frac{dv}{dt}
\end{aligned}\right\} \quad \dots\dots\dots\dots (66),$$

$$\frac{du}{dx} + \frac{dv}{dy} = 0 \quad \dots\dots\dots \quad \dots\dots\dots (67).$$

By virtue of (67), $udy - vdx$ is an exact differential. Let then

$$udy - vdx = d\chi \dots\dots\dots\dots\dots (68).$$

Eliminating p by differentiation from the two equations (66), and expressing u and v in terms of χ in the resulting equation, we get

$$\left(\frac{d^2}{dx^2} + \frac{d^2}{dy^2} - \frac{1}{\mu'}\frac{d}{dt}\right)\left(\frac{d^2}{dx^2} + \frac{d^2}{dy^2}\right)\chi = 0 \dots\dots\dots(69),$$

and, as before

$$\chi = \chi_1 + \chi_2 \dots\dots\dots \dots\dots\dots(70),$$

where

$$\left(\frac{d^2}{dx^2} + \frac{d^2}{dy^2}\right)\chi_1 = 0 \dots\dots\dots\dots(71),$$

$$\left(\frac{d^2}{dx^2} + \frac{d^2}{dy^2} - \frac{1}{\mu'}\frac{d}{dt}\right)\chi_2 = 0\dots\dots\dots \dots\dots(72).$$

We get from (66) and (68)

$$dp = \mu'\rho dx \cdot \frac{d}{dy}\left(\frac{d^2}{dx^2} + \frac{d^2}{dy^2} - \frac{1}{\mu'}\frac{d}{dt}\right)\chi$$
$$- \mu'\rho dy \cdot \frac{d}{dx}\left(\frac{d^2}{dx^2} + \frac{d^2}{dy^2} - \frac{1}{\mu'}\frac{d}{dt}\right)\chi,$$

which becomes by means of (70), (71), and (72)

$$dp = \rho\left(\frac{d^2\chi_1}{dt\,dx}\,dy - \frac{d^2\chi_1}{dt\,dy}\,dx\right) \dots\dots\dots\dots(73).$$

26. Passing to polar co-ordinates r, θ, where θ is supposed to be measured from the axis of x, we get from (68), (71), (72), and (73)

$$Rrd\theta - \Theta dr = d\chi \dots\dots\dots\dots\dots(74),$$

$$\left(\frac{d^2}{dr^2} + \frac{1}{r}\frac{d}{dr} + \frac{1}{r^2}\frac{d^2}{d\theta^2}\right)\chi_1 = 0 \dots\dots\dots\dots(75),$$

$$\left(\frac{d^2}{dr^2} + \frac{1}{r}\frac{d}{dr} + \frac{1}{r^2}\frac{d^2}{d\theta^2} - \frac{1}{\mu'}\frac{d}{dt}\right)\chi_2 = 0 \dots\dots\dots(76),$$

$$dp = \rho\frac{d}{dt}\left(\frac{d\chi_1}{dr}rd\theta - \frac{d\chi_1}{rd\theta}\,dr\right)\dots\dots\dots\dots(77),$$

R, Θ in (74) being the velocities along and perpendicular to the radius vector.

27.   Let a be the radius of the cylinder; and as before let the cylinder's motion be defined by the equation

$$\frac{d\xi}{dt} = c\epsilon^{\sqrt{-1}nt} = c\epsilon^{\mu'm^2t} \quad\quad\dotsc\dotsc\dotsc(78);$$

then we have for the equations of condition which relate to the surface of the cylinder

$$\left.\begin{aligned} R = \frac{d\chi}{rd\theta} = \cos\theta\,\frac{d\xi}{dt} = c\cos\theta\,\epsilon^{\mu'm^2t},\\[2mm] \Theta = -\frac{d\chi}{dr} = -\sin\theta\,\frac{d\xi}{dt} = -c\sin\theta\,\epsilon^{\mu'm^2t} \end{aligned}\right\} \text{ when } r = a\dotsc(79).$$

The general equations (75), (76), as well as the equations of condition (79), may be satisfied by taking

$$\chi_1 = \epsilon^{\mu'm^2t}\sin\theta\, F_1(r), \quad \chi_2 = \epsilon^{\mu'm^2t}\sin\theta\, F_2(r)\dotsc\dotsc\dotsc(80).$$

Substituting in (75), (76), and (79), we get

$$F_1''(r) + \frac{1}{r}F_1'(r) - \frac{1}{r^2}F_1(r) = 0\dotsc\dotsc\dotsc\dotsc(81),$$

$$F_2''(r) + \frac{1}{r}F_2'(r) - \frac{1}{r^2}F_2(r) - m^2 F_2(r) = 0\dotsc\dotsc\dotsc(82),$$

$$F_1(a) + F_2(a) = ac, \quad F_1'(a) + F_2'(a) = c\dotsc\dotsc\dotsc\dotsc(83),$$

besides which we have the condition that the velocity shall vanish at an infinite distance.

28.   The integral of (81) is

$$F_1(r) = \frac{A}{r} + Br \quad\quad\dotsc\dotsc\dotsc\dotsc\dotsc(84).$$

The integral of (82) cannot be obtained in finite terms.

To simplify the latter equation, assume $F_2(r) = F_3'(r)$. Substituting in (82), and integrating once, we get

$$F_3''(r) + \frac{1}{r}F_3'(r) - m^2 F_3(r) = 0 \quad\quad\dotsc\dotsc\dotsc\dotsc(85).$$

It is unnecessary to add an arbitrary constant, because such a constant, if introduced, might be got rid of by writing $F_3(r) + C$ for $F_3(r)$.

To integrate (85) by series according to ascending powers of r, let us first, instead of (85), take the equation formed from it by multiplying the second term by $1 - \delta$. Assuming in this new equation $F_3(r) = A_{\prime}x^\alpha + B_{\prime}x^\beta + ...$, and determining the arbitrary indices $\alpha, \beta ...$ and the arbitrary constants $A_{\prime}, B_{\prime}...$ so as to satisfy the equation, we get

$$F_3(r) = A_{\prime}\left\{1 + \frac{m^2r^2}{2(2-\delta)} + \frac{m^4r^4}{2.4(2-\delta)(4-\delta)} + ...\right\}$$

$$+ A_{\prime\prime}r^\delta\left\{1 + \frac{m^2r^2}{2(2+\delta)} + \frac{m^4r^4}{2.4(2+\delta)(4+\delta)} + ...\right\}$$

$$= (A_{\prime} + A_{\prime\prime} + A_{\prime\prime}\delta\log r)\left\{1 + \frac{m^2r^2}{2^2} + \frac{m^4r^4}{2^2.4^2} + ...\right\}$$

$$+ \tfrac{1}{2}(A_{\prime} - A_{\prime\prime})\delta\left\{\frac{m^2r^2}{2^2}S_1 + \frac{m^4r^4}{2^2.4^2}S_2 + \frac{m^6r^6}{2^2.4^2.6^2}S_3 + ...\right\}$$

$$+ \text{terms involving } \delta^2, \delta^3 ....$$

In this expression
$$S_i = 1^{-1} + 2^{-1} + 3^{-1} ... + i^{-1} ....(86).$$

Putting now
$$A_{\prime} = C - A_{\prime\prime}, \quad A_{\prime\prime} = D\delta^{-1},$$

substituting in the above equation, and then making δ vanish, we get

$$F_3(r) = (C + D\log r)\left(1 + \frac{m^2r^2}{2^2} + \frac{m^4r^4}{2^2.4^2} + ...\right)$$

$$- D\left(\frac{m^2r^2}{2^2}S_1 + \frac{m^4r^4}{2^2.4^2}S_2 + \frac{m^6r^6}{2^2.4^2.6^2}S_3 + ...\right) ......(87).$$

The series in this equation are evidently convergent for all values of r, however great; but, nevertheless, they give us no information as to what becomes of $F_3(r)$ when r becomes infinite, and yet one relation between C and D has to be determined by the condition that $F_3(r)$ shall not become infinite with r.

The equation (85) may be integrated by means of descending series combined with exponentials, by assuming
$$F_3(r) = \epsilon^{\pm mr}(A_{\prime}r^\alpha + B_{\prime}r^\beta ...).$$

I have already given the integral in this form in a paper, *On the numerical calculation of a class of definite integrals and infinite series**. The result is

$$F_3(r) = C' \epsilon^{-mr} r^{-\frac{1}{2}} \left\{ 1 - \frac{1^2}{2 \cdot 4mr} + \frac{1^2 \cdot 3^2}{2 \cdot 4 \, (4mr)^2} - \frac{1^2 \cdot 3^2 \cdot 5^2}{2 \cdot 4 \cdot 6 \, (4mr)^3} + \ldots \right\}$$

$$+ D' \, \epsilon^{mr} r^{-\frac{1}{2}} \left\{ 1 + \frac{1^2}{2 \cdot 4mr} + \frac{1^2 \cdot 3^2}{2 \cdot 4 \, (4mr)^2} + \frac{1^2 \cdot 3^2 \cdot 5^2}{2 \cdot 4 \cdot 6 \, (4mr)^3} + \ldots \right\} \ldots (88).$$

These series, although ultimately divergent in all cases, are very convenient for numerical calculation when the modulus of mr is large. Moreover they give at once $D' = 0$ for the condition that $F_3(r)$ shall not become infinite with r, and therefore we shall be able to obtain the required relation between C and D, provided we can express D' as a function of C and D.

29. This may be effected by means of the integral of (85) expressed by different integrals. This form of the integral is already known. It becomes, by a slight transformation,

$$F_3(r) = \int_0^{\frac{\pi}{2}} \{C'' + D'' \log (r \sin^2 \omega)\} \, (\epsilon^{mr \cos \omega} + \epsilon^{-mr \cos \omega}) \, d\omega \ldots (89),$$

C'', D'' being the two arbitrary constants. If we expand the exponentials in (89), and integrate the terms separately, we obtain, in fact, an expression of the same form as (87). This transformation requires the reduction of the definite integral

$$P_i = \int_0^{\frac{\pi}{2}} \cos^{2i} \omega \log \sin \omega \, d\omega.$$

If we integrate by parts, integrating $\cos \omega \log \sin \omega \, d\omega$, and differentiating $\cos^{2i-1} \omega$, we shall make P_i depend on P_{i-1}. Assuming $P_0 = Q_0$, $P_1 = \frac{1}{2}Q_1 \ldots$, and generally

$$P_i = \frac{1 \cdot 3 \ldots (2i - 1)}{2 \cdot 4 \ldots 2i} \, Q_i,$$

we get

$$Q_i = Q_0 - \{2^{-1} + 4^{-1} \ldots + (2i)^{-1}\} \frac{\pi}{2} = \frac{\pi}{2} \log (\tfrac{1}{2}) \, - \frac{\pi}{4} \, S_i. \dagger$$

---

* See *Camb. Phil. Trans.* Vol. IX. p. 182. [*Ante*, Vol. II. p. 349.]

† A demonstration by Mr Ellis of the theorem

$$\int_0^{\frac{\pi}{2}} \log \sin \theta \, d\theta = \frac{\pi}{2} \log (\tfrac{1}{2}),$$

The equivalence of the expressions (87) and (89) having been ascertained, in order to find the relations between C, D and C'', D'', it will be sufficient to write down the two leading terms in (87) and (89), and equate the results. We thus get

$$C + D \log r = \pi C'' + \pi D'' \log r + 2\pi D'' \log (\tfrac{1}{2}),$$

whence

$$C = \pi C'' + 2\pi \log (\tfrac{1}{2}) \cdot D'', \quad D = \pi D'' \dots\dots\dots(90).$$

There remains the more difficult step of finding the relation between D' and C'', D''. For this purpose let us seek the ultimate value of the second member of equation (89) when r increases indefinitely. In the first place we may observe that if Ω, Ω' be two imaginary quantities having their real parts positive, if the real part of Ω be greater than that of Ω', and if r be supposed to increase indefinitely, $\epsilon^{\Omega r}$ will ultimately be incomparably greater than $\epsilon^{\Omega' r}$, or even than $\log r \cdot \epsilon^{\Omega' r}$, or, to speak more precisely, the modulus of the former expression will ultimately be incomparably greater than the modulus of either of the latter. Hence, in finding the ultimate value of the expression for $F_3(r)$ in (89), we may replace the limits 0 and $\tfrac{1}{2}\pi$ of ω by 0 and ω_1, where ω_1 is a positive quantity as small as we please, which we may suppose to vanish after r has become infinite. We may also, for the same reason, omit the second of the exponentials. Let $\cos \omega = 1 - \lambda$, so that

$$\sin^2 \omega = 2\lambda \left(1 - \frac{\lambda}{2}\right), \quad d\omega = \frac{d\lambda}{\sqrt{2\lambda - \lambda^2}} = \left(1 + \frac{\lambda}{4} + \dots\right) \frac{d\lambda}{\sqrt{(2\lambda)}};$$

then the limits of λ will be 0 and λ_1, where $\lambda_1 = 1 - \cos \omega_1$. Since $\log \left(1 - \dfrac{\lambda}{2}\right)$ ultimately vanishes, and $1 + \dfrac{\lambda}{4} + \dots$ becomes ultimately 1, we get from (89)

$$\text{limit of } F_3(r) = \epsilon^{mr} \times \text{limit of } \int_0^{\lambda_1} (C'' + D'' \log 2\lambda r)\, \epsilon^{-m\lambda r} \frac{d\lambda}{\sqrt{(2\lambda)}} \, *.$$

due to Euler, will be found in the second volume of the *Cambridge Mathematical Journal*, p. 282, or in Gregory's *Examples*, p. 484.

* The word *limit* is here used in the sense in which $f(r)$ may be called the limit of $\phi(r)$ when the ratio of $\phi(r)$ to $f(r)$ is ultimately a ratio of equality, though $f(r)$ and $\phi(r)$ may vanish or become infinite together, in which case the limit of $\phi(r)$, according to the usual sense of the word *limit*, would be said to be zero or infinity.

If now we put $\lambda = \lambda' r^{-1}$, we shall have 0 and $\lambda_1 r$ for the limits of λ', and the second of these becomes infinite with r.   Hence

$$\text{limit of } F_3(r) = (2r)^{-\frac{1}{2}} \epsilon^{mr} \int_0^\infty (C'' + D'' \log 2\lambda') \, \epsilon^{-m\lambda'} \lambda'^{-\frac{1}{2}} \, d\lambda' \ \dots\dots(91).$$

Now $\displaystyle\int_0^\infty \epsilon^{-x} x^{-\frac{1}{2}} dx = \pi^{\frac{1}{2}}$, and if we differentiate both sides of the equation

$$\int_0^\infty \epsilon^{-x} x^{s-1} \, dx = \Gamma(s)$$

with respect to s, and after differentiation put $s = \frac{1}{2}$, we get

$$\int_0^\infty \epsilon^{-x} x^{-\frac{1}{2}} \log x \, dx = \Gamma'\left(\tfrac{1}{2}\right).$$

Putting $x = m\lambda'$ in these equations we get

$$\int_0^\infty \epsilon^{-m\lambda'} \lambda'^{-\frac{1}{2}} d\lambda' = \pi^{\frac{1}{2}} m^{-\frac{1}{2}},$$

$$\int_0^\infty \epsilon^{-m\lambda'} \lambda'^{-\frac{1}{2}} \log \lambda' \, d\lambda' = m^{-\frac{1}{2}} \left\{ \Gamma'\left(\tfrac{1}{2}\right) - \pi^{\frac{1}{2}} \log m \right\},$$

where that value of $m^{-\frac{1}{2}}$ is to be taken which has its real part positive.   Substituting in (91) we get

$$\text{limit of } F_3(r) = \left(\frac{\pi}{2mr}\right)^{\frac{1}{2}} \epsilon^{mr} \left\{ C'' + \left(\pi^{-\frac{1}{2}} \Gamma'\tfrac{1}{2} - \log \frac{m}{2}\right) D'' \right\}.$$

Comparing with (88) we get

$$D' = \left(\frac{\pi}{2m}\right)^{\frac{1}{2}} \left\{ C'' + \left(\pi^{-\frac{1}{2}} \Gamma'\tfrac{1}{2} - \log \frac{m}{2}\right) D'' \right\} \ \dots\dots\dots(92).$$

30.   We are now enabled to find the relation between C and D arising from the condition that the motion of the fluid shall not become infinitely great at an infinite distance from the cylinder. The determination of the arbitrary constants A, B, C, D will present no further difficulty.   We must have $B = 0$, since otherwise the velocity would be finite at an infinite distance, and then the two equations (83), combined with the relation above mentioned, will serve to determine A, C, D. The motion of the fluid will thus be completely determined, the functions $F_1(r)$, $F_3(r)$ being given by (84) and (87).   When the modulus of mr is large, the series in (87), though ultimately hypergeometrically conver-

gent, are at first rapidly divergent, and in calculating the numerical value of $F_3(r)$ in such a case it would be far more convenient to employ equation (88). The employment of this equation for the purpose would require the previous determination of the constant C'. It will be found however that in calculating the resultant pressure of the fluid on the cylinder, which it is the main object of the present investigation to determine, a knowledge of the value of C' will not be required, and that, even though the equation (88) be employed *.

Putting $D' = 0$ in (92), and eliminating C'' and D'' between the resulting equation and the two equations (90), we get

$$C = \left(\log \frac{m}{8} - \pi^{-\frac{1}{2}} \Gamma' \tfrac{1}{2} \right) D \dots \dots \dots (93) ;$$

and we get from (83) and (84), observing that $F_2(r) = F_3'(r)$, and that $B = 0$,

$$\frac{A}{a} + F_3'(a) = ac, \quad -\frac{A}{a} + aF_3''(a) = ac \dots \dots \dots (94),$$

whence

$$\frac{a^2 c + A}{a^2 c - A} = \frac{aF_3''(a)}{F_3'(a)} \dots \dots \dots \dots (95).$$

This equation will determine A, because if $F_3(a)$ be expressed by (87) the second member of (95) will only contain the ratio of C to D, which is given by (93), and if $F_3(a)$ be expressed by (88) C' will disappear, inasmuch as $D' = 0$.

31. Let us now form the expression for the resultant of the forces which the fluid exerts on the cylinder. Let F be the resultant of the pressures acting on a length dl of the cylinder, which will evidently be a force acting in the direction of the axis of x; then we get in the same way as the expression (47) was obtained

$$F = adl \int_0^{2\pi} (- P_r \cos \theta + T_\theta \sin \theta)_a \, d\theta \dots \dots \dots (96),$$

and P_r, T_θ are given in terms of R and Θ by the same formulæ (46) as before. When the right-hand members of these equations are expressed in terms of χ, there will be only one term in which

---

* [C' as subsequently determined will be given at the end of the paper.]

the differentiation with respect to r rises to the second order, and we get from (70), (75), and (76)

$$\frac{d^2\chi}{dr^2} = -\frac{1}{r}\frac{d\chi}{dr} - \frac{1}{r^2}\frac{d^2\chi}{d\theta^2} + \frac{1}{\mu'}\frac{d\chi_2}{dt}.$$

We get from this equation and the equations of condition (79)

$$\left(\frac{dR}{dr}\right)_a = \frac{1}{a}\left(\frac{d\chi}{d\theta}\right)_a - \frac{1}{a^2}\left(\frac{d^2\chi}{drd\theta}\right)_a = 0,$$

$$\left(\frac{dR}{rd\theta}\right)_a = \frac{1}{a^2}\left(\frac{d^2\chi}{d\theta^2}\right)_a = -\frac{\sin\theta}{a}\frac{d\xi}{dt} = \frac{\Theta}{a},$$

$$\left(\frac{d\Theta}{dr}\right)_a = -\left(\frac{d^2\chi}{dr^2}\right)_a = \frac{1}{a}\left(\frac{d\chi}{dr}\right)_a + \frac{1}{a^2}\left(\frac{d^2\chi}{d\theta^2}\right)_a - \frac{1}{\mu'}\left(\frac{d\chi_2}{dt}\right)_a = -\frac{1}{\mu'}\left(\frac{d\chi_2}{dt}\right)_a.$$

Hence

$$F = adl\int_0^{2\pi}\left\{-p_a\cos\theta + \rho\left(\frac{d\chi_2}{dt}\right)_a\sin\theta\right\}d\theta\ldots\ldots\ldots(97).$$

We get by integration by parts

$$\int p_a\cos\theta\,d\theta = p_a\sin\theta - \int\left(\frac{dp}{d\theta}\right)_a\sin\theta\,d\theta.$$

The first term vanishes at both limits; and putting for $dp/d\theta$ its value given by (77), and substituting in (97), we get

$$F = \rho adl\frac{d}{dt}\int_0^{2\pi}\left\{a\left(\frac{d\chi_1}{dr}\right)_a + (\chi_2)_a\right\}\sin\theta\,d\theta,$$

or

$$F = \pi\rho adl.n\sqrt{-1}\,\{aF_1'(a) + F_3'(a)\}\,\epsilon^{\sqrt{-1}nt}.$$

Observing that $F_3'(a)$ or $F_2'(a) = ac - F_1(a)$ from (83), and that $F_1(a) = Aa^{-1}$, where A is given by (95), and putting M' for $\pi\rho a^2dl$, the mass of the fluid displaced, we get

$$F = M'cn\sqrt{-1}\left\{1 - 2\frac{aF_3''(a) - F_3'(a)}{aF_3''(a) + F_3'(a)}\right\}\epsilon^{\sqrt{-1}}{}_{nt},$$

which becomes by means of the differential equation (85) which F_3 satisfies

$$F = -M'cn\sqrt{-1}\left\{1 - \frac{4F_3'(a)}{m^2aF_3(a)}\right\}\epsilon^{\sqrt{-1}nt}\quad\ldots\ldots\ldots(98).$$

Let

$$1 - \frac{4F_3'(a)}{m^2aF_3(a)} = k - \sqrt{-1}\,k'\ldots\ldots\ldots\ldots(99),$$

where k and k' are real, then, as before, $kM' \, d^2\xi/dt^2$ will be the part of F which alters the time of oscillation, and $k'M'n \, d\xi/dt$ the part which produces a diminution in the arc of oscillation.

When μ' vanishes, m becomes infinite, and we get from (88) and (99), remembering that $D' = 0$; $k = 1$, $k' = 0$, a result which follows directly and very simply from the ordinary equations of hydrodynamics*.

32. Every thing is now reduced to the numerical calculation of the quantities k, k', of which the analytical expressions are given. The series (87) being always convergent might be employed in all cases, but when the modulus of ma is large, it will be far more convenient to employ a series according to descending powers of a. Let us consider the ascending series first.

Let $2\mathfrak{m}$ be the modulus of ma; then

$$ma = 2\mathfrak{m}\, \epsilon^{\frac{\pi}{4}\sqrt{-1}}, \quad \mathfrak{m} = \frac{a}{2}\sqrt{\frac{n}{\mu'}} = \frac{a}{2}\sqrt{\frac{\pi}{\mu'\tau}} \quad \dots\dots\dots(100),$$

τ being as before the time of oscillation from rest to rest. Substituting in (99) the above expression for ma, we get

$$k - \sqrt{-1}\, k' = 1 + \frac{\sqrt{-1}\, a F_3'(a)}{\mathfrak{m}^2 F_3(a)} \quad \dots\dots\dots\dots(101).$$

Putting for shortness

$$\log_e 4 + \pi^{-\frac{1}{2}}\, \Gamma'\left(\tfrac{1}{2}\right) = -\Lambda \quad \dots\dots\dots\dots\dots(102),$$

we get from (87) and (93)

$$\frac{1}{D} F_3(a) = \left(\log \mathfrak{m} + \Lambda + \frac{\pi}{4}\sqrt{-1}\right)$$

$$\left(1 + \frac{\mathfrak{m}^2}{1^2}\sqrt{-1} - \frac{\mathfrak{m}^4}{1^2 \cdot 2^2} - \frac{\mathfrak{m}^6}{1^2 \cdot 2^2 \cdot 3^2}\sqrt{-1} + \dots\right)$$

$$- \left(\frac{\mathfrak{m}^2}{1^2} S_1 \sqrt{-1} - \frac{\mathfrak{m}^4}{1^2 \cdot 2^2} S_2 - \frac{\mathfrak{m}^6}{1^2 \cdot 2^2 \cdot 3^2} S_3 \sqrt{-1} + \dots\right),$$

$$\frac{1}{D} a F_3'(a) = 1 + \frac{\mathfrak{m}^2}{1^2}\sqrt{-1} - \frac{\mathfrak{m}^4}{1^2 \cdot 2^2} - \cdots$$

$$+ 2\left(\log \mathfrak{m} + \Lambda + \frac{\pi}{4}\sqrt{-1}\right)\left(\frac{\mathfrak{m}^2}{1}\sqrt{-1} - \frac{\mathfrak{m}^4}{1^2 \cdot 2} - \cdots\right)$$

$$- 2\left(\frac{\mathfrak{m}^2}{1} S_1 \sqrt{-1} - \frac{\mathfrak{m}^4}{1^2 \cdot 2} S_2 - \frac{\mathfrak{m}^6}{1^2 \cdot 2^2 \cdot 3} S_3 \sqrt{-1} + \dots\right).$$

---

* See *Camb. Phil. Trans.* Vol. VIII. p. 116. [*Ante*, Vol. I. p. 37.]

Let

$$\left.\begin{array}{ll}
\dfrac{\mathfrak{m}^2}{1} - \dfrac{\mathfrak{m}^6}{1^2 . 2^2 . 3} + \ldots = M_0, & \dfrac{\mathfrak{m}^4}{1^2 . 2} - \dfrac{\mathfrak{m}^8}{1^2 . 2^2 . 3^2 . 4} + \ldots = M_e \\[3mm]
\dfrac{\mathfrak{m}^2}{1^2} - \dfrac{\mathfrak{m}^6}{1^2 . 2^2 . 3^2} + \ldots = M_0', & \dfrac{\mathfrak{m}^4}{1^2 . 2^2} - \dfrac{\mathfrak{m}^8}{1^2 . 2^2 . 3^2 . 4^2} + \ldots = M_e' \\[3mm]
\dfrac{\mathfrak{m}^2}{1} S_1 - \dfrac{\mathfrak{m}^6}{1^2 . 2^2 . 3} S_2 + \ldots = N_0, & \dfrac{\mathfrak{m}^4}{1^2 . 2} S_2 - \dfrac{\mathfrak{m}^8}{1^2 . 2^2 . 3^2 . 4} S_4 + \ldots = N_e \\[3mm]
\dfrac{\mathfrak{m}^2}{1^2} S_1 - \dfrac{\mathfrak{m}^6}{1^2 . 2^2 . 3^2} S_3 + \ldots = N_0', & \dfrac{\mathfrak{m}^4}{1^2 . 2} S_2 - \dfrac{\mathfrak{m}^8}{1^2 . 2^2 . 3^2 . 4^2} S_4 + \ldots = N_e'
\end{array}\right\}. \quad (103),$$

$$\log_e \mathfrak{m} + \Lambda = L \quad\ldots\ldots\ldots\ldots\ldots\ldots\ldots(104):$$

then substituting in (101), changing the sign of $\sqrt{-1}$, and arranging the terms, we get

$$k + \sqrt{-1}\, k'$$

$$= 1 + \frac{2}{\mathfrak{m}^2} \cdot \frac{-LM_0 + \frac{\pi}{4} M_e - \frac{1}{2} M_0' + N_0 + \left\{\frac{\pi}{4} M_0 + L M_e - \frac{1}{2}(1 - M_e') - N_e\right\}\sqrt{-1}}{-\frac{\pi}{4} M_0' + L(1 - M_e') + N_e' + \left\{- LM_0' - \frac{\pi}{4}(1 - M_e') + N_0'\right\}\sqrt{-1}}$$

$$\ldots\ldots\ldots(105).$$

33. Before going on with the calculation, it will be requisite to know the numerical value of the transcendental quantity Λ. Now

$$\pi^{-\frac{1}{2}} \Gamma'(\tfrac{1}{2}) = (\Gamma\tfrac{1}{2})^{-1} \Gamma'(\tfrac{1}{2}) = \frac{d}{ds} \log \Gamma(s)$$

$$= \frac{d}{ds} \log \Gamma(1+s) - \frac{1}{s}, \text{ for } s = \tfrac{1}{2},$$

and the value of $d/ds . \log \Gamma(1+s)$ may be got at once from Legendre's table of the common logarithms of $\Gamma(1+s)$, in which the interval of s is ·001. Putting l_s for the tabular number corresponding to s, we have

$$\frac{d}{ds} \log \Gamma(1+s) = 1000 \log_e 10 \left\{\Delta l_s - \tfrac{1}{2} \Delta^2 l_s + \tfrac{1}{3} \Delta^3 l_s - \tfrac{1}{4} \Delta^4 l_s + \ldots\right\}.$$

For $s = \tfrac{1}{2}$

$$\Delta l_s = +16050324, \quad \Delta^2 l_s = +405620, \quad \Delta^3 l_s = -359, \quad \Delta^4 l_s = +6*,$$

* These numbers are copied from De Morgan's *Differential and Integral Calculus*, p. 588.

the last figure being in each case in the 12th place of decimals.
We thus get

$$\pi^{-\frac{1}{2}}\Gamma'\left(\tfrac{1}{2}\right)=-1\cdot9635102,\quad \Lambda=+\cdot5772158*\ldots\ldots\ldots(106).$$

34. When $\mathfrak{m}$ is large, it will be more convenient to employ
series according to descending powers of a. Observing that the
general term of $F_3(a)$ as given by (88), in which $D'=0$, is

$$(-1)^i\,C'\,\epsilon^{-ma}\,\frac{[1\,.\,3\ldots(2i-1)]^2}{2\,.\,4\ldots2i\,(4ma)^i\,a^{\frac{1}{2}}},$$

we get for the general term of $F_3'(a)$

$$(-1)^{i-1}\,C'\epsilon^{-ma}\,\frac{[1\,.\,3\ldots(2i-3)]^2}{2\,.\,4\ldots(2i-2)\,(4ma)^{i-1}\,a^{\frac{1}{2}}}\left\{m\,\frac{(2i-1)^2}{2i\,.\,4ma}-\frac{2i-1}{2a}\right\},$$

and the expression within brackets is equivalent to

$$-\frac{(2i-1)\,(2i+1)}{8ia},$$

whence

$$aF_3'(a)=C'\epsilon^{-ma}\,ma^{\frac{1}{2}}\left\{-1-\frac{1\,.\,3}{2\,.\,4ma}+\frac{1^2\,.\,3\,.\,5}{2\,.\,4\,(4ma)^2}-\ldots\right\},$$

and we find by actual division

$$\frac{aF_3'(a)}{F_3(a)}=-ma-\tfrac{1}{2}+\tfrac{1}{8}(ma)^{-1}\ldots.$$

35. When many terms are required, the calculation of the
coefficients may be facilitated in the following manner.

Assuming $aF_3'(a)=v(a)\,F_3(a)$, we have

$$F_3'(a)=a^{-1}v(a)\,F_3(a),$$
$$F_3''(a)=\{a^{-1}v'(a)-a^{-2}v(a)+a^{-2}(va)^2\}\,F_3(a).$$

Substituting in the differential equation (85) which F_3 has to
satisfy, we get

$$av'(a)+\{v(a)\}^2-m^2a^2=0\ldots\ldots\ldots(107).$$

Assuming

$$v(a)=-ma+A_0+A_1(ma)^{-1}+A_2(ma)^{-2}+\ldots\ldots(108),$$

<hr/>

* [Λ is in fact the well-known transcendent called *Euler's Constant*, the value of
which is $\cdot5772156649$ &c. This, which I ought to have known, was pointed out to
me just after the publication of the paper by my friend Prof. F. Newman.]

and substituting in the above equation, we get

$$- ma - 1A_1 (ma)^{-1} - 2A_2 (ma)^{-2} - 3A_3 (ma)^{-3} \ldots$$

$$+ \{- 2ma + A_0 + A_1 (ma)^{-1} + \ldots\} \{A_0 + A_1 (ma)^{-1} + \ldots\} = 0,$$

which gives on equating coefficients, $A_0 = - \frac{1}{2}$, and for $i > 0$

$$2A_{i+1} = - iA_i + A_0A_i + A_1A_{i-1} \ldots + A_iA_0,$$

or, assuming to avoid fractions,

$$A_i = 2^{-2i-1} B_i \ldots\ldots\ldots\ldots\ldots\ldots\ldots(109),$$

$$B_{i+1} = - 2iB_i + B_0B_i + B_1B_{i-1} \ldots + B_iB_0 \ldots\ldots(110),$$

a formula by means of which the coefficients B_1, B_2, $B_3 \ldots$ may be readily calculated one after another. We get

$$\left.\begin{array}{llll} B_0 = -1, & B_1 = +1, & B_2 = -4, & B_3 = +25, \\ B_4 = -208, & B_5 = +2146, & B_6 = -26368, \\ B_7 = +375733, & B_8 = -6092032. \end{array}\right\} \quad (111).$$

We get now from (100), (101), (108), and (109)

$$k - \sqrt{-1}\,k'$$

$$= 1 + 2\epsilon^{-\frac{\pi}{4}\sqrt{-1}}\,\mathfrak{m}^{-1} - \tfrac{1}{2}B_0\epsilon^{-2\frac{\pi}{4}\sqrt{-1}}\,\mathfrak{m}^{-2} - \frac{1}{2^4}B_1\epsilon^{-3\frac{\pi}{4}\sqrt{-1}}\,\mathfrak{m}^{-3} \ldots \quad (112),$$

whence if we calculate

$$u_1 = 2\mathfrak{m}^{-1}, \quad u_2 = - \tfrac{1}{2}B_0\mathfrak{m}^{-2}, \quad u_3 = \tfrac{1}{16}B_1\mathfrak{m}^{-3} \ldots,$$

$$u_i = (-1)^{i+1}\tfrac{1}{2}B_{i-2}\,8^{-i+2}\,\mathfrak{m}^{-i},$$

we shall have, changing the sign of $\sqrt{-1}$ in (112), and writing $\eth$ for $\epsilon^{\frac{\pi}{4}\sqrt{-1}}$,

$$\left.\begin{array}{l} k + \sqrt{-1}\,k' = 1 + u_1\eth + u_2\eth^2 - u_3\eth^3 + u_4\eth^4 - u_5\eth^5 + \ldots \\ k = 1 + \sqrt{\tfrac{1}{2}}u_1 + \sqrt{\tfrac{1}{2}}u_3 - u_4 + \sqrt{\tfrac{1}{2}}u_5 - \sqrt{\tfrac{1}{2}}u_7 + u_8 - \sqrt{\tfrac{1}{2}}u_9 \ldots \\ k' = \sqrt{\tfrac{1}{2}}u_1 + u_2 - \sqrt{\tfrac{1}{2}}u_3 + \sqrt{\tfrac{1}{2}}u_5 - u_6 + \sqrt{\tfrac{1}{2}}u_7 - \sqrt{\tfrac{1}{2}}u_9 \ldots \end{array}\right\} \quad (113).$$

If l_1, $l_2 \ldots$ be the common logarithms of the coefficients of $\mathfrak{m}^{-1}$, $\mathfrak{m}^{-2} \ldots$ in the last two of the formulæ (113),

$$l_1 = \cdot 1505150; \quad l_4 = \bar{2}\cdot 4948500; \quad l_7 = \bar{2}\cdot 3646348;$$

$$l_2 = \bar{1}\cdot 6989700; \quad l_5 = \bar{2}\cdot 2371251; \quad l_8 = \bar{2}\cdot 7019316;$$

$$l_3 = \bar{2}\cdot 6453650; \quad l_6 = \bar{2}\cdot 4046734; \quad l_9 = \bar{2}\cdot 6017045;$$

and if the logarithms of the coefficients of $\mathfrak{m}^{-1}$, $\mathfrak{m}^{-2}$... in u_1, u_2... be required, it will be sufficient to add ·1505150 to the 1st, 3rd, 5th, &c. of the logarithms above given.

36. It will be found that when $\mathfrak{m}$ is at all large, the series (113) are at first convergent, and afterwards divergent, and in passing from convergent to divergent the quantities u_i become nearly equal for several successive terms. If after having calculated i terms of the first of the series (113) we wish to complete the series by a formula involving the differences of u_i, we have

$$u_i 8^i - u_{i+1} 8^{i+1} + u_{i+2} 8^{i+2} - \ldots = 8^i \left\{ 1 - 8(1+\Delta) + 8^2 (1+\Delta)^2 - \ldots \right\} u_i$$

$$= 8^i \left\{ 1 + 8(1+\Delta) \right\}^{-1} u_i$$

$$= \frac{8^i}{1+8} \left\{ 1 - \frac{8}{1+8} \Delta + \left(\frac{8}{1+8} \right)^2 \Delta^2 - \ldots \right\} u_i,$$

and

$$1 + 8 = 1 + \cos \frac{\pi}{4} + \sqrt{-1} \sin \frac{\pi}{4} = 2 \cos \frac{\pi}{8} \epsilon^{\frac{\pi}{8} \sqrt{-1}},$$

$$8 (1 + 8)^{-1} = \tfrac{1}{2} \sec \frac{\pi}{8} \cdot \epsilon^{\frac{\pi}{8} \sqrt{-1}},$$

so that the quantities to be added to k, k', are

$$\left.\begin{array}{l}
\text{to } k, \quad (-1)^i \tfrac{1}{2} \sec \dfrac{\pi}{8} \left\{ \cos \dfrac{2i-1}{8} \pi . u_i - \tfrac{1}{2} \sec \dfrac{\pi}{8} \cos \dfrac{2i}{8} \pi . \Delta u_i \right. \\[2mm]
\qquad\qquad \left. + \left(\tfrac{1}{2} \sec \dfrac{\pi}{8} \right)^2 \cos \dfrac{2i+1}{8} \pi . \Delta^2 u_i \ldots \right\} \\[4mm]
\text{to } k', \quad (-1)^i \tfrac{1}{2} \sec \dfrac{\pi}{8} \left\{ \sin \dfrac{2i-1}{8} \pi . u_i - \tfrac{1}{2} \sec \dfrac{\pi}{8} \sin \dfrac{2i}{8} \pi . \Delta u_i \right. \\[2mm]
\qquad\qquad \left. + \left(\tfrac{1}{2} \sec \dfrac{\pi}{8} \right)^2 \sin \dfrac{2i+1}{8} \pi . \Delta^2 u_i \ldots \right\}
\end{array}\right\} \quad (114).$$

37. The following table contains the values of the functions k and k' calculated for 40 different values of $\mathfrak{m}$. From $\mathfrak{m} = \cdot 1$ to $\mathfrak{m} = 1 \cdot 5$ the calculation was performed by means of the formula (105); the rest of the table was calculated by means of the series (113). In the former part of the calculation, six places of decimals were employed in calculating the functions M_0, &c. given by (103). The last figure was then struck out, and five-figure logarithms were employed in multiplying the four functions M_0,

M'_0, M_e, and $1 - M'_e$ by $\pi/4$, and by L, as well as in reducing the right-hand member of (105) to the form $k + \sqrt{-1}\,k'$. The terms of the series (113) were calculated to five places of decimals. That these series are sufficiently convergent to be employed when $\mathfrak{m} = 1\cdot5$, might be presumed from the numerical values of the terms, and is confirmed by finding that they give $k = 1\cdot952$, and $k' = 1\cdot153$. For $\mathfrak{m} = 1\cdot5$ and a few of the succeeding values, the second and third of the series (113) were summed directly as far as $\mathfrak{m}^{-5}$ inclusively, and the remainders were calculated from the formulæ (114). Two columns are annexed, which give the values of $\mathfrak{m}^2 k$ and $\mathfrak{m}^2 k'$, and exhibit the law of the variation of the two parts of the force F, when the radius of the cylinder varies, the nature of the fluid and time of oscillation remaining unchanged. Four significant figures are retained in all the results.

$\mathfrak{m}$	k	k'	$\mathfrak{m}^2 k$	$\mathfrak{m}^2 k'$	$\mathfrak{m}$	k	k'	$\mathfrak{m}^2 k$	$\mathfrak{m}^2 k'$
0	∞	∞	0	0	2·1	1·677	·7822	7·395	3·450
·1	19·70	48·63	·1970	·4863	2·2	1·646	·7421	7·966	3·592
·2	9·166	16·73	·3666	·6691	2·3	1·618	·7059	8·557	3·734
·3	6·166	9·258	·5549	·8332	2·4	1·592	·6730	9·168	3·877
·4	4·771	6·185	·7633	·9896	2·5	1·568	·6430	9·799	4·019
·5	3·968	4·567	·9920	1·142	2·6	1·546	·6154	10·45	4·160
·6	3·445	3·589	1·240	1·292	2·7	1·526	·5902	11·12	4·303
·7	3·082	2·936	1·510	1·439	2·8	1·507	·5669	11·81	4·444
·8	2·812	2·477	1·800	1·585	2·9	1·489	·5453	12·52	4·586
·9	2·604	2·137	2·110	1·731	3·0	1·473	·5253	13·25	4·728
1·0	2·439	1·876	2·439	1·876	3·1	1·457	·5068	14·01	4·870
1·1	2·306	1·678	2·790	2·021	3·2	1·443	·4895	14·78	5·012
1·2	2·194	1·503	3·160	2·164	3·3	1·430	·4732	15·57	5·154
1·3	2·102	1·365	3·552	2·307	3·4	1·417	·4581	16·38	5·296
1·4	2·021	1·250	3·961	2·450	3·5	1·405	·4439	17·21	5·437
1·5	1·951	1·153	4·389	2·595	3·6	1·394	·4305	18·06	5·580
1·6	1·891	1·069	4·841	2·739	3·7	1·383	·4179	18·93	5·721
1·7	1·838	·9965	5·312	2·880	3·8	1·373	·4060	19·82	5·863
1·8	1·791	·9332	5·804	3·024	3·9	1·363	·3948	20·73	6·005
1·9	1·749	·8767	6·314	3·165	4·0	1·354	·3841	21·67	6·145
2·0	1·711	·8268	6·845	3·307	∞	1	0	∞	∞

The numerical calculation by means of the formulæ (103), (104), (105) becomes very laborious when many values of the functions are required. The difficulty of the calculation increases with the value of $\mathfrak{m}$ for two reasons, first, the calculation of the functions M_0, &c. becomes longer, and secondly, the moduli of the numerator and denominator of the fraction in the right-hand member of (105) go on decreasing, so that greater and greater accuracy is required in the calculation of the functions M_0, &c., and of the products LM_0, &c., in order to ensure a given degree of accuracy in the result. The calculation by the descending series (113) is on the contrary very easy.

It will be found that the first differences of $\mathfrak{m}^2 k'$ and of $\mathfrak{m}^2(k-1)$ are nearly constant, except near the very beginning of the table. Hence in the earlier part of the table the value of k or k' for a value of $\mathfrak{m}$ not found in the table will be best got by finding $\mathfrak{m}^2 k - \mathfrak{m}^2$ or $\mathfrak{m}^2 k'$ by interpolation, and thence passing to the value of k or k'. Very near the beginning of the table, interpolation would not succeed, but in such a case recourse may be had to the formulæ (103), (104), (105), the calculation of which is comparatively easy when $\mathfrak{m}$ is small. It did not seem worth while to extend the table beyond $\mathfrak{m} = 4$, because where $\mathfrak{m}$ is greater than 4, the series (113) are so rapidly convergent that k and k' may be calculated to a sufficient degree of accuracy with extreme facility.

38. Let us now examine the progress of the functions k and k'.

When $\mathfrak{m}$ is very small, we may neglect the powers of $\mathfrak{m}$ in the numerator and denominator of the fraction in the right-hand member of equation (105), retaining only the logarithms and the constant terms. We thus get

$$k + \sqrt{-1}\, k' = 1 - \frac{\mathfrak{m}^{-2} \sqrt{-1}}{L - \frac{1}{4}\pi \sqrt{-1}},$$

whence

$$\mathfrak{m}^2 (k-1) = \frac{\frac{1}{4}\pi}{L^2 + (\frac{1}{4}\pi)^2}, \qquad \mathfrak{m}^2 k' = \frac{-L}{L^2 + (\frac{1}{4}\pi)^2} \ ...(115),$$

L being given by (102) and (104), or (104) and (106). When $\mathfrak{m}$ vanishes, L, which involves the logarithm of $\mathfrak{m}^{-1}$, becomes infinite, but ultimately increases more slowly than if it varied as $\mathfrak{m}$ affected with any negative index however small. Hence it appears from (115), that $k-1$ and k' are expressed by $\mathfrak{m}^{-2}$ multiplied by two functions of $\mathfrak{m}$ which, though they ultimately vanish with $\mathfrak{m}$, decrease very slowly, so that a considerable change in $\mathfrak{m}$ makes but a small change in these functions. Now when the radius a of the cylinder varies, everything else remaining the same, $\mathfrak{m}$ varies as a,

and in general the parts of the force F on which depend the alteration in the time of vibration, and the diminution in the arc of
oscillation, vary as a^2k, a^2k', respectively.   Hence in the case of a
cylinder of small radius, such as the wire used to support a sphere
in a pendulum experiment, a considerable change in the radius of
the cylinder produces a comparatively small change in the part of
the alteration in the time and arc of vibration which is due to the
resistance experienced by the wire.   The simple formulæ (115)
are accurate enough for the fine wires usually employed in such
experiments if the theory itself be applicable; but reasons will
presently be given for regarding the application of the theory to
such fine wires as extremely questionable.

From $\mathfrak{m} = {\cdot}3$ or ${\cdot}4$ to the end of the table, the first differences
of each of the functions $\mathfrak{m}^2(k-1)$ and $\mathfrak{m}^2k'$ remain nearly constant.
Hence for a considerable range of values of $\mathfrak{m}$, each of the functions may be expressed pretty accurately by $A + B\mathfrak{m}$.   When $\mathfrak{m}$ is
at all large, the first two terms in the 2nd and 3rd of the formulæ
(113) will give k and k' with considerable accuracy, because, independently of the decrease of the successive quantities $\mathfrak{m}^{-1}$, $\mathfrak{m}^{-2}$,
$\mathfrak{m}^{-3} \dots$, the coefficients of $\mathfrak{m}^{-1}$ and $\mathfrak{m}^{-2}$ are considerably larger than
those of several of the succeeding powers.   If we neglect in these
formulæ the terms after u_2, we get

$$k = 1 + \sqrt{2} \cdot \mathfrak{m}^{-1}, \quad k' = \sqrt{2} \cdot \mathfrak{m}^{-1} + \tfrac{1}{2}\mathfrak{m}^{-2}.$$

It may be remarked that these approximate expressions, regarded
as functions of the radius a, have precisely the same form as the
exact expressions obtained for a sphere, the coefficients only being
different.

Section IV.

Determination of the motion of a fluid about a sphere which moves uniformly with a small velocity. Justification of the application of the solutions obtained in Sections II. and III. to cases in which the extent of oscillation is not small in comparison with the radius of the sphere or cylinder. Discussion of a difficulty which presents itself with reference to the uniform motion of a cylinder in a fluid.

39. Let a sphere move in a fluid with a uniform velocity V, its centre moving in a right line; and let the rest of the notation be the same as in Section II. Conceive a velocity equal and opposite to that of the sphere impressed both on the sphere and on the fluid, which will not affect the relative motion of the sphere and fluid, and will reduce the determination of the motion of the fluid to a problem of steady motion. Then we have for the equations of condition

$$R = 0, \qquad \Theta = 0, \text{ when } r = a \ldots\ldots\ldots\ldots(116);$$
$$R = -V\cos\theta, \ \ \Theta = V\sin\theta, \text{ when } r = \infty \ \ \ldots\ldots(117).$$

The equations of condition, as well as the equations of motion, may be satisfied by supposing ψ to have the form $\sin^2\theta f(r)$. We get from (20′), by the same process as that by which (33), (34) were obtained,

$$\left(\frac{d^2}{dr^2} - \frac{2}{r^2}\right)^2 f(r) = 0 \ldots\ldots\ldots\ldots\ldots\ldots(118),$$

the only difference being that in the present case the equation (20′) cannot be replaced by the two (22), (23), which become identical, inasmuch as the velocity of the fluid is independent of the time.

The integral of (118) is

$$f(r) = Ar^{-1} + Br + Cr^2 + Dr^4 \ldots\ldots\ldots\ldots(119),$$

which gives

$$R = \frac{1}{r^2\sin\theta}\frac{d\psi}{d\theta} = 2\cos\theta\,(Ar^{-3} + Br^{-1} + C + Dr^2),$$

$$\Theta = -\frac{1}{r\sin\theta}\frac{d\psi}{dr} = \sin\theta\,(Ar^{-3} - Br^{-1} - 2C - 4Dr^2).$$

The first of the equations of condition (117) requires that

$$D = 0, \quad C = -\tfrac{1}{2} V \quad \ldots\ldots\ldots\ldots\ldots(120).$$

It is particularly to be remarked that inasmuch as the two arbitrary constants C, D are determined by the first of the conditions (117), none remain whereby to satisfy the second. Nevertheless it happens that the second of these conditions leads to precisely the same equations (120) as the first. The equations of condition (116) give

$$A = -\tfrac{1}{4} Va^3, \quad B = \tfrac{3}{4} Va ;$$

whence

$$\psi = \tfrac{1}{4} Va^2 \left(-\frac{2r^2}{a^3} + \frac{3r}{a} - \frac{a}{r} \right) \sin^2 \theta * \quad \ldots\ldots(121),$$

$$R = -V \left(1 - \frac{3a}{2r} + \frac{a^3}{2r^3} \right) \cos \theta \quad \ldots\ldots\ldots\ldots(122),$$

$$\Theta = V \left(1 - \frac{3a}{4r} - \frac{a^3}{4r^3} \right) \sin \theta \ldots\ldots\ldots\ldots\ldots(123).$$

If now we wish to obtain the solution of the problem in its original shape, in which the sphere is in motion and the fluid at rest, except so far as it is disturbed by the sphere, we have merely to add $V \cos \theta$, $-V \sin \theta$, $\tfrac{1}{2} Vr^2 \sin^2 \theta$ to the expressions for R, Θ, ψ. We get from (121)

$$\psi = \tfrac{1}{4} Va^2 \left(\frac{3r}{a} - \frac{a}{r} \right) \sin^2 \theta \ldots\ldots\ldots\ldots(124).$$

40. Let us now return to the problem of Section II.; let us suppose the time of oscillation to increase indefinitely, and examine what equation (40) becomes in the limit.

When σ becomes infinite, n, and therefore m, vanishes; the expression within brackets in (40) takes the form $\infty - \infty$, and its limiting value is easily found by the ordinary methods. We must retain the m^2 in the coefficient of t, because t is susceptible of unlimited increase. We get in the limit

$$\psi = \tfrac{1}{4} a^2 c e^{\mu'm^2 t} \left(\frac{3r}{a} - \frac{a}{r} \right) \sin^2 \theta \ldots\ldots\ldots\ldots(125).$$

---

* I have already had occasion, in treating of another subject, to publish the solution expressed by this equation, which I had obtained as a limiting case of the problem of a ball pendulum. See *Philosophical Magazine* for May, 1848, p. 343. [*Ante*, Vol. II. p. 10.]

If now we put V for $d\xi/dt$, the velocity of the sphere, we get from (39), $ce^{\mu'm^2t} = V$. After substituting in (125), the equation will remain unchanged when we pass from the symbolical to the real values of ψ and V, and thus (125) will be reduced to (124).

41. It appears then that by supposing the rate of alteration of the velocity of the sphere to decrease indefinitely, we obtain from the solution of the problem of Section II. the same result as was obtained in Art. 39, by treating the motion as steady. As yet, however, the method of Art. 40 is subject to a limitation from which that of Art. 39 is free. In obtaining equation (40), it was supposed that the maximum excursion of the sphere was small in comparison with its radius. Retaining this restriction while we suppose τ to become very large, we are obliged to suppose c to become very small, so that the velocity of the sphere is not merely so small that we may neglect terms depending upon its square, a restriction to which Art. 39 is also subject, but so extremely small that the space passed over by the sphere in even a long time is small in comparison with its radius.

We have seen, however, that on supposing τ very large in (40) we obtain a result identical with (124), not merely a result with which (124) becomes identical when the restriction above mentioned is introduced. This leads to the supposition that the solution expressed by (40) is in fact more general than would appear from the way in which it was obtained. That such is really the case may be shewn by a slight modification of the analysis. Instead of referring the fluid to axes fixed in space, refer it to axes originating at the centre of the sphere, and moveable with it. In the general equations of motion, the terms which contain differential coefficients taken with respect to the co-ordinates will remain unchanged, inasmuch as they represent the very same limiting ratios as before: it is only those in which differentiation with respect to t occurs that will be altered. If d'/dt be the symbol of differentiation with respect to t when the fluid is referred to the moveable axes, we shall have

$$\frac{d}{dt} = \frac{d'}{dt} - \frac{d\xi}{dt}\frac{d}{dx};$$

but the terms arising from $d\xi/dt \cdot d/dx$ are of the order of the square of the velocity, and are therefore to be neglected. Hence the general equations have the same form whether the fluid be referred to the fixed or moveable axes. But on the latter supposition the equations of condition (30) become rigorously exact. Hence equation (40) gives correctly the solution of the problem, independently of the restriction that the maximum excursion of the sphere be small compared with its radius, provided we suppose the polar co-ordinates r, θ measured from the centre of the sphere in its actual, not its mean position. Similar remarks apply to the problem of the cylinder. Moreover, in the case of a sphere oscillating within a concentric spherical envelope, it is not necessary, in order to employ the solution obtained in Section II., that the maximum excursion of the sphere be small compared with its radius; it is sufficient that it be small compared with the radius of the envelope.

These are points of great importance, because the excursions of an oscillating sphere in a pendulum experiment are not by any means extremely small compared with the radius of the sphere; and in the case of a narrow cylinder, such as the suspending wire, so far from the maximum excursion being small compared with the radius of the cylinder, it is, on the contrary, the radius which is small compared with the maximum excursion.

42.   Let us now return to the case of the uniform motion of a sphere. In order to obtain directly the expression for the resistance of the fluid, it would be requisite first to find p, then to get P_r and T_θ from (46), or at least to get the values of these functions for $r = a$, and lastly to substitute in (47) and perform the integration. We should obtain p by integrating the expression for dp got from (16) and (17). It would be requisite first to express u and q in terms of ψ, then to transform the expression for dp so as to involve polar co-ordinates, and then substitute for ψ its value given by (121); or else to express the right-hand member of (121) by the co-ordinates x, ϖ, and substitute in the expression for dp^*. We

---

* The equations (16), (17) give, after a troublesome transformation to polar co-ordinates,

$$\frac{dp}{dr} = \frac{\mu}{r^2 \sin\theta} \frac{d}{d\theta} \left(\frac{d^2}{dr^2} + \frac{\sin\theta}{r^2} \frac{d}{d\theta} \frac{1}{\sin\theta} \frac{d}{d\theta} - \frac{\rho}{\mu} \frac{d}{dt} \right) \psi \dots\dots\dots\dots(a),$$

$$\frac{dp}{d\theta} = - \frac{\mu}{\sin\theta} \frac{d}{dr} \left(\frac{d^2}{dr^2} + \frac{\sin\theta}{r^2} \frac{d}{d\theta} \frac{1}{\sin\theta} \frac{d}{d\theta} - \frac{\rho}{\mu} \frac{d}{dt} \right) \psi \dots\dots\dots\dots(b).$$

The expression for dp got from these equations is an exact differential by virtue of the equation which determines ψ; and in the problems considered in Section II.

have seen, however, that the results applicable to uniform motion may be deduced as limiting cases of those which relate to oscillatory motion, and consequently, we may make use of the expression for F already worked out. Writing V for $ce^{\sqrt{-1}nt}$ in the first equation of Art. 20, expressing m in terms of n, and then making n vanish, we get

$$- F = 6\pi\mu'\rho a V \dots\dots\dots (126),$$

and $- F$ is the resistance required.

This equation may be employed to determine the terminal velocity of a sphere ascending or descending in a fluid, provided the motion be so slow that the square of the velocity may be neglected. It has been shewn experimentally by Coulomb [*], that in the case of very slow motions, the resistance of a fluid depends partly on the square and partly on the first power of the velocity. The formula (126) determines, in the particular case of a sphere, that part of the whole resistance which depends on the first power of the velocity, even though the part which depends on the square of the velocity be not wholly insensible.

It is particularly to be remarked, that according to the formula (126), the resistance varies not as the surface but as the radius of the sphere, and consequently the quotient of the resistance divided by the mass increases in a higher ratio, as the radius diminishes, than if the resistance varied as the surface. Accordingly, fine powders remain nearly suspended in a fluid of widely different specific gravity.

and in the present Section ψ has the form $\Psi \sin^2\theta$, where Ψ is independent of θ. Hence we get from (b), by integrating partially with respect to θ,

$$p = \mu \cos\theta \frac{d}{dr}\left(\frac{d^2}{dr^2} - \frac{2}{r^2} - \frac{\rho}{\mu}\frac{d}{dt}\right)\Psi \dots\dots (c).$$

It is unnecessary to add an arbitrary function of r, because if $\lambda(r)$ be such a function which we suppose added to the right-hand member of (c), we must determine λ by substituting in (a). The resulting expression for $\lambda'(r)$ cannot contain θ, inasmuch as the expression for dp is an exact differential, but it is composed of terms which all involve $\cos\theta$ as a factor, and therefore we know, without working out, that these terms must destroy one another. Hence $\lambda(r)$ must be constant, or at most be a function of t, which we may suppose included in Π. $\lambda(r)$ will in fact be equal to zero if Π be the equilibrium pressure at the depth at which $\int gdz'$ vanishes.

[*] *Mémoires de l'Institut*, Tom. III. p. 246.

43. When the motion is so slow that the part of the resistance which depends on the square of the velocity may be neglected, we have, supposing V to be the terminal velocity, $-F = \frac{4}{3}\pi g (\sigma - \rho) a^3$, where g is the force of gravity, and σ, which is supposed greater than ρ, the density of the sphere. Substituting in (126) we get

$$V = \frac{2g}{9\mu'} \left(\frac{\sigma}{\rho} - 1 \right) a^2 \dots\dots\dots\dots\dots(127).$$

Let us apply this equation to determine the terminal velocity of a globule of water forming part of a cloud. Putting $g = 386$, $\mu' = (\cdot 116)^2$, an inch being the unit of length, and supposing $\sigma \rho^{-1} - 1 = 1000$, in order to allow a little for the rarity of the air at the height of the cloud, we get $V = 6372 \times 1000a^2$. Thus, for a globule the one thousandth of an inch in diameter, we have $V = 1\cdot593$ inch per second. For a globule the one ten thousandth of an inch in diameter, the terminal velocity would be a hundred times smaller, so as not to amount to the one sixtieth part of an inch per second.

We may form a very good judgment of the magnitude of that part of the resistance which varies as the square of the velocity, and which is the only kind of resistance that could exist if the pressure were equal in all directions, by calculating the numerical value of the resistance according to the common theory, imperfect though it be. It follows from this theory that if h be the height due to the velocity V, the resistance is to the weight as $3\rho h$ to $8\sigma a$. For $V = 1\cdot593$ inch per second, the resistance is not quite the one four hundredth part of the weight; and for a sphere only the one ten thousandth of an inch in diameter, moving with the velocity calculated from the formula (127), the ratio of the resistance to the weight would be ten times as small. The terminal velocities of the globules calculated from the common theory would be $32\cdot07$ and $10\cdot14$ inches per second, instead of only $1\cdot593$ and $\cdot01593$ inch. It appears then that the apparent suspension of the clouds is mainly due to the internal friction of air.

44. The resistance to the globule has here been determined as if the globule were a solid sphere. In strictness, account ought to be taken of the relative motion of the fluid particles forming the globule itself. Although it may readily be imagined that no material change would thus be made in the numerical result, it may be worth while to point out the mode of solution of the problem. Suppose the globule preserved in a strictly spherical shape by capillary attraction, which will very nearly indeed be the case. Conceive a velocity equal and opposite to that of the globule impressed both on the globule and on the surrounding fluid, which will reduce the problem to one of steady motion. Let ψ_1, &c. refer to the fluid forming the globule, and assume $\psi_1 = f_1(r)\sin^2\theta$. Then we get on changing the constants in (119)

$$f_1(r) = A_1 r^{-1} + B_1 r + C_1 r^2 + D_1 r^4.$$

The arbitrary constants, A_1, B_1 vanish by the condition that the velocity shall not become infinite at the centre. There remain the two arbitrary constants C_1, D_1 to be determined, in addition to those which appeared in the former problem. But we have now four instead of two equations of condition which have to be satisfied at the surface of the sphere, which are that

$$R = 0, \quad R_1 = 0, \quad \Theta = \Theta_1, \quad T_\theta = T_{1\theta}, \quad \text{when } r = a \quad \ldots\ldots(128).$$

We shall thus have the same number of arbitrary constants as conditions to be satisfied. Now $T_{1\theta}$ will involve μ_1 as a coefficient, just as T_θ involves $\mu'\rho$ or μ; and μ_1, which refers to water, is much larger than μ, which refers to air, although μ' is larger than μ_1'. Hence the results will be nearly the same as if we had taken $\mu_1 = \infty$, or regarded the sphere as solid.

If, however, instead of a globule of liquid descending in a gas we have a very small bubble ascending in a liquid, we must not treat the bubble as a solid sphere. We may in this case also neglect the motion of the fluid forming the sphere, but we have now arrived at the other extreme case of the general problem, and the two equations of condition which have to be satisfied at the surface of the sphere are that $R = 0$ and $T_\theta = 0$ when $r = a$, instead of $R = 0$ and $\Theta = 0$, when $r = a$.

The equation of condition $T_\theta = 0$ which applies to a bubble, as well as the fourth of equations (128), will not be the true equations, if forces arising from internal friction exist in the superficial film of a fluid which are of a different order of magnitude from those which exist throughout the mass. At the end of the memoir already referred to, Coulomb states that in very slow motions the resistance of bodies not completely immersed in a liquid is much greater than that of bodies wholly immersed, and promises to communicate a second memoir in continuation of the first. This memoir, so far as I can find out, has never appeared. Should the existence of such forces in the superficial film of a liquid be made out, the results deduced from the theory of internal friction will be modified in a manner analogous to that in which the results deduced from the common principles of hydrostatics are modified by capillary attraction. It may be remarked that we have nothing to do with forces of this kind in considering the motion of pendulums in air, or even in considering the oscillations of a sphere in water, except as regards the very minute fraction of the whole effect which relates to the resistance experienced by the suspending wire in the immediate neighbourhood of the free surface.

It may readily be seen that the effect of a set of forces in the superficial film of a liquid offering a peculiar resistance to the relative motion of the particles would be, to make the resistance of a gas to a descending globule agree still more closely with the result obtained by regarding the globule as solid, while the resistance experienced by an ascending bubble would be materially increased, and made to approach to that which would belong to a solid sphere of the same size without mass, or more strictly, with a mass only equal to that of the gas forming the bubble. Possibly the determination of the velocity of ascent of very small bubbles may turn out to be a good mode of measuring the amount of friction in the superficial film of a liquid, if it be true that forces of this kind have any existence. But any investigation relating to such a subject would at present be premature.

45. Let us now attempt to determine the uniform motion of a fluid about an infinite cylinder. Employing the notation of Section III., and reducing the problem to one of steady motion as in Art. 39, we obtain the same equations of condition (116), (117), as in the case of the sphere. Assuming $\chi = \sin \theta \, F(r)$, and substituting in the equation obtained from (69) by transforming to polar coordinates and leaving out the terms which involve d/dt, we get

$$\left(\frac{d^2}{dr^2} + \frac{1}{r} \frac{d}{dr} - \frac{1}{r^2} \right)^2 F(r) = 0 \dots\dots\dots (129).$$

The integral of this equation may readily be obtained by multiplying the last term of the operating factor by $(1 + \delta)^2$, integrating the transformed equation, and then making δ vanish. It is

$$F(r) = A r^{-1} + Br + Cr \log r + Dr^3 \dots\dots (130)$$

which gives

$$R = \frac{d\chi}{r d\theta} = (Ar^{-2} + B + C \log r + Dr^{2}) \cos \theta$$

$$\Theta = -\frac{d\chi}{dr} = (Ar^{-2} - B - C - C \log r - 3Dr^{2}) \sin \theta.$$

The first of the equations of condition (117) requires that

$$C = 0, \quad D = 0, \quad B = -V,$$

which also satisfies the second. We have thus only one arbitrary constant left whereby to satisfy the two equations of condition (116), and the same value of A will not satisfy these two equations.

46. It appears then that the supposition of steady motion is inadmissible. It will be remembered that, in the case of the sphere, the solution of the problem was only possible because it so happened that the values of two arbitrary constants determined by satisfying the first of the equations of condition (117) satisfied also the second, which indicates that the solution was to a certain extent tentative. We have evidently a right to conceive a sphere or infinite cylinder to exist at rest in an infinite mass of fluid also at rest, to suppose the sphere or cylinder to be then moved with a uniform velocity V, and to propose for determination the motion of the fluid at the end of the time t. But we have no right to assume that the motion approaches a permanent state as t increases indefinitely. We may follow either of two courses. We may proceed to solve the general problem in which the sphere or cylinder is supposed to move from rest, and then examine what results we obtain by supposing t to increase indefinitely, or else we may assume *for trial* that the motion is steady, and proceed to inquire whether we can satisfy all the conditions of the problem on this supposition. The former course would have the disadvantage of requiring a complicated analysis for the sake of obtaining a comparatively simple result, and it is even possible that the solution of the problem might baffle us altogether; but if we adopt the latter course, we must not forget that the equations with which we work are only provisional.

It might be objected that the impossibility of satisfying the conditions of the problem on the hypothesis of steady motion arose from our assumption that $\sin \theta$ was a factor of χ, the other factor being independent of θ. This however is not the case. For, for given values of r and t, χ is a finite function of θ from $\theta = 0$ to $\theta = \pi$. We have a right to suppose χ to vanish at any point of the axis of x positive that we please; and if we suppose χ to vanish at one such point, it may be shewn as in the note to Art. 15, that χ will vanish at all points of the axis of x positive or x negative. Hence χ may be expanded in a convergent series of sines of θ and its multiples; and since χ and its derivatives with respect to θ alter continuously with θ, the expansions of the derivatives will be got by direct differentiation*. This being true for all other pairs of values of r and t, χ can in general be expanded in a convergent series of sines of θ and its multiples; but the coefficients, instead of being constant, will be functions of r and t, or in the particular case of steady motion, functions of r alone. Now a very slight examination of the general equations will suffice to shew that the coefficients of the sines of the different multiples of θ remain perfectly independent throughout the whole process, and consequently had we employed the general expansion, we should have been led to the very same conclusions which have been deduced from the assumed form of χ.

47.   If we take the impossibility of the existence of a limiting state of motion, which has just been established, in connexion with the results obtained in Section III., we shall be able to understand the general nature of the motion of the fluid around an infinite cylinder which is at first at rest, and is then moved on indefinitely with a uniform velocity.

The fluid being treated as incompressible, the first motion which takes place is impulsive. Since the terms depending on the internal friction will not appear in the calculation of this motion, we may employ the ordinary équations of hydrodynamics. The result, which is easily obtained, is

$$R dr + \Theta r d\theta = d\phi, \quad \text{where } \phi = - \frac{V a^2}{r} \cos \theta \dagger \quad \ldots\ldots (131).$$

* See a paper "On the Critical Values of the Sums of Periodic Series," *Camb. Phil. Trans.* Vol. VIII. p. 533. [*Ante*, Vol. I. p. 236.]

† According to these equations, the fluid flows past the surface of the cylinder with a finite velocity. At the end of the small time t' after the impact, the friction has reduced the velocity of the fluid in contact with the cylinder to that of the

As the cylinder moves on, it carries more and more of the fluid with it, in consequence of friction. For the sake of precision, let the quantity carried by the element dl of the cylinder be defined to be that which, moving with the velocity V, would have the same momentum in the direction of the motion that is actually possessed by the elementary portion of fluid which is contained between two parallel infinite planes drawn perpendicular to the axis of the cylinder, at an interval dl, the particles composing which are moving with velocities that vary from V to zero in passing from the surface outwards. The pressure of the cylinder on the fluid continually tends to increase the quantity of fluid which it carries with it, while the friction of the fluid at a distance from the cylinder continually tends to diminish it. In the case of a sphere, these two causes eventually counteract each other, and the motion becomes uniform. But in the case of a cylinder, the increase in the quantity of fluid carried continually gains on the decrease due to the friction of the surrounding fluid, and the quantity carried increases indefinitely as the cylinder moves on. The rate at which the quantity carried is increased decreases continually, because the motion of the fluid in the neighbourhood of the cylinder becomes more and more nearly a simple motion of translation equal to that of the cylinder itself, and therefore the rate at which the quantity of fluid carried is increased would become smaller and smaller, even were no resistance offered by the surrounding fluid.

The correctness of this explanation is confirmed by the following considerations. Suppose that $F(r)$ had been given by the equation

$$F(r) = Ar^{-1} + Br + Cr^{1-\delta} + Dr^3$$

instead of (130), δ being a small positive quantity. On this supposition it would have been possible to satisfy all the equations of

cylinder itself, and the tangential velocity alters very rapidly in passing from the surface outwards. At a small distance s from the surface of the cylinder, the relative velocity of the fluid and the cylinder, in a direction tangential to the surface, is a function of the independent variables t', s, which vanishes with s for any given value of t', however small, but which for any given value of s, however small, approaches indefinitely to the quantity determined by (131) as t vanishes. The communication of lateral motion is similar to the communication of temperature when the surface of a body has its temperature instantaneously raised or lowered by a finite quantity.

condition, and therefore steady motion would have been possible. By determining the arbitrary constants, and substituting in χ, we should have obtained

$$\psi = aV\left\{-\frac{\delta}{2-}\frac{a}{r}-\frac{r}{a}+\frac{2}{2-\delta}\left(\frac{r}{a}\right)^{1-\delta}\right\}\sin\theta,$$

$$R = V\left\{-\frac{\delta}{2-\delta}\left(\frac{a}{r}\right)^{2}-1+\frac{2}{2-\delta}\left(\frac{r}{a}\right)^{-\delta}\right\}\cos\theta,$$

$$\Theta = V\left\{-\frac{\delta}{2-\delta}\left(\frac{a}{r}\right)^{2}+1-\frac{2(1-\delta)}{2-\delta}\left(\frac{r}{a}\right)^{-\delta}\right\}\sin\theta.$$

Since δ is supposed to be extremely small, it follows from these expressions that when r is not greater than a moderate multiple of a, the velocities R, Θ are extremely small; but, however small be δ, we have only to go far enough from the cylinder in order to find velocities as nearly equal to $-V\cos\theta$, $+V\sin\theta$ as we please. But the distance from the cylinder to which we must proceed in order to find velocities R, Θ which do not differ from their limiting values $-V\cos\theta$, $+V\sin\theta$ by more than certain given quantities, increases indefinitely as δ decreases. Hence, restoring to the fluid and the cylinder the velocity V, we see that in the neighbourhood of the cylinder the motion of the fluid does not sensibly differ from a motion of translation, the same as that of the cylinder itself, while the distance to which the cylinder exerts a sensible influence in disturbing the motion of the fluid increases indefinitely as δ decreases.

48. When we have formed the equations of motion of a fluid on any particular dynamical hypothesis, it becomes a perfectly definite mathematical problem to determine the motion of the fluid when a given solid, initially at rest as well as the fluid, is moved in a given manner, or to discuss the character of the analytical solution in any extreme case proposed. It is quite another thing to enquire how far the principles which furnished the mathematical data of the problem may be modified in extreme cases, or what will be the nature of the actual motion in such cases. Let us regard in this point of view the case considered in the preceding article as a mathematical problem. When the quantity of fluid carried with the cylinder· becomes considerable compared with the quantity displaced, it would seem that the motion must become unstable, in the sense in which the motion of a sphere

rolling down the highest generating line of an inclined cylinder may be said to be unstable. But besides the instability, it may not be safe in such an extreme case to neglect the terms depending on the square of the velocity, not that they become unusually large in themselves, but only unusually large compared with the terms retained, because when the relative motions of neighbouring portions of the fluid become very small, the tangential pressures which arise from friction become very small likewise.

Now the general character of the motion must be nearly the same whether the velocity of the cylinder be constant, or vary slowly with the time, so that it does not vary materially when the cylinder passes through a space equal to a small multiple of its radius. To return to the problem considered in Section III., it would seem that when the radius of the cylinder is very small, the motion which would be expressed by the formulæ of that Section would be unstable. This might very well be the case with the fine wires used in supporting the spheres employed in pendulum experiments. If so, the quantity of fluid carried by the wire would be diminished, portions being continually left behind and forming eddies. The resistance to the wire would on the whole be increased, and would moreover approximate to a resistance which would be a function of the velocity. Hence, so far as depends on the wire, the arc of oscillation would be more affected by the resistance of the air than would follow from the formulæ of Section III. Whether the effect on the time of oscillation would be greater or less than that expressed by the formulæ is difficult to say, because the increase of resistance would tend to increase the effect on the time of vibration, while on the other hand the approximation of the law of resistance to that of a function of the velocity would tend to diminish it.

SECTION V.

On the effect of internal friction in causing the motion of a fluid to subside. Application to the case of oscillatory waves.

49. We have already had instances of the effect of friction in causing a gradual subsidence in the motion of a solid oscillating in a fluid; but a result may easily be obtained from the equations of

motion in their most general shape, which shews very clearly the effect of friction in continually consuming a portion of the work of the forces acting on the fluid.

Let P_1, P_2, P_3 be the three normal, and T_1, T_2, T_3 the three tangential pressures in the direction of three rectangular planes parallel to the co-ordinate planes, and let D be the symbol of differentiation with respect to t when the particle and not the point of space remains the same. Then the general equations applicable to a heterogeneous fluid (the equations (10) of my former paper) are

$$\rho \left(\frac{Du}{Dt} - X \right) + \frac{dP_1}{dx} + \frac{dT_3}{dy} + \frac{dT_2}{dz} = 0 \ldots\ldots\ldots(132),$$

with the two other equations which may be written down from symmetry. The pressures P_1, &c. are given by the equations

$$P_1 = p - 2\mu \left(\frac{du}{dx} - \delta \right), \qquad T_1 = -\mu \left(\frac{dv}{dz} + \frac{dw}{dy} \right) \ldots(133)$$

and four other similar equations. In these equations

$$3\delta = \frac{du}{dx} + \frac{dv}{dy} + \frac{dw}{dz} \ldots\ldots\ldots\ldots\ldots(134).$$

At the end of the time t let V be the *vis viva* of a limited portion of the fluid, occupying the space which lies inside the closed surface S, and let $V + DV$ be the *vis viva* of the same mass at the end of the time $t + Dt$. Then

$$V = \iiint \rho \left(u^2 + v^2 + w^2 \right) dx\, dy\, dz,$$

$$DV = 2Dt \iiint \rho \left(u \frac{Du}{Dt} + v \frac{Dv}{Dt} + w \frac{Dw}{Dt} \right) dx\, dy\, dz \ldots..(135),$$

the triple integrals extending throughout the space bounded by S. Substituting now for $\frac{Du}{Dt}$, &c. their values given by the equations of the system (132), we get

$$DV = 2Dt \iiint \rho \left(uX + vY + wZ \right) dx\, dy\, dz$$

$$- 2Dt \iiint \left\{ u \left(\frac{dP_1}{dx} + \frac{dT_3}{dy} + \frac{dT_2}{dz} \right) + v \left(\frac{dP_2}{dy} + \frac{dT_1}{dz} + \frac{dT_3}{dx} \right) \right.$$

$$\left. + w \left(\frac{dP_3}{dz} + \frac{dT_2}{dx} + \frac{dT_1}{dy} \right) \right\} dx\, dy\, dz \ldots\ldots\ldots(136).$$

The first part of this expression is evidently twice the work, during the time Dt, of the external forces which act all over the mass. The second part becomes after integration by parts

$$- 2Dt \iint (uP_1 + vT_3 + wT_2)\, dy\, dz - 2Dt \iint (vP_2 + wT_1 + uT_3)\, dz\, dx$$
$$- 2Dt \iint (wP_3 + uT_2 + vT_1)\, dx\, dy$$
$$+ 2Dt \iiint \left\{ \frac{du}{dx} P_1 + \frac{dv}{dy} P_2 + \frac{dw}{dz} P_3 + \left(\frac{dv}{dz} + \frac{dw}{dy} \right) T_1 + \left(\frac{dw}{dx} + \frac{du}{dz} \right) T_2 \right.$$
$$\left. + \left(\frac{du}{dy} + \frac{dv}{dx} \right) T_3 \right\} dx\, dy\, dz.$$

The double integrals in this expression are to be extended over the whole surface S. If dS be an element of this surface, l', m', n' the direction-cosines of the normal drawn outwards at dS, we may write $l'dS$, $m'dS$, $n'dS$ for $dy\, dz$, $dz\, dx$, $dx\, dy$. The second part of DV thus becomes

$$- 2Dt \iint \{ u\, (l'P_1 + m'T_3 + n'T_2) + v\, (m'P_2 + n'T_1 + l'T_3)$$
$$+ w\, (n'P_3 + l'T_2 + m'T_1) \}\, dS.$$

The coefficients of u, v, w in this expression are the resolved parts, in the direction of x, y, z, of the pressure on a plane in the direction of the elementary surface dS, whence it appears that the expression itself denotes twice the work of the pressures applied to the surface of the portion of fluid that we are considering.

On substituting for P_1, &c. their values given by the equations (133), (134), we get for the last part of DV

$$+ 2Dt \iiint p \left(\frac{du}{dx} + \frac{dv}{dy} + \frac{dw}{dz} \right) dx\, dy\, dz$$
$$- 2Dt \iiint \mu \left\{ 2 \left(\frac{du}{dx} \right)^2 + 2 \left(\frac{dv}{dy} \right)^2 + 2 \left(\frac{dw}{dz} \right)^2 - \tfrac{2}{3} \left(\frac{du}{dx} + \frac{dv}{dy} + \frac{dw}{dz} \right)^2 \right.$$
$$\left. + \left(\frac{dv}{dz} + \frac{dw}{dy} \right)^2 + \left(\frac{dw}{dx} + \frac{du}{dz} \right)^2 + \left(\frac{du}{dy} + \frac{dv}{dx} \right)^2 \right\} dx\, dy\, dz.$$

In this expression p denotes, in the case of an elastic fluid, the pressure statically corresponding to the density which actually exists about the point whose co-ordinates are x, y, z, and the part of the expression which contains p denotes twice the work converted into *vis viva* in consequence of internal expansions, and arising from the forces on which the elasticity depends. The last part of the expression is essentially negative, or at least cannot be positive, and can only vanish in one very particular case. It denotes the *vis viva* consumed, or twice the work lost in the system during the time dt, in consequence of internal friction. According to the very important theory of Mr Joule, which is founded on a set of most striking and satisfactory experiments, the work thus apparently lost is in fact converted into heat, at such a rate, that the work expressed by the descent of 772 lbs through one foot, supplies the quantity of heat required to raise 1 lb. of water through 1° of Fahrenheit's thermometer.

50.   The triple integral containing μ can only vanish when the differential coefficients of u, v, w satisfy the five following equations,

$$\left. \begin{array}{c} \dfrac{du}{dx} = \dfrac{dv}{dy} = \dfrac{dw}{dz}\,, \\[2mm] \dfrac{dv}{dz} + \dfrac{dw}{dy} = 0, \quad \dfrac{dw}{dx} + \dfrac{du}{dz} = 0, \quad \dfrac{du}{dy} + \dfrac{dv}{dx} = 0 \end{array} \right\} \dots(137).$$

These equations give immediately the following expressions for the differentials of u, v, w, in which the co-ordinates alone are supposed to vary, the time being constant :

$$\left. \begin{array}{l} du = \delta dx - \omega''' dy + \omega'' dz \\ dv = \delta dy - \omega' dz + \omega''' dx \\ dw = \delta dz - \omega'' dx + \omega' dy \end{array} \right\} \dots\dots\dots\dots(138).$$

In these equations δ, ω', ω'', ω''' are certain functions of which the forms are defined by the equations (138), but need not at present be considered. It follows from equations (138) that the motion of each element of the fluid within the surface S is compounded of a motion of translation, a motion of rotation, and a motion of dilatation alike in all directions. So far as regards the first two kinds of motion, the fluid element moves like a solid, and of course there is nothing to call internal friction into play. For the reasons stated in my former paper, I was led to assume that a motion of dilatation alike in all directions (which of course can only exist in the case of an elastic fluid) has no effect in causing the pressure to differ from the statical pressure corresponding to the actual density, that is, in occasioning a violation of the functional relation commonly supposed to exist between the pressure, density, and temperature. The reader will observe that this is a totally different thing from assuming that a motion of dilatation has no effect on the pressure at all.

When the fluid is incompressible $\delta = 0$, and it may be proved without difficulty that ω', ω'', ω''' are constant, that is to say, constant so far as the co-ordinates are concerned. In this case we get by integrating equations (137)

$$\left. \begin{aligned} u &= a - \omega''' y + \omega'' z \\ v &= b - \omega' z + \omega''' x \\ w &= c - \omega'' x + \omega' y \end{aligned} \right\} \quad \ldots\ldots\ldots\ldots\ldots(139).$$

Hence, in the case of an incompressible fluid, unless the whole mass comprised within the surface S move together like a solid, there cannot fail to be a certain portion of *vis viva* lost by internal friction. In the case of an elastic fluid, the motion which may take place without causing a loss of *vis viva* in consequence of friction is somewhat more general, and corresponds to velocities $u + \Delta u$, $v + \Delta v$, $w + \Delta w$, where u, v, w are the same as in (139), and

$$\Delta u = \delta x + 2 (\alpha x + \beta y + \gamma z) x - \alpha (x^2 + y^2 + z^2),$$

with similar expressions for Δv and Δw. In these expressions $\alpha,\, \beta,\, \gamma$ are three constants symmetrically related to x, y, z, and δ is a constant which has the same relation to each of the co-ordinates*.

51. By means of the expression given in Art. 49, for the loss of *vis viva* due to internal friction, we may readily obtain a very approximate solution of the problem: To determine the rate at which the motion subsides, in consequence of internal friction, in the case of a series of oscillatory waves propagated along the surface of a liquid.

Let the vertical plane of xy be parallel to the plane of motion, and let y be measured vertically downwards from the mean surface; and for simplicity's sake suppose the depth of the fluid very great compared with the length of a wave, and the motion so small that the square of the velocity may be neglected. In the case of motion which we are considering, $u\,dx + v\,dy$ is an exact differential $d\phi$ when friction is neglected, and

$$\phi = c\epsilon^{-my} \sin (mx - nt) \ldots\ldots\ldots\ldots\ldots(140),$$

* See note C. at the end.

where c, m, n are three constants, of which the last two are connected by a relation which it is not necessary to write down. We may continue to employ this equation as a near approximation when friction is taken into account, provided we suppose c, instead of being constant, to be a parameter which varies slowly with the time.   Let V be the *vis viva* of a given portion of the fluid at the end of the time t, then

$$V = \rho c^2 m^2 \iiint \epsilon^{-2my} \, dx \, dy \, dz \ \ldots\ldots\ldots\ldots(141).$$

But by means of the expression given in Art. 49, we get for the loss of *vis viva* * during the time dt, observing that in the present case μ is constant, $w = 0$, $\delta = 0$, and $udx + vdy = d\phi$, where ϕ is independent of z,

$$4\mu \, dt \iiint \left\{ \left(\frac{d^2\phi}{dx^2}\right)^2 + \left(\frac{d^2\phi}{dy^2}\right)^2 + 2\left(\frac{d^2\phi}{dx\,dy}\right)^2 \right\} dx \, dy \, dz,$$

which becomes, on substituting for ϕ its value,

$$8\mu c^2 m^4 \, dt \iiint \epsilon^{-2my} \, dx \, dy \, dz.$$

But we get from (141) for the decrement of *vis viva* of the same mass arising from the variation of the parameter c

$$- 2\rho m^2 c \frac{dc}{dt} \, dt \iiint \epsilon^{-2my} \, dx \, dy \, dz.$$

Equating the two expressions for the decrement of *vis viva*, putting for m its value $2\pi\lambda^{-1}$, where λ is the length of a wave, replacing μ by $\mu'\rho$, integrating, and supposing c_0 to be the initial value of c, we get

$$c = c_0 \epsilon^{-\frac{16\pi^2\mu't}{\lambda^2}}.$$

It will presently appear that the value of $\sqrt{\mu'}$ for water is about 0·0564, an inch and a second being the units of space and

---

* [There is an oversight here, which M. Boussinesq has pointed out (*Mémoires des Savans Étrangers*, Tome xxiv. No. 2, p. 34). I should have said "the loss of *energy*." Now in a series of waves of small disturbance the total energy is half kinetic and half potential. Hence the reduction of energy consequent upon a reduction in the amplitude falls half on the kinetic and half on the potential energy. Hence the reduction of the *kinetic* energy or *vis viva* is only half of that given by the formula in the text, and therefore the expression for dc/dt is twice what it ought to be. Hence the numerical coefficient in the index of the exponential should be 8 instead of 16; and retaining the same numerical data as in the examples, we should have for the ripples $c : c_0 :: 1 : 0\cdot5337$, and the height of the long waves would be reduced in a day by little more than the one four-hundredth part.]

time. Suppose first that λ is two inches, and t ten seconds. Then $16\pi^2\mu't\lambda^{-2} = 1\cdot256$, and $c : c_0 :: 1 : 0\cdot2848$, so that the height of the waves, which varies as c, is only about a quarter of what it was. Accordingly, the ripples excited on a small pool by a puff of wind rapidly subside when the exciting cause ceases to act.

Now suppose that λ is 40 fathoms or 2880 inches, and that t is 86400 seconds or a whole day. In this case $16\pi^2\mu't\lambda^{-2}$ is equal to only $0\cdot005232$, so that by the end of an entire day, in which time waves of this length would travel 574 English miles, the height would be diminished by little more than the one two-hundredth part in consequence of friction. Accordingly, the long swells of the ocean are but little allayed by friction, and at last break on some shore situated at the distance of perhaps hundreds of miles from the region where they were first excited.

52. It is worthy of remark, that in the case of a homogeneous incompressible fluid, whenever $u\,dx + v\,dy + w\,dz$ is an exact differential, not only are the ordinary equations of fluid motion satisfied*, but the equations obtained when friction is taken into account are satisfied likewise. It is only the equations of condition which belong to the boundaries of the fluid that are violated. Hence any kind of motion which is possible according to the ordinary equations, and which is such that $u\,dx + v\,dy + w\,dz$ is an exact differential, is possible likewise when friction is taken into account, provided we suppose a certain system of normal and tangential pressures to act at the boundaries of the fluid, so as to satisfy the equations of condition. The requisite system of pressures is given by the system of equations (133). Since μ disappears from the general equations (1), it follows that p is the same function as before. But in the first case the system of pressures at the surface was $P_1 = P_2 = P_3 = p$, $T_1 = T_2 = T_3 = 0$. Hence if ΔP_1 &c. be the additional pressures arising from friction, we get from (133), observing that $\delta = 0$, and that $u\,dx + v\,dy + w\,dz$ is an exact differential $d\phi$,

$$\Delta P_1 = -2\mu\,\frac{d^2\phi}{dx^2}, \qquad \Delta P_2 = -2\mu\,\frac{d^2\phi}{dy^2},$$

$$\Delta P_3 = -2\mu\,\frac{d^2\phi}{dz^2} \quad\dots\dots\dots\dots\dots\dots(142),$$

---

* It is here supposed that the forces X, Y, Z are such that $X\,dx + Y\,dy + Z\,dz$ is an exact differential.

$$\Delta T_1 = - 2\mu \frac{d^2\phi}{dy\,dz}, \qquad \Delta T_2 = - 2\mu \frac{d^2\phi}{dz\,dx},$$

$$\Delta T_3 = - 2\mu \frac{d^2\phi}{dx\,dy} \quad\ldots\ldots\ldots\ldots\ldots(143).$$

Let dS be an element of the bounding surface, l', m', n' the direction-cosines of the normal drawn outwards, ΔP, ΔQ, ΔR the components in the direction of x, y, z of the additional pressure on a plane in the direction of dS. Then by the formulæ (9) of my former paper applied to the equations (142), (143) we get

$$\Delta P = - 2\mu \left\{ l' \frac{d^2\phi}{dx^2} + m' \frac{d^2\phi}{dx\,dy} + n' \frac{d^2\phi}{dx\,dz} \right\} \quad \ldots(144),$$

with similar expressions for ΔQ and ΔR, and ΔP, ΔQ, ΔR are the components of the pressure which must be applied at the surface, in order to preserve the original motion unaltered by friction.

53. Let us apply this method to the case of oscillatory waves, considered in Art. 51. In this case the bounding surface is nearly horizontal, and its vertical ordinates are very small, and since the squares of small quantities are neglected, we may suppose the surface to coincide with the plane of xz in calculating the system of pressures which must be supplied, in order to keep up the motion. Moreover, since the motion is symmetrical with respect to the plane of xy, there will be no tangential pressure in the direction of z, so that the only pressures we have to calculate are ΔP_2 and ΔT_3. We get from (140), (142), and (143), putting $y = 0$ after differentiation,

$$\Delta P_2 = - 2\mu m^2 c \sin\,(mx - nt), \quad \Delta T_3 = 2\mu m^2 c \cos\,(mx - nt)\ldots(145).$$

If u_1, v_1 be the velocities of the surface, we get from (140), putting $y = 0$ after differentiation,

$$u_1 = mc \cos\,(mx - nt), \quad v_1 = - mc \sin\,(mx - nt)\ldots(146).$$

It appears from (145) and (146) that the oblique pressure which must be supplied at the surface in order to keep up the motion is constant in magnitude, and always acts in the direction in which the particles are moving.

The work of this pressure during the time dt corresponding to the element of surface $dx\,dz$, is equal to

$$dx\,dz(\Delta T_3 \,.\, u_1 dt + \Delta P_1 \,.\, v_1 dt).$$

Hence the work exerted over a given portion of the surface is equal to

$$2\mu m^3 c^2 dt \iint dx\,dz.$$

In the absence of pressures ΔP_2, ΔT_3 at the surface, this work must be supplied at the expense of *vis viva*. Hence $4\mu m^3 c^2 dt \iint dx\,dz$ is the *vis viva* lost by friction, which agrees with the expression obtained in Art. 51, as will be seen on performing in the latter the integration with respect to y, the limits being $y = 0$ to $y = \infty$.

PART II.

COMPARISON OF THEORY AND EXPERIMENT.

SECTION I.

Discussion of the Experiments of Baily, Bessel, Coulomb, and Dubuat*.

54. THE experiments discussed in this Section will be taken in the order which is most convenient for discussion, which happens to be almost exactly the reverse of the chronological order. I commence with the experiments of the late Mr Baily, which are described in the *Philosophical Transactions* for 1832, in a memoir entitled "On the Correction of a Pendulum for the Reduction to a Vacuum: together with Remarks on some anomalies observed in Pendulum experiments."

---

* [At the time when this paper was read, the relation between μ and ρ cannot be said to have been known. It is true that it may be inferred (at least for air, and thence presumably for other gases) from certain of Graham's experiments on the transpiration of gases. These however had been but recently published, having appeared in the *Philosophical Transactions* for 1846; and it was not till many years afterwards, about 1859, that Maxwell first inferred from the kinetic theory of gases the law that bears his name, namely that the coefficient of viscosity μ is independent of the density.

In the comparison of theory and experiment as regards the effect of the presence of air on the motion of pendulums, I relied mainly on the experiments of Baily, which were made by a direct method, while at the same time they were conducted with all the accuracy of modern physical research, and embraced a great variety of forms of pendulum, many of them such as to admit of comparison with theory.

These experiments were strictly differential, giving the difference between the time of vibration at atmospheric pressure and in rarefied air. Had the vacuum been absolutely perfect, the difference would have given at once the effect of air at the atmospheric pressure. Had it merely been very high, the effect of the residual air on the time of vibration would have been insensible, and the result as regards the time would still have been the same. It is true that the *whole* effect of the rarefied air would not thus disappear; as a result of Maxwell's law it would tend, as the exhaustion proceeded, to fall wholly on the arc of vibration, and to approach a finite limit; and this limit would not begin to break down till an exhaustion was reached comparable with the highest we have to deal with in radiometer vacua.

But in Baily's experiments no high exhaustions were aimed at; the air was merely pumped out till the pressure was reduced to about one inch of mercury, and

The object of these experiments was, to determine by actual observation the correction to the time of vibration due to the presence of the air in the case of a great number of pendulums of various forms. This was effected by placing each pendulum in succession in a vacuum apparatus, by which means the pendulum, without being dismounted, could be swung alternately under the full atmospheric pressure, and in air so highly rarefied as nearly to approach to a vacuum. The paper, as originally presented to the Royal Society, contained the results obtained with 41 pendulums, the same body being counted as a different pendulum when swung in a different manner. Out of these, 14 are of such forms as to admit of comparison with theory. An addition to the paper contains the results obtained with 45 pendulums more, of which 24 admit of comparison with theory. The details of these additional experiments are omitted, the results only being given.

Baily has exhibited the results obtained with the several pendulums in each of two ways, first, by the value of the factor n by which the correction for buoyancy must be multiplied in order to amount to the whole effect of the air as given by observation, and, secondly, by the weight of air which must be conceived to be

the effect of the air was supposed to be arrived at by increasing the observed difference in the times of vibration in the ratio of the difference of densities to the atmospheric density. As the effect of the air at the lower density was too large to be neglected, it was necessary, in order to compare with sufficient accuracy the results of experiment with the formulæ of this paper, to know the relation between μ and ρ. As already mentioned, I assumed in accordance with what appeared to be indicated by a single experiment of Sabine's that μ varies as ρ, or in other words that μ' is independent of the density. The results of the experiments when thus reduced seemed to indicate a most remarkable accord with theory.

When it became known that the law of nature is that μ and not μ' is independent of ρ, it seemed very strange that the experiments when reduced on the assumption of a wrong law as to the relation of μ to ρ should have led to such a remarkable agreement with theory. I contemplated at one time undertaking the re-computation of the whole series of Baily's experiments here discussed in accordance with Maxwell's law, and it was this that delayed the reprinting of the present paper. The value of the result at the present time would however hardly repay the labour of the calculation, more especially as the remarkable agreement between theory and observation notwithstanding the employment of a wrong law as to the relation between μ and ρ admits of being readily explained, and the value of μ obtained as in the text of being very approximately corrected, in a very simple manner. As however this would be too long for a footnote, I must reserve it for an addition to be made at the end of the paper.]

attached to the centre of gyration of the pendulum, adding to its inertia without adding to its weight, in order that the increased inertia, combined with the buoyancy of the air, may account for the whole effect observed. I shall uniformly write $\mathfrak{n}$ for Baily's n, in order to distinguish it from the n of Part I. of the present paper, which has a totally different meaning. In the case of a pendulum oscillating in air, it will be sufficient, unless the pendulum be composed of extremely light materials, to add together the effects of buoyancy and inertia. Hence if the pendulum consist of a sphere attached to a fine wire of which the effect is neglected, or else of a uniform cylindrical rod, we may suppose $\mathfrak{n} = 1 + k$, where k is the factor so denoted in Part I.; so that if M' be the mass of air displaced, kM' will be the mass which we must suppose collected at the centre of the sphere, or distributed uniformly along the axis of the cylinder, in order to express the effect of the inertia of the air. The second mode of exhibiting the effect of the air was suggested by Mr Airy, and is better adapted than the former for investigating the effect of the several pieces of which a pendulum of complicated form is composed. Since the value of the factor $\mathfrak{n}$ and that of the weight of air are merely two different expressions for the result of the same experiment, it would be sufficient to compare either with the result calculated from theory. In some cases, however, I have computed both. In almost all the calculations I have employed 4-figure logarithms. The experimental result is sometimes exhibited to four figures, but no reliance can be placed on the last. In fact, in the best observations, the mean error in different determinations of $\mathfrak{n}$ for the same pendulum appears to have been about the one-hundredth part of the whole, and that it should be so small, is a proof of the extreme care with which the experiments must have been performed.

55. I commence with the 13th set of experiments—*Results with plain cylindrical rods*—page 441. This set contains three pendulums, each consisting of a long rod attached to a knife-edge apparatus. The result obtained with each pendulum furnishes an equation for the determination of μ', and the theory is to be tested by the accordance or discordance of the values so obtained. The principal steps of the calculation are contained in the following table.

Determination of $\sqrt{\mu'}$ *by means of* Baily's *experiments with plain cylindrical rods.*

Pendulum rod	No.	Diameter 2 a	Time of vibration τ	n by experiment
Copper, 58·8 inches long	21	0·410	1·0136	2·932
Brass, 56·4	43	0·185	0·9933	4·083
Steel, 56·4	44	0·072	0·9933	7·530

No.	Correction for confined space (by theory)	Deduced value of k by experiment	Corresponding value of m	Resulting value of $\sqrt{\mu'}$
21	−0·009	1·923	1·5445	0·1166
43	−0·002	3·081	0·7000	0·1175
44		6·530	0·2822	0·1134

In this table the first column explains itself. The next contains the reference number. In the case of the copper rod I have replaced 42 by 21, under which number the details of the experiment will be found. The diameters of the rods are expressed in decimals of an inch. The time of vibration of the pendulum No. 21 may be got from the tables at the end of Baily's memoir, which contain the details of the experiments. Nos. 43 and 44 belong to the "additional experiments," of which all the details are suppressed. Baily has not even given the times of vibration, not having been aware of the circumstance, indicated by the theory of this paper, that the factor n and the weight of air which must be conceived as dragged by the pendulum are functions of the time of vibration. Accordingly, in the cases of the pendulums Nos. 43 and 44, and in all similar cases, I have calculated the time of vibration by the ordinary formulæ of dynamics. In calculating τ, I have added 1·55 inch, the length of the shank of the knife-edge apparatus, to the length of the rods. The result so obtained is abundantly accurate enough for my purpose. Had the rod, retaining its actual length, been supposed to begin directly at the knife-edge, the error thence resulting in the value of τ, or rather the corresponding error in the calculated value of $\mathfrak{n}$ or k, might just have been sensible. The fifth column in the above table is copied from Baily's table. The next contains a small correction necessary to reduce the value of $\mathfrak{n}$ got from observation to what would have been got from observations made in an unlimited mass of fluid. It is calculated from the formula $2a^2 (b^2 - a^2)^{-1}$ or $2a^2 b^{-2}$ nearly, which is obtained from the ordinary equations of hydrodynamics, and therefore it cannot be regarded as more than a rude approximation. It will be useful, however, as affording an estimate of the magnitude of the effect produced by confining the air. The diameter of the vacuum tube (whether external or internal is not specified) is stated to have been six inches and a half, whence $2b = 6 \cdot 5$. The values of k given in the next column are obtained by applying the correction for confined space to Baily's values of $\mathfrak{n}$, and subtracting unity. The value of $\mathfrak{m}$ corresponding to each value of k was got by interpolation from the table near the end of Section III. of the former part of this paper. For $k = 1 \cdot 923$ the interpolation is easy. The value $3 \cdot 081$ happens to be almost exactly found in the table. For $k = 6 \cdot 530$, a remark already made will be found to be of importance, namely, that the first differences of $\mathfrak{m}^2 (k - 1)$ are nearly constant. The last column contains the value of $\sqrt{\mu'}$ obtained from the equation

$$\mathfrak{m} = \frac{a}{2} \sqrt{\frac{\pi}{\mu' \tau}} \quad \ldots\ldots\ldots\ldots\ldots\ldots (147),$$

which contains the definition of $\mathfrak{m}$.

It will be observed that the three values of $\sqrt{\mu'}$ are nearly identical. Of course any theory professing to account for a set of experiments by means of a particular value of a disposable constant, when applied to the experiments would lead to nearly the same numerical value of the constant if the experiments were made under nearly the same circumstances. But in the present case the circumstances of the experiments are widely different. The diameter of the steel rod is little more than the sixth part of that of the copper rod, and the value of k obtained by experiment for the steel rod is more than three times as great as that obtained for the copper rod. It is a simple consequence of the ordinary theory of hydrodynamics that in the case of a long rod oscillating

in an unlimited fluid $k = 1$, and we see that this value of k must
be multiplied, in round numbers, by 2, by 3, and by $6\frac{1}{2}$, in order
to account for the observed effect. The value $1\cdot5445$ of $\mathfrak{m}$ is so
large that the descending series comes into play in the calculation
of the function k, while $0\cdot2822$ is so small that the ascending
series are rapidly convergent. Hence the near agreement of the
values of $\sqrt{\mu'}$ deduced from the three experiments is a striking
confirmation of the theory. The mean of the three is $0\cdot1158$, but
of course the last figure cannot be trusted. I shall accordingly
assume as the value of the square root of the index of friction of
air in its average state of pressure, temperature, and moisture

$$\sqrt{\mu'} = 0\cdot116.$$

It is to be remembered that $\sqrt{\mu'}$ expresses a length divided by
the square root of a time, and that the numerical value above
given is adapted to an English inch as the unit of length, and a
second of mean solar time as the unit of time.

56. I now proceed to compare the observed values of $\mathfrak{n}$ with
those calculated from theory with the assumed value of $\sqrt{\mu}$. I
begin with the same cylindrical rods as before, together with the
long brass tubes Nos. 35 to 38. The diameter of this tube was
$1\cdot5$ inch, and its length 56 inches. The ends were open, but as
the included air was treated by Mr Baily in the reduction of his
experiments as if it formed part of the pendulum, we may regard
the pendulum as a solid rod. The tube was furnished with six
agate planes, represented in the wood-cut at page 417, which
rested on fixed knife-edges. The pendulums Nos. 35, 36, 37, and
38 consisted of the same tube swung on the planes marked A, C, a, c.
In air the pendulum swung at the rate of about 90080 vibrations
in a day, so that $\tau = 0\cdot9596$ nearly. The values of $\mathfrak{n}$ obtained
with the end planes A, c were slightly though sensibly greater
than the values obtained with the mean planes C, a. I shall
suppose the mean of the four values of $\mathfrak{n}$, namely $2\cdot290$, to be the
result of the experiments. In the following table the difference
between the theoretical and experimental values of $\mathfrak{n}$ is exhibited
both by decimals and as a fractional part of the former of these
values.

Baily's results with a long brass tube and with long cylindrical rods.

No.	2 a.	m	k	Add for confined space.	Total n by theory.	n by experiment.	Difference.
35 to 38	1·5	5·849	1·242	0·122	2·364	2·290	$-0·074$, or $-\frac{1}{31}$
21 or 42	0·410	1·555	1·917	0·009	2·926	2·932	$+0·006$, or $+\frac{1}{489}$
43	0·185	0·7089	3·055	0·002	4·057	4·083	$+0·026$, or $+\frac{1}{157}$
44	0·072	0·2759	6·670		7·670	7·530	$-0·140$, or $-\frac{1}{54}$

It will be seen at once how closely the experiments are represented by theory. The largest proportionate difference occurs in the case of the brass tube, and even that is less than one-thirtieth. A glance at Baily's wood-cut at page 417 will shew that the six planes with which the tube was furnished caused the whole figure to deviate sensibly from the cylindrical form. Moreover the resistance experienced by each element of the cylinder has been calculated by supposing the element in question to belong to an infinite cylinder oscillating with the same linear velocity, and the resistance thus determined must be a little too great in the immediate neighbourhood of the ends of the cylinder, where the free motion of the air is less impeded than it would be if the cylinder were prolonged. Lastly, the correction for confined space is calculated according to the ordinary equations of hydrodynamics, and on that account, as well as on account of the abrupt termination of the cylinder, will be only approximate. The small discrepancy between theory and observation, as well as the small difference (amounting to about the 1-83rd of the whole) detected by experiment between the results obtained with the extreme planes and those obtained with the mean planes, may reasonably be attributed to some such causes as those just mentioned. In the case of the steel rod or wire, the difference between theory and observation may be altogether removed by supposing a very small error to have existed in the measurement of the diameter of the rod. Since, as we have seen, the observation is satisfied by $m = ·02822$, and (147) gives $a \propto m$ when μ' and τ are constant, it is sufficient, in order to satisfy the experiment, to increase the diameter of the rod in the ratio of 0·2759 to 0·2822, or to suppose an error of only 0·0017 inch in defect to have existed in the measurement of the diameter.

57. I proceed next to the experiments on spheres attached to fine wires. The pendulums of this construction comprise four 1½-inch spheres, Nos. 1, 2, 3, and 4; three 2-inch spheres, Nos. 5, 6, and 7; and one 3-inch sphere, No. 66. Nos. 8 and 9 are the same spheres as Nos. 5 and 7 respectively, swung by suspending the wire over a cylinder instead of attaching it to a knife-edge apparatus. As this mode of suspension was not found very satisfactory, and the results are marked by Baily as doubtful cases, I shall omit the pendulums Nos. 8 and 9, more especially as with reference to the present inquiry they are merely repetitions of Nos. 5 and 7.

In the case of a sphere attached to a fine wire of which the effect is neglected, and swung in an unconfined mass of fluid, we have by the formulæ (52)

$$k = \tfrac{1}{2} + \frac{9}{2a} \sqrt{\frac{\mu' \tau}{2\pi}} \quad \dots\dots\dots\dots\dots(148),$$

$2a$ being in this case the diameter of the sphere. Before employing this formula in the comparison of theory and experiment, it will be requisite to consider two corrections, one for the effect of the wire, the other for the effect of the confinement of the air by the sides of the vacuum tube.

I have already remarked at the end of Section IV., Part I., that the application of the formulæ of Section III. to the case of such fine wires as those used in pendulum experiments is not quite safe. Be that as it may, these formulæ will at any rate afford us a good estimate of the probable magnitude of the correction.

Let l be the length, a_1 the radius, V_1 the volume of the wire, V the volume of the sphere, I the moment of inertia of the pendulum, I' that of the air which we may conceive dragged by it, H the sum of the elements of the mass of the pendulum multiplied by their respective vertical distances below the axis of suspension, H' the same for the air displaced, σ the density of the air. Then the length of the isochronous simple pendulum is IH^{-1} in vacuum, and $(I+I')(H-H')^{-1}$ in air, and the time of vibration is increased by the air in the ratio of $I^{\frac{1}{2}}H^{-\frac{1}{2}}$ to

$(I + I')^{\frac{1}{2}} (H - H')^{-\frac{1}{2}}$, or, on account of the smallness of σ, in the ratio of 1 to $1 + \frac{1}{2} (I'I^{-1} + H'H^{-1})$ nearly. Now $\frac{1}{2} H'H^{-1}$ is the correction for buoyancy, and therefore

$$\mathfrak{n} - 1 = \frac{I'}{H'} \cdot \frac{H}{I} \quad\ldots\ldots\ldots\ldots\ldots(149).$$

We have also, if k_1 be the value of the function k of Section III., Part I.,

$$I' = k\sigma V (l + a)^2 + \tfrac{1}{3} k_1 \sigma V_1 l^2, \quad H' = \sigma V (l + a) + \tfrac{1}{2} \sigma V_1 l \ldots (150),$$

and $HI^{-1} = (l + a)^{-1}$ very nearly. Substituting in (149), expanding the denominator, and neglecting V_1^2, we get

$$\mathfrak{n} - 1 = k + \tfrac{1}{3} \frac{V_1}{V} k_1 \left(\frac{l}{l + a} \right)^2 - \tfrac{1}{2} \frac{V_1}{V} k \frac{l}{l + a}.$$

Now V_1 is very small compared with V, and it is only by being multiplied by the large factor k_1 that it becomes important. We may then, without any material error, replace the last term in the above equation by $\tfrac{1}{3} V_1 V^{-1} l^2 (l + a)^{-2}$, and if λ be the length of the isochronous simple pendulum, we may suppose $l + a = \lambda$, and replace $l^2 (l + a)^{-2}$ by $1 - 2a\lambda^{-1}$, since a is small compared with λ. We thus get, putting $\Delta\mathfrak{n}$ for the correction due to the wire,

$$\Delta\mathfrak{n} = \tfrac{1}{3} \frac{V_1}{V} \left(1 - \frac{2a}{\lambda} \right) (k_1 - 1).$$

Substituting for $k_1 - 1$ from (115), and for $\mathfrak{m}$ from (147), in which equations, however, k_1, a_1 must be supposed to be written for k, a, expressing V_1, V in terms of the diameters of the wire and sphere, and neglecting as before a^2 in comparison with λ^2, we get

$$\Delta\mathfrak{n} = \frac{(2\lambda - 3 \times 2a) \mu' \tau}{\left\{ L^2 + \left(\frac{\pi}{4} \right)^2 \right\} (2a)^3} \quad\ldots\ldots\ldots\ldots(151),$$

where

$$-L = \log_e \frac{4}{2a_1} \sqrt{\frac{\mu' \tau}{\pi}} - 0\cdot5772 \ldots\ldots\ldots\ldots(152).$$

It is by these formulæ that I have computed the correction for the wire in the following table. In the experiments, the time of oscillation was so nearly one second that it is sufficient in the formulæ (148), (151), and (152) to put $\tau = 1$, and take λ for the length of the seconds' pendulum, or 39·14 inches.

With respect to the correction for confined space, it seems evident that the vacuum tube must have impeded the free motion of the air, and consequently increased the resistance experienced by the pendulum when it was swung in air, and that the increase of resistance caused by the cylindrical tube must have been somewhat less than that which would have been produced by a spherical envelope of the same radius surrounding the sphere. The effect of a spherical envelope has been investigated in Section II., Part I.; but as we are obliged at last to have recourse to estimation, it is needless to be very precise in calculating the increase of resistance due to such an envelope, and we may accordingly employ the expression obtained from the ordinary theory of hydrodynamics. According to this theory, the increase of the factor k, which is due to the envelope, is equal to $\frac{3}{2} a^3 (b^3 - a^3)^{-1}$, or $\frac{3}{2} a^3 b^{-3}$ nearly, when b is large compared with a. The increase due to a cylindrical envelope whose axis is vertical, and consequently perpendicular to the direction of oscillation of the sphere, may be estimated at about two-thirds of the increase due to a spherical envelope of the same diameter. I have accordingly taken $+ a^3 b^{-3}$ for the correction for confined space, and have supposed $2b = 6.5$ inches.

The diameter of the wire employed in the pendulums Nos. 1, 2, 3, 5, 6, and 7, is stated to have been about the $\frac{1}{70}$th of an inch, and that of the wire employed with the heavy brass sphere No. 66, about 0.023 inch. The ivory sphere No. 4 was swung with a fine wire weighing rather more than half a grain. Taking the weight at half a grain, and the specific gravity of silver at 10.5, we have for this wire $2a_1 = 0.00251$ nearly. The diameters of the three brass spheres in the following table are taken from page 447 of Baily's memoir. The several parts of which, according to theory, n is composed, are exhibited separately.

The mean error in different determinations of n for the same sphere was about 0.01 or 0.02, and this does not include errors arising from small errors in specific gravities, &c. Hence, if we except the spheres Nos. 1, 2, and 4, the discrepancies between theory and experiment are altogether insignificant. In considering the confirmation thence arising to the theory, it must be borne in mind that the theory did not furnish a single disposable constant, inasmuch as $\sqrt{\mu'}$ was already determined from the experiments

Baily's results with spheres suspended by fine wires.

No. and kind	Diameter of sphere $2a$	Diameter of wire. $2a_1$	n By theory		
			For buoyancy	For inertia on common theory	Additional for inertia on account of internal friction
$1\frac{1}{2}$-INCH SPHERES					
No. 1, Platina	1·44	0·01429	1	0·5	0·289
No. 2, Lead	1·46	0·01429	1	0·5	0·285
No. 3, Brass	1·465	0·01429	1	0·5	0·284
No. 4, Ivory	1·46	0·00251	1	0·5	0·285
2-INCH SPHERES					
No. 5, Lead	2·06	0·01429	1	0·5	0·202
No. 6, Brass	2·065	0·01429	1	0·5	0·202
No. 7, Ivory	2·06	0·01429	1	0·5	0·202
3-INCH SPHERE					
No. 66, Brass	3·030	0·023	1	0·5	0·137

No.	n By theory (continued)			n By experiment	Difference
	Correction for wire	Correction for confined space	Total		
1	0·035	0·011	1·835	1·881	$+0\cdot046$, or $+\frac{1}{40}$
2	0·035	0·011	1·831	1·871	$+0\cdot040$, or $+\frac{1}{46}$
3	0·035	0·011	1·830	1·834	$+0\cdot004$, or $+\frac{1}{457}$
4	0·016	0·011	1·812	1·872	$+0\cdot060$, or $+\frac{1}{30}$
5	0·012	0·032	1·746	1·738	$-0\cdot008$, or $-\frac{1}{218}$
6	0·012	0·032	1·746	1·751	$+0\cdot005$, or $+\frac{1}{349}$
7	0·012	0·032	1·746	1·755	$+0\cdot009$, or $+\frac{1}{194}$
66	0·005	0·101	1·743	1·748	$+0\cdot005$, or $+\frac{1}{349}$

with cylindrical rods. The result obtained with the brass sphere
No. 3 happens to agree almost exactly with theory. However, as
the results obtained with this sphere exhibited some anomalies, it
seems best to exclude it from consideration. The value of n, then,
which belongs to a $1\frac{1}{2}$ inch sphere, appears to exceed by a minute
quantity the value deduced from theory. The difference is indeed
so small that it might well be attributed to errors of observation,
were it not that all the spheres tell the same tale. Thus the
error $+ 0\cdot046$ in the case of the platina sphere corresponds to an

error of less than the fortieth part of a second in the observation of an interval of time amounting to 4½ hours. If the apparent defect, amounting to about 0·04 or 0·05, in the theoretical result be real, it may be attributed with probability to an error in the correction for the wire. This would be no objection to the theory, for it will be remembered that the theory itself indicated the probable failure of the formulæ generally applicable to a long cylinder when the cylinder comes to be of such extreme fineness as the wires employed in pendulum experiments.

58. The preceding experiments of Baily's are the most important for the purposes of the present paper, inasmuch as they were performed on pendulums of simple and very different forms; but there still remain three sets of experiments, the fourteenth, fifteenth, and sixteenth, in which the pendulum consisted of a combination of a sphere and a rod, so that the results can be compared with theory. The details of these experiments being suppressed, I have been obliged to calculate the time of oscillation from the ordinary formulæ of dynamics, but the results will no doubt be accurate enough for the purpose required. In all the calculations I have supposed the rod to reach up to the axis of suspension, and have consequently added 1·55 inch (the length of the shank of the knife-edge apparatus) to the length of the rod, and have added to the weight of the rod a quantity bearing to the whole weight the ratio of 1·55 inch to the whole length.

In the case of the spheres attached to the ends of the rods (sets 14 and 16) the process of calculation is as follows. Let l be the length of the rod increased by 1·55 inch, W_1 its weight, increased as above explained, a the radius and W the weight of the sphere, λ the length of the isochronous simple pendulum. Then supposing the masses of the rod and sphere to be respectively distributed along the axis, and collected at the centre, which will be quite accurate enough for the present purpose, and putting α for the ratio of a to l, we have by the ordinary formula

$$\lambda = \frac{\frac13 W_1 + (1 + \alpha)^2 W}{\frac12 W_1 + (1 + \alpha) W} l \quad \ldots\ldots\ldots\ldots(153),$$

whence τ, the time of vibration, is known. The formula (148) then gives k, which applies to the sphere, and (147) gives $\mathfrak{m}$, the a

in this formula being the radius of the rod, from whence k_1, which applies to the rod, may be got by interpolation from the table in Part I. Let Δk, Δk_1 be the corrections which must be applied to k, k_1 on account of the confined space of the vacuum apparatus, and let S_1, S be the specific gravities of the rod and sphere respectively; then we get by means of the formulæ (149), (150)

$$n - 1 = \frac{\frac{1}{3}(k_1 + \Delta k_1)\dfrac{W_1}{S_1} + (1 + \alpha)^2 (k + \Delta k)\dfrac{W}{S}}{\frac{1}{3} W_1 + (1 + \alpha)^3 W}$$

$$\times \frac{\frac{1}{2} W_1 + (1 + \alpha)\, W}{\frac{1}{2}\dfrac{W_1}{S_1} + (1 + \alpha)\dfrac{W}{S}} \quad \ldots\ldots (154).$$

The first of the two factors connected by the sign $\times$ in this equation is equal to $\sigma^{-1} I' I^{-1}$, and if we want to calculate the weight of air which we must conceive attached to the centre of gyration of the pendulum in order to allow for the inertia of the air, we have only to multiply the factor just mentioned by σ and by the weight of the whole pendulum. The following table contains the comparison of theory and experiment in the case of the 14th set. The rods here mentioned are the same as those which composed the pendulums Nos. 21, 43, and 44, and the spheres are the three brass spheres of Nos. 3, 5, and 66. It appears from p. 432 of Baily's paper that his results are all reduced to a standard pressure and temperature, on the supposition that the effect of the air on the time of vibration is proportional to its density. The theory of the present paper shews that this will only be the case if μ' be constant, which however there is reason for supposing it to be when the pressure alone varies. Be that as it may, no material error can be produced by reducing the observations in this way, because the difference of density in any pair of experiments did not much differ from the density of air at the standard pressure and temperature. The standard pressure and temperature taken were 29·9218 inches of mercury and 32° F., and the assumed specific gravity of air at this pressure and temperature was the 1-770th of that of water, so that in the calculations from theory it is to be supposed that $\sigma^{-1} = 770$.

If w be the weight of the whole pendulum, w' that of the air which we must suppose attached to the pendulum at its centre of gyration in order to express the effect of the inertia of the air, S the *vibrating* specific gravity of the pendulum, the effects of buoyancy and inertia are as σS^{-1} to $w'w^{-1}$; but they are also as 1 to $\mathfrak{n} - 1$, according to the definition of the factor $\mathfrak{n}$, and therefore

$$w' = (\mathfrak{n} - 1)\frac{\sigma}{S}w \quad\ldots\ldots\ldots\ldots\ldots(155),$$

a formula which may be employed to calculate w' when $\mathfrak{n}$ is known.

Baily's results with spheres at the end of long rods.

No. 45 — 1½-inch sphere with copper rod.
No. 46 — 2-inch sphere with ditto.
No. 47 — 3-inch sphere with ditto.
No. 48 — 1½-inch sphere with brass rod.
No. 49 — 2-inch sphere with ditto.
No. 50 — 3-inch sphere with ditto.
No. 51 — 1½-inch sphere with steel rod.
No. 52 — 2-inch sphere with ditto.
No. 53 — 3-inch sphere with ditto.

No.	Value of $\mathfrak{n}$			Weight of adhesive air, in grains		
	By theory	By experiment	Difference	By theory	By experiment	Difference
45	2·525	2·458	− 0·067, or − $\frac{1}{38}$	4·863	4·564	− 0·299, or − $\frac{1}{16}$
46	2·202	2·234	+ 0·032, or + $\frac{1}{69}$	5·005	5·076	+ 0·071, or + $\frac{1}{70}$
47	1·957	1·873	− 0·084, or − $\frac{1}{23}$	7·071	6·425	− 0·646, or − $\frac{1}{11}$
48	2·375	2·356	− 0·019, or − $\frac{1}{125}$	1·447	1·417	− 0·030, or − $\frac{1}{48}$
49	2·060	1·982	− 0·078, or − $\frac{1}{26}$	2·135	1·973	− 0·162, or − $\frac{1}{13}$
50	1·631	1·933 ?	+ 0·302 ?	4·411	4·868 ?	+ 0·457 ?
51	2·099	2·344 ?	+ 0·245 ?	0·682	0·834 ?	+ 0·152 ?
52	1·920	1·793	− 0·127, or − $\frac{1}{15}$	1·457	1·259	− 0·198, or − $\frac{1}{7}$
53	1·781	1·759	− 0·022, or − $\frac{1}{81}$	3·742	3·670	− 0·072, or − $\frac{1}{52}$

With respect to the two experiments marked ? Baily remarks, "These two experiments (with the pendulums Nos. 50 and 51) are very unsatisfactory; and are marked as such in my journal. It was consequently my intention to have repeated them; but the subject was overlooked till it was too late. I should propose their being rejected altogether." If these two experiments be struck out, it will be seen that the differences between theory and experiment are very small, especially when the difficulty of this set of experiments is considered, arising from the frequency of the coincidences with the mean solar clock.

59. On account of the difficulty which Baily experienced in obtaining accurate results with the long rods and spheres attached, he divided the brass and steel rods near the centre of oscillation, and after having cut off an inch from each portion inserted the spheres where the rods had been divided. The results thus obtained constitute the 15th set of experiments. He afterwards removed the lower segments of the rods, and obtained the results contained in the 16th set. I shall give the computation of the latter set first, inasmuch as the formulæ to be employed are exactly the same as those required for the 14th set. The experiments belonging to this set in which the spheres were swung with iron wires have already been computed under the head of spheres attached to fine wires.

Baily's results with the spheres at the end of the short rods.

No. 60 $-1\frac{1}{2}$-inch sphere with brass rod.
No. 61 -2-inch sphere with ditto.
No. 62 -3-inch sphere with ditto.
No. 63 $-1\frac{1}{2}$-inch sphere with steel rod.
No. 64 -2-inch sphere with ditto.
No. 65 -3-inch sphere with ditto.

No.	Value of n			Weight of adhesive air		
	By theory	By experiment	Difference	By theory	By experiment	Difference
60	2·149	2·198	$+0·049$, or $+\frac{1}{44}$	1·011	1·047	$+0·036$, or $+\frac{1}{28}$
61	1·879	1·901	$+0·022$, or $+\frac{1}{85}$	1·619	1·513	$-0·106$, or $-\frac{1}{15}$
62	1·787	1·830	$+0·043$, or $+\frac{1}{42}$	3·970	4·202	$+0·232$, or $+\frac{1}{17}$
63	1·960	1·904	$-0·056$, or $-\frac{1}{35}$	0·570	0·537	$-0·033$, or $-\frac{1}{16}$
64	1·796	1·785	$-0·011$, or $-\frac{1}{163}$	1·239	1·227	$-0·012$, or $-\frac{1}{103}$
65	1·758	1·779	$+0·021$, or $+\frac{1}{84}$	3·609	3·720	$+0·111$, or $+\frac{1}{32}$

Here again the differences between theory and experiment are extremely small. In the case of the pendulum No. 61, Baily's two results 1·901 and 1·513 appear to be inconsistent, as not agreeing with the formula (155).

60. The following table contains the values of τ, k, and k_1 deduced from the given data, and employed in the calculations of which the results are contained in the two preceding tables. It is added, partly to facilitate a comparison of the circumstances of the different experiments, partly to assist in the re-computation

of any of the experiments, or the detection of any numerical error which I may have committed. I may here observe that I have not, generally speaking, re-examined the calculations, except where an error was apparent, but that each step requiring addition, subtraction, multiplication, or division, was checked immediately after it was performed. I have not thought it requisite to check in this manner the taking of logarithms or antilogarithms out of a table.

Values of τ, k, *and* k_1 *employed in the calculation of the theoretical results employed in the two preceding tables.*

(For the description of the pendulums Nos. 45 to 53, see p. 89.)

Long rods				Short rods			
No.	τ	k	k_1	No.	τ	k	k_1
45	1·090	0·7968	1·951				
46	1·158	0·7170	1·981				
47	1·227	0·6523	2·010				
48	1·155	0·8055	3·222	60	0·9517	0·7772	3·012
49	1·198	0·7207	3·264	61	0·9806	0·7005	3·042
50	1·222	0·6520	3·288	62	0·9982	0·6373	3·062
51	1·190	0·8099	7·272	63	0·9868	0·7824	6·649
52	1·199	0·7208	7·299	64	0·9954	0·7021	6·679
53	1·231	0·6525	7·396	65	1·0030	0·6377	6·714

The corrections for confined space employed are, for the spheres, (Δk), 0·0115, 0·0321, 0·1013; and for the rods, (Δk_1), 0·009, 0·002, 0·000. These corrections are to be added to the values of k, k_1 given in the preceding table before going on with the calculation.

61. In the 14th set of experiments, the weight of adhesive air due to the spheres alone has been computed by Baily by subtracting from the whole weight, as given by observation, the weight due to the rods as given by the 13th set of experiments, taking account of the change of weight corresponding to the change in the position of the centre of gyration, the point at which the air is supposed to be attached. According to theory, this process is not legitimate, inasmuch as the weight dragged by a rod is a function of the time of vibration, which is altered when a sphere is attached to the end of the rod. But in the 15th set

of experiments the spheres did not materially affect the time of vibration, inasmuch as they were inserted nearly at the centre of oscillation of the rods, and therefore in this case the process is legitimate. Accordingly, I think it is a sufficient comparison between theory and experiment in the case of the 15th set, to compare the weights of air due to the spheres alone, as calculated by Baily, with the weights calculated according to the theory of this paper with the assumed value of $\sqrt{\mu'}$. I have exhibited separately the weight corresponding to the correction for confined space, in order to enable the reader to form an estimate of the extent to which the results may be affected by the uncertainty relating to the amount of this correction.

Weights of air dragged by the spheres alone, as deduced from Baily's results with the spheres at the centre of oscillation of the long rods.

By Theory

	1½-inch sphere	2-inch sphere	3-inch sphere
In free air	0·431	1·060	3·002
Additional for confined space	0·006	0·048	0·476
Total	0·437	1·108	3·478
Difference, theory and experiment, as decimal	− 0·012	+ 0·001	− 0·101

By Experiment

	1½-inch sphere	2-inch sphere	3-inch sphere
From experiments with brass rod	0·446	1·180	3·382
From experiments with steel rod	0·405	1·039	3·371
Mean	0·425	1·109	3·377
Difference, as fraction of the whole	$-\frac{1}{36}$	$+\frac{1}{1108}$	$-\frac{1}{34}$

62. I pass now to Bessel's experiments described in his memoir entitled *Untersuchungen über die Länge des einfachen Sekundenpendels*, which is printed among the memoirs of the Academy of Sciences of Berlin for the year 1826. The object of

this memoir was to determine the length of the seconds' pendulum by a new method, which consisted in swinging the same sphere with wires of two different lengths, the difference of lengths being measured with extreme precision. In the calculation, the absolute length of the simple pendulum isochronous with either the long or the short compound pendulum was regarded as unknown, but the difference of the two as known, and this difference, combined with the observed times of oscillation, is sufficient for the determination of the quantity sought. Nothing more would have been required if the pendulums had been swung in a vacuum; but inasmuch as they were swung in air, a further correction was necessary to reduce the observations to a vacuum. Since it is necessary to take into account the inertia of the air, as well as its buoyancy, in reducing the observations to a vacuum, Bessel sought to determine by experiment the value of the factor k, of which the meaning has been already explained. The value of this factor, as Bessel remarked, will depend upon the form of the body; but he does not seem, at least in his first memoir, to have contemplated the possibility of its depending on the time of oscillation, and consequently he supposed it to have the same value for the long as for the short pendulum. When the factor k is introduced, the equation obtained from the known difference of length of the two simple pendulums contains two unknown quantities, namely k, and the length of the seconds' pendulum. To obtain a second equation, Bessel made another set of experiments, in which the brass sphere was replaced by an ivory sphere, having as nearly as possible the same diameter. The results obtained with the ivory sphere furnished a second equation, in which k appeared with a much larger coefficient, on account of the lightness of ivory compared with brass. The two equations determined the two unknown quantities.

Let λ be the length of the seconds' pendulum, t_1, t_2 the times of oscillation of the brass sphere when swung with the short wire and long wire respectively, l_1, l_2 the lengths of the corresponding simple pendulums, corrected for everything except the inertia of the air, m the mass of the sphere, m_1 the mass of the fluid displaced; then

$$\lambda t_1^{\,2} \left(1 + \frac{m_1}{m}\, k\right)^{-1} = l_1 \, ;$$

or, since m_1 is so small that we may neglect m_1^2,

$$\lambda t_1^2 \left(1 - \frac{m_1}{m} k\right) = l_1.$$

The long pendulum furnishes a similar equation, and the result obtained from the brass sphere is

$$\lambda (t_2^2 - t_1^2)\left(1 - \frac{m_1}{m} k\right) = l_2 - l_1 \quad \ldots\ldots\ldots\ldots(156),$$

since $l_2 - l_1$ is the quantity which is regarded as accurately known. The ivory sphere in like manner furnishes the equation

$$\lambda (t_2'^2 - t_1'^2)\left(1 - \frac{m_1}{m'} k\right) = l_2' - l_1' \quad \ldots\ldots\ldots\ldots(157),$$

where the accented letters refer to that sphere. The equation for the determination of k results from the elimination of λ between the equations (156) and (157).

Now, according to the theory of this paper, the factor k has really different values for the long and short pendulums. Let k_1 refer to the short, and k_2 to the long pendulum with the brass sphere, k_1' to the short, and k_2' to the long pendulum with the ivory sphere. Then

$$\lambda t_1^2 \left(1 - \frac{m_1}{m} k_1\right) = l_1, \quad \lambda t_2^2 \left(1 - \frac{m_1}{m} k_2\right) = l_2,$$

and therefore

$$l_2 - l_1 = \lambda t_2^2 \left(1 - \frac{m_1}{m} k_2\right) - \lambda t_1^2 \left(1 - \frac{m_1}{m} k_1\right) \quad \ldots\ldots(158).$$

In the equation resulting from the elimination of λ between (156) and (157), let the values of $l_2 - l_1$ and $l_2' - l_1'$ got from (158) and the similar equation relating to the ivory sphere be substituted. The result is

$$(t_2^2 - t_1^2)\left(1 - \frac{m_1}{m} k\right)\left\{t_2'^2\left(1 - \frac{m_1}{m'} k_2'\right) - t_1'^2\left(1 - \frac{m_1}{m'} k_1'\right)\right\}$$

$$= (t_2'^2 - t_1'^2)\left(1 - \frac{m_1}{m'} k\right)\left\{t_2^2\left(1 - \frac{m_1}{m} k_2\right) - t_1^2\left(1 - \frac{m_1}{m} k_1\right)\right\}.$$

This equation is of the form

$$P + Qm_1 + Rm_1^2 = P' + Q'm_1 + R'm_1^2,$$

and $P = P'$, and Rm_1^2, $R'm_1^2$ may be neglected, so that the equation is reduced to $Q = Q'$. It is now no longer necessary to distinguish between t_2 and $t_2{}'$, and between t_1 and $t_1{}'$, which may be supposed equal. Also $m : m' :: S : S'$, where S, S' are the specific gravities of the brass and ivory spheres respectively. Substituting in the equation $Q = Q'$, and solving with respect to k, we get

$$k = \frac{t_2{}^2 (Sk'_2 - S'k_2) - t_1{}^2 (Sk'_1 - S'k_1)}{(t_2{}^2 - t_1{}^2)(S - S')} \quad \ldots\ldots\ldots(159).$$

This equation contains the algebraical definition of that function k of which the numerical value is determined by combining, in Bessel's manner, the results obtained with the four pendulums. Since the equation is linear so far as regards k, k_1, &c., we may consider separately the different parts of which these quantities are composed, and add the results. For the part which relates to the spheres, regarded as suspended by infinitely fine wires, we have $k'_2 = k_2$ and $k'_1 = k_1$, since the radii of the two spheres were equal, or at least so nearly equal that the difference is insensible in the present enquiry. We get then from (159)

$$k = \frac{t_2{}^2 k_2 - t_1{}^2 k_1}{t_2{}^2 - t_1{}^2} \quad \ldots\ldots\ldots\ldots\ldots(160),$$

which gives

$$\frac{k - k_1}{t_2{}^2} = \frac{k - k_2}{t_1{}^2} = \frac{k_2 - k_1}{t_2{}^2 - t_1{}^2} \quad \ldots\ldots\ldots\ldots(161).$$

Since $t_2 > t_1$ and $k_2 > k_1$, the equations (161) shew that the value of k determined by Bessel's method is greater than the factor which relates to the short pendulum, which was a seconds' pendulum nearly, and even greater than that which relates to the long pendulum, as has been already remarked in Art. 6.

If k_s be the factor relating to either sphere oscillating once in a second, and if the effect of the confinement of the air be neglected, we have from the formula (148)

$$k_1 - \tfrac{1}{2} : k_2 - \tfrac{1}{2} : k_s - \tfrac{1}{2} :: t_1{}^{\frac{1}{2}} : t_2{}^{\frac{1}{2}} : 1,$$

and in Bessel's experiments $t_1 = 1\cdot001$, $t_2 = 1\cdot721$, $2a = 2\cdot143$ in English inches. We thus get from either of the equations (160) or (161), on substituting $0\cdot116$ for $\sqrt{\mu'}$, $k = 0\cdot786$. The value of the factor k_s, which relates to the sphere of the same size, swung as a seconds' pendulum, is only $0\cdot694$, and k_1 may be regarded as equal to k_s. The formula (148) gives $k_2 = 0\cdot755$.

63. We have next to investigate the correction for the wire. The effect of the inertia of the air set in motion by the wire was altogether neglected by Bessel, and indeed it would have been quite insensible had the parts of the correction for inertia due to the wire and to the sphere, respectively, been to each other in nearly the same ratio as the parts of the correction for buoyancy. Baily, however, was led to conclude from his experiments that the effect of the wire was probably not altogether insignificant, and the theory of this paper leads, as we have seen, to the result that the factor $\mathfrak{n}$ is very large in the case of a very fine wire.

The ivory sphere in Bessel's experiments was swung with a finer wire than the brass sphere. It was for this reason that I did not from the first suppose $k_1' = k_1$ and $k_2' = k_2$. Let Δk, Δk_1 &c. be the corrections due to the wire. The values of Δk_1, Δk_2, $\Delta k_1'$, $\Delta k_2'$, may be got from the formula (151), in which it is to be remembered that λ denotes the length of the isochronous simple pendulum, not, as in Bessel's notation, the length of the seconds' pendulum. It is stated by Bessel (p. 131), that the wire used with the brass sphere weighed 10·95 Prussian grains in the case of the long pendulum, and 3·58 grains in the case of the short. This gives 7·37 grains for the weight of one toise or 72 French inches. The weight of one toise of the wire employed with the ivory sphere was 6·28 − 2·04 or 4·24 grains (p. 141). The specific gravity of the wire was 7·6 (p. 40), and the weight of a cubic line (French) of water is about 0·1885 grain. From these data it results that the radii of the wires were 0·003867 and 0·002933 inch English. The formula (147) gives $\mathfrak{m}$, whence L is known from (152). The lengths of the isochronous simple pendulums were about 39·20 inches for the short pendulum, and 116·94 for the long. On substituting the numerical values we get from (151), since $k_1 = \mathfrak{n}_1 - 1$ and $k_2 = \mathfrak{n}_2 - 1$,

$$\Delta k_1 = 0·0107, \quad \Delta k_2 = 0·0286, \quad \Delta k_1' = 0·0090, \quad \Delta k_2' = 0·0244.$$

The specific gravities of the two spheres were about 8·190 and 1·794, whence we get from (159) $\Delta k = 0·0308$, or 0·031 nearly.

The value of k deduced by Bessel from his experiments was 0·9459 or 0·946 nearly, which in a subsequent paper he increased to 0·956. In this paper he contemplates the possibility of its being different in the cases of the long and of the short pendulum, and remarks with justice that no sensible error would thence result in the length of the seconds' pendulum, as determined by his method, but that the factor k would belong to the system of the two pendulums.

The following is the result of the comparison of theory and experiment in the case of Bessel's experiments on the oscillations of spheres in air.

Value of k belonging to the system of a long and a short pendulum, as determined experimentally by Bessel .... 0·956

Value deduced from theory, including the correction for the wire, but not the correction for confined space .... 0·817

difference　+0·139

I cannot find that Bessel has stated exactly the distance of the centre of the sphere from the back of the frame within which it was swung, but if we may judge by the sketch of the whole apparatus which is given in Plate I., and by a comparison of figs. 2 and 3, Plate II., it must have been very small, that is to say, a small fraction of the radius of the sphere*. If so, although the exact calculation of the correction for confined space would form a problem of extreme difficulty, it may be shewn from theoretical considerations that the correction would be by no means insensible, so that it might wholly or in part account for the difference +0·139 between the results of theory and observation. It is, however, not improbable, for a reason which has been already mentioned, that the theoretical correction for the wire is not quite exact.

64. The experiments performed by Bessel on a sphere vibrating in water will be more conveniently considered after the discussion of some experiment of Coulomb's, to which I now proceed. These experiments are contained in a memoir entitled

---

* The measurement of either of Bessel's figures, figs. 5 or 6, Plate II. gives 1·53 inch for the distance of the centre of the sphere from the surface of the broad iron bar which formed the back of the frame, the surface of the bar being supposed truly vertical; and the measurement of fig. 2 giving 2·06 inches for the diameter of the sphere, it appears that the distance of the surface of the sphere from the surface of the bar was barely equal to half the radius of the sphere.

Expériences destinées à déterminer la cohérence des fluides et les lois de leur résistance dans les mouvements très-lents, which will be found in the 3rd Volume of the *Mémoires de l'Institut*, p. 246. The experiments which I shall first consider are those which relate to the oscillations of disks suspended in water with their planes horizontal. In these experiments the disk operated upon was attached to the lower extremity of a vertical cylinder of copper, not quite half an inch in diameter, the axis of which passed through the centre of the disk. The cylinder was suspended by a fine wire attached to its upper extremity. The under portion of the cylinder, together with the attached disk, were immersed in water, the disk at the bottom of the cylinder being immersed to the depth of 4 or 5 centimetres below the surface. The upper portion carried a horizontal metallic graduated disk, by means of which the arc of oscillation could be read off, and which, on account of its size and weight, mainly determined the inertia of the system, so that the time of oscillation in the different experiments was nearly the same. The observations were taken as follows. The whole system was turned very slowly round by applying the hands of the graduated disk, taking care not to derange the vertical position of the suspending wire. The arc through which the system had been turned was read by means of the graduation, or rather the system was turned through an arc previously fixed on ; the system was then left to itself, and the arc again read off to a certain number of oscillations. Thus it was the decrement of the arc of oscillation that was observed; the time of oscillation was indeed also observed, but only approximately, for the sake of determining a subsidiary quantity required in the calculation. Indeed, it will be easily seen that the experiments were not adapted to determine the effect of the fluid on the time of oscillation. The decrement of arc so determined had to be corrected for the effect of the imperfect elasticity of the wire, and of the resistance of the air against the graduated disk, and of the water against the portion of the copper cylinder immersed. The amount of the correction was determined by repeating the observation when the lower disk had been removed.

It appeared from the experiments, *first*, that with the same disk immersed, the successive amplitudes of oscillation decreased in geometric progression ; *secondly*, that with different disks the moment of the resisting force was proportional to the fourth power of the radius. From these laws Coulomb concluded that each small element of any one of the disks experienced a resistance varying as the area of the element multiplied by its linear velocity. It should be observed that Coulomb was only authorized by his experiments to assert this law to be true in the case of oscillations of given period, inasmuch as the time of oscillation was nearly the same in all the experiments.

Let a be the radius of the disk in the fluid, τ the time of oscillation, θ the angular displacement of the disk, measured from its mean position, I the moment of inertia of the whole system; and let $1 : 1 - m$ be the ratio in which the arc of oscillation is diminished in one oscillation. According to the formula (15) we have

$$\epsilon^{-n\beta t}$$

for the factor which expresses the ratio of the arc of oscillation at the end of the time t to the initial arc. At the end of one oscillation $t = \tau$, and the value of the above factor is $1 - m$, which is given by observation. Putting for β its value, in which $M\gamma^2 = I$, and $n\tau = \pi$, we get

$$\log_e (1 - m) = -\frac{\rho a^4}{I} \sqrt{\frac{\pi^3 \mu' \tau}{8}} \dots\dots\dots(162).$$

Let T be the time of oscillation, and I_0 the moment of inertia, when the under disk is removed: then $I = I_0 \tau^2 T^{-2}$. Also if M be the mass and R the radius of the large graduated disk, we have $I_0 = \frac{1}{2} MR^2$, neglecting, as Coulomb did, the rotatory inertia of the copper cylinder. Substituting in (162), we get

$$\log_e (1 - m)^{-1} = 2^{-\frac{1}{2}} \pi^{\frac{3}{2}} \rho \mu'^{\frac{1}{2}} \tau^{-\frac{3}{2}} T^2 a^4 R^{-2} M^{-1} \dots\dots(163).$$

Let W be the weight of the disk in grammes. Then the mass of the disk is equal to that of W cubic centimetres or $1000W$ cubic millimetres of water. Hence $M = 1000\rho W$, a millimetre being the unit of length. Substituting in (163), and solving with respect to $\sqrt{\mu'}$, we get

$$\sqrt{\mu'} = 1000 \times 2^{\frac{1}{2}} \log_e 10 . \pi^{-\frac{3}{2}} WR^2 T^{-2} a^{-4} \tau^{\frac{3}{2}} \log_{10} (1 - m)^{-1} \dots(164),$$

and the same value of $\sqrt{\mu'}$ ought to result from different experiments.

The weight of the disk is stated to have been 1003 grammes, and its diameter 271 millimetres, and it made 4 oscillations in 91 seconds.   Hence $W = 1003$, $R = 135·5$, $T = 22·75$.   The last three factors in (164) vary from one experiment to another. After making experiments with three disks of different radii attached to the copper cylinder, Coulomb made another set with nothing attached, for the purpose of eliminating the effect of the imperfect elasticity of the wire.   The following table contains the data furnished by experiment, together with the value of $\sqrt{\mu'}$ deduced from the several experiments.   The latter is reduced to the decimal of an English inch, by including $\overline{2}·5952$ (the logarithm of the ratio of a millimetre to an inch) in the logarithm of the constant part of the 2nd member of equation (164).

Determination of the value of $\sqrt{\mu'}$ for water from Coulomb's experiments on the decrement of the arc of oscillation of disks, oscillating in their own plane by the force of torsion.

No.	Diameter of disk $2a$ in millimetres	Time of four oscillations 4τ	$\log_{10}(1-m)-1$	Resulting value of $\sqrt{\mu'}$ in inches
1	195	97	0·0568	0·05519
2	140	92	0·021	0·05716
3	119	91	0·0135	0·05436
4	0	91	0·0058	

In correcting the results of the first three experiments for the imperfect elasticity of the wire, Coulomb calculated the values of m given by the four experiments, and subtracted the value given by the fourth from each of the others.   But it is at the same time easier and more exact to subtract the value of $\log(1-m)^{-1}$ given by the fourth experiment from that given by each of the others. For if

$$-2c\,\frac{d\theta}{dt}, \qquad -2c'\,\frac{d\theta}{dt}$$

be the moments of two forces, each varying as the velocity, divided by the moment of inertia, the factors by which the initial arc of oscillation must be multiplied to get the arc at the end of the time t, first, when the two forces act together, secondly, when the second force acts alone, are

$$\epsilon^{-(c+c')\,t}, \qquad \epsilon^{-c'\,t},$$

respectively, and that, whether the time t be great or small. Hence if we subtract the logarithm of the second factor from that of the first we shall get the logarithm of the factor due to the action of the first force alone. But if we put each factor under the form $1-m$, and subtract the m of the second factor from the m of the first, we shall not get the m due to the first force alone, unless t be small enough to allow of our neglecting the squares of ct and $c't$, or at least the product $ct \cdot c't$. In truth, when $t = \tau$, the quantities m, are sufficiently small to be treated in Coulomb's manner without any material error, since the corrected values of $\log(1-m)$, obtained in the two ways, would only differ in the 4th place of decimals.

The numbers given in the last column of the above table were calculated from the formula (164), on substituting for $\log(1-m)^{-1}$ the numbers found in the first three lines of the 4th column, corrected by subtracting $0\cdot0058$. The mean of the three results is $0\cdot05557$, but the three experiments are not equally valuable for the determination of $\sqrt{\mu'}$. For the three numbers from which $\sqrt{\mu'}$ was deduced are $0\cdot0510$, $0\cdot0152$, $0\cdot0077$, and a given error in the first of these numbers would produce a smaller error in $\sqrt{\mu'}$ than that which would be produced by the same error in the second, still more, than that which would be produced by the same error in the third. If we multiply the three values of $\sqrt{\mu'}$ by 510, 152, and 77, respectively, and divide the sum of the products by $510 + 152 + 77$ or 739, we get $0\cdot05551$. We may then take $0\cdot555$ as the result of the experiments. Assuming $\sqrt{\mu'} = 0\cdot0555$ we have

$\log(1-m)^{-1}$ from experiment	$0\cdot0568$ in No. 1,	$0\cdot021$ in No. 2,	$0\cdot0135$ in No. 3,
.... from theory	$0\cdot0571$	$0\cdot0206$	$0\cdot0137$
difference	$-0\cdot0003$	$+0\cdot0004$	$-0\cdot0002$

65. So far the accordance of the theoretical and observed results is no very searching test of the truth of the theory. For, in fact, the theory is involved in the result only so far as this, that it shews that the resistance experienced by a given small element of a disk oscillating in a given period varies as the linear velocity; since the difference of periods in Coulomb's experiments was so small that the effects thence arising would be mixed up with errors of observation. This law is so simple that it might

very well result from theories differing in some essential particulars from the theory of this paper. But should the numerical value of $\sqrt{\mu'}$ determined by Coulomb's experiments on disks be found to give results in accordance with theory in totally different cases, then the theory will receive a striking confirmation. Before proceeding to the discussion of other experiments, there are one or two minute corrections to be applied to the value of $\sqrt{\mu'}$ given above, which it will be convenient to consider.

In the first place, the result obtained in Art. 8 is only approximate, the approximation depending upon the circumstance that the diameter of the revolving body is large compared with a certain line determined by the values of μ' and τ. In the particular case in which the revolving solid is a circular disk, it happens that the approximate solution satisfies the general equations exactly, except so far as relates to the abrupt termination of the disk at its edge*. In consequence of this abrupt termination, the fluid annuli in the immediate neighbourhood of the edge are more retarded by the action of the surrounding fluid than they would have been were the disk continued, and consequently the resistance experienced by the disk in the immediate neighbourhood of its edge is actually a little greater than that given by the formula. I have not investigated the correction due to this cause, but it would doubtless be very small.

In the second place, the formula (15) is adapted to an indefinite succession of oscillations, whereas Coulomb did not turn the disk through an angle greater than the largest intended to be observed, and suffer one or two oscillations to pass before the observation commenced, but took for the initial arc that at which the disk had been set by the hand. Probably the disk was held in this position for a short time, so that the fluid came nearly to rest. If so, the resulting value of $\sqrt{\mu'}$, as may readily be shewn, would be a little too small. For in the course of an indefinite series of oscillations, the disk, in its forward motion, carries a certain quantity of fluid with it, and this fluid, in consequence of its inertia, tends to preserve its motion. Hence, when the disk, having attained its maximum displacement in the positive direction, begins to return, it finds the fluid moving in such a manner as to oppose its return, and therefore it experiences a greater

* See note A. at the end.

resistance than if it had started from the same position with the
fluid at rest. In fact, it appears from the expression for G in
Art. 8, that the moment of the resistance vanishes, in passing
from negative to positive, not when the disk has reached the end
of its excursion in the positive direction, but the eighth part of a
period earlier. Hence, had the observation commenced during a
series of oscillations, a larger initial arc would have been necessary,
to overcome the greater resistance, in order to produce, after a
given number of oscillations, the same final arc as that actually
observed. I have investigated the correction to be applied on
account of this cause, and find it to be about $+ 0.009$, but I must
refer to a note for the demonstration, in order not to interrupt
the present discussion*. I shall assume then, in the following
comparisons, that for water

$$\sqrt{\mu'} = 0.0564,$$

the units being the same as before, namely, an English inch and a
second. That μ' is independent of the pressure of the fluid, or at
least very nearly so, appears from an experiment of Coulomb's, in
which it was found that the decrement of the arc of oscillation of
a disk oscillating in water was the same in an exhausted receiver
as under the full atmospheric pressure.

I will here mention another experiment of Coulomb's which
bears directly on one part of the theory. On covering the disk
with a thin coating of tallow, the resistance was found to be the
same as before; and even when the tallow was sprinkled with
powdered sandstone, by means of a sieve, the increase of resistance
was barely sensible. This strikingly confirms the correctness of
the equations of condition assumed to hold good at the surface
of a solid.

66. I will now compare the formula (148) with the results
obtained by Bessel for the oscillations of the brass sphere in water,
which will be found at page 65 of his memoir. This sphere was
suspended so as to be immersed in the water contained in a large
vessel, and was swung with two different lengths of wire, the
same as those employed for the experiments in air. The times of
oscillation were 1·9085 second for the long pendulum, and 1·1078
for the short. The results are

* See note B. at the end.

	Long pendulum.	Short pendulum,
k, by experiment	0·648	0·602
k, by theory	0·631	0·600
difference	+ 0·017	+ 0·002

The depth to which the spheres were immersed is not stated, but it was probably sufficient to render the effect of the free surface small, if not insensible. The vessel was three feet in diameter, and the water 10 inches deep, so that unless the spheres were suspended near the bottom, which is not likely to have been the case, the effect of the limitation of the fluid by the sides of the vessel must have been but trifling. The agreement of theory and observation, as will be seen, is very close.

67. In the same memoir which contains the experiments on disks, Coulomb has given the results of some experiments in which the disk immersed in the fluid was replaced by a long narrow cylinder, placed with its axis horizontal and its middle point in the prolongation of the axis of the vertical copper cylinder. In these experiments, the arcs did not decrease in geometric progression, as would have been the case if the resistance had varied as the velocity; but it was found that the results of observation could be satisfied by supposing the resistance to vary partly as the first power, and partly as the square of the velocity. In Coulomb's notation, $1 : 1 - m$ denotes the ratio in which the arc of oscillation would be altered after one oscillation, if the part of the resistance varying as the square of the velocity were destroyed. The several experiments performed with the same cylinder were found to be sufficiently satisfied by the formula deduced from the above-mentioned hypothesis respecting the resistance, when suitable numerical values were assigned to two disposable constants m and p, of which p related to the part of the resistance varying as the square of the velocity.

Conceive the cylinder divided into elementary slices by planes perpendicular to its axis. Let r be the distance of any slice from the middle point, θ the angle between the actual and the mean positions of the axis, dF that part of the resistance experienced by the slice which varies as the first power of the velocity. Then calculating the resistance as if the element in question belonged to an infinite cylinder moving with the same linear velocity, we have by the formulæ of Art. 31

$$dF = k'M'n\frac{d\xi}{dt}, \quad \text{where } M' = \pi\rho a^2 dr, \quad \frac{d\xi}{dt} = r\frac{d\theta}{dt}.$$

If G be the moment of the resistance, l the whole length of the cylinder, we have, putting $n = \pi\tau^{-1}$,

$$G = \frac{\pi^2 k' \rho a^2 l^3}{12\tau} \frac{d\theta}{dt};$$

whence

$$\log_e (1 - m)^{-1} = \frac{\pi^2 k' \rho a^2 l^3}{24I} \quad \text{...............(165)},$$

I being the moment of inertia.

Expressing I in terms of the same quantities as in the case of the disk, we get from (147) and (165)

$$\log_{10} (1 - m)^{-1} = \log_{10} \epsilon \cdot \frac{\pi\mu' Tl^3}{3R^2} \cdot \frac{g\rho}{W} \cdot \frac{T}{\tau} \cdot \mathfrak{m}^2 k' \quad \text{......(166)},$$

and $g\rho$ is the weight of a cubic millimetre of water, or the 1000th part of a gramme. The numerical values of μ', T, R, W have been already given, but μ' must be reduced from square inches to square millimetres. The cylinders, of which three were tried in succession, had all the same length, namely, 249 millimetres. Their circumferences, calculated from their weights and expressed in millimetres, were 21·1, 11·2, and 0·87, and the time of four oscillations was 92^s, 91^s, 91^s. The values of $\mathfrak{m}$ calculated from these data by means of the formula (147) are 0·4332, 0·2312, and 0·01796. For the first and second of these values, $\mathfrak{m}^2 k'$ may be obtained by interpolation from the table given in Part I.; for the third it will be sufficient to employ the second of the formulæ (115).

The following are the results:

	Cylinder, No. 1.	No. 2.	No. 3.
m, by experiment	0·0400	0·0260	0·0136
m, by theory	0·0413	0·0291	0·0113
Difference	− 0·0013	− 0·0031	+ 0·0023

The differences between the results of theory and experiment are perhaps as small as could reasonably be expected, when it is considered that, notwithstanding the delicate nature of the experiments, the numerical values of two constants, m and p, had to be deduced from their results.

68. This memoir of Coulomb's contains also a notice of a set of experiments with disks and cylinders in which the water was replaced by oil. The experiments with disks shewed that with a given disk the arc of oscillation decreased in geometric progression, and that with different disks the moments of the resistances were as the fourth powers of the diameters. The absolute resistances were greater than in the case of water in the ratio of about 17·5 to 1. The details of Coulomb's experiments on cylinders oscillating in oil are entirely omitted. It is merely stated that on making the same cylinders as before, or shorter cylinders when the resistance was too great, oscillate in oil, it was found, conformably with the results obtained with planes, that the coherence of oil was to that of water as 17 to 1. The coherence is here supposed to be measured by that part of the resistance which is proportional to the first power of the velocity. On making a rough calculation of the ratio of the resistances to cylinders oscillating in oil and in water, on the supposition that $\sqrt{\mu'}$ for oil is to $\sqrt{\mu'}$ for water as 17·5 to 1, as would follow from the experiments on disks if the difference of the specific gravities of the two fluids be neglected, I found that the ratio in question ought to have been somewhere about 100 to 1, instead of only 17 to 1. It would seem from this that the theory of the present paper is not applicable to oil; but fresh experiments would be required before this point can be considered as established, on account of the theoretical doubt respecting the application of the formulæ of Section III. Part I., to extremely fine cylinders, especially in cases in which μ' is large, so that $\mathfrak{m}$ is very small. It would be interesting to make out whether what I have called internal friction is or is not of the same nature as viscosity. Coulomb and Dubuat apply the term *viscosity* to that property of water by virtue of which certain effects are produced which have been shewn in this paper to be perfectly explicable on the theory of internal friction ; whereas Poisson, in one of his memoirs, expressly asserts that the terms in the equations of motion which result from what has been called in this paper internal friction belong to perfect fluids, and

have nothing to do with viscosity*. Poisson does not give the slightest hint as to the grounds on which he rested his opinion.

69. I come now to the experiments of Dubuat, which are contained in an excellent work of his entitled *Principes d'Hydraulique*, of which the second edition was published in 1786. The first edition does not contain the experiments in question. Dubuat justly remarked that the time of oscillation of a pendulum oscillating in a fluid is greater than it would be in vacuum, not only on account of the buoyancy of the fluid, which diminishes the moving force, but also on account of the mass of fluid which must be regarded as accompanying the pendulum in its motion; and even determined experimentally the mass of fluid which must be regarded as carried by the oscillating body in the case of spheres and of several other solids. Thus Dubuat anticipated by about forty years the discovery of Bessel; but it was not until after the appearance of Bessel's memoir that Dubuat's labours relating to the same subject attracted attention.

Dubuat's method was as follows. Imagine a body suspended by a fine thread or wire and swung in vacuum, and let a be the length of the pendulum, reckoned from the centre of suspension to the centre of oscillation. Now imagine the same body swung in a fluid, in which its apparent weight is p, so that if P denote the weight of fluid displaced, the true weight of the body will be $p + P$. Since the moving force is diminished in the ratio of $p + P$ to p, if the inertia of the body were all that had to be overcome, it would be necessary to diminish the length of the pendulum in the same ratio, in order to preserve the same time of oscillation. But since the mass in motion consists not only of the mass of the body itself, but also of that of the fluid which it carries with it, the pendulum must be shortened still more, in order that the time of oscillation may be unaltered. Let l be the length of the pendulum so shortened, and $\mathfrak{n}$ (which for the same reason as before I write instead of Dubuat's n,) a factor greater than unity, such that $p + \mathfrak{n}P$ is the weight of the mass in motion; then

$$l = \frac{ap}{p + \mathfrak{n}P}, \quad \text{whence } \mathfrak{n} = \frac{p}{P}\left(\frac{a}{l} - 1\right) \quad \ldots\ldots\ldots(167).$$

* *Journal de l'École Polytechnique*, Tom. XIII. p. 95.

Dubuat's experiments on this subject consist of 44 experiments on spheres oscillating in water (Tom. II. p. 236); 31 experiments on other solids oscillating in water (p. 246); and 3 experiments on spheres oscillating in air (p. 283). The following table contains a comparison of the formula (148) with Dubuat's results for spheres oscillating in water. The value of $\sqrt{\mu'}$ employed in the calculation is 0·0564 inch English, or 0·05291 inch French.

Dubuat's experiments on spheres oscillating in water.

	τ	n calc.	n obs.	n diff.
Sphere of lead Diameter 1·0113 inches Weight in water 2102 grains	$\frac{1}{2}$	1·633	1·502	− ·131
	1	1·687	1·502	− ·185
	2	1·766	1·522	− ·244
	3	1·825	1·620	− ·205
Sphere of glass Diameter 2·645 inches Weight in water 574 grains	2	1·602	1·518	− ·084
	4	1·644	1·569	− ·075
	6	1·676	1·598	− ·078
Same sphere weighing in water 2102 grains	1	1·572	1·515	− ·057
	2	1·602	1·516	− ·086
	3	1·624	1·523	− ·101
	4	1·644	1·546	− ·098
Same sphere weighing in water 4204 grains	1	1·572	1·537	− ·035
	2	1·602	1·523	− ·079
	3	1·624	1·524	− ·100
	4	1·644	1·538	− ·106
Same sphere weighing in water 9216 grains	$\frac{1}{2}$	1·551	1·449	− ·102
	1	1·572	1·372	− ·200
	2	1·602	1·494	− ·108
	3	1·624	1·494	− ·130
Sphere of wood Diameter 4·076 inches Weight in water 2102 grains	2	1·566	1·507	− ·059
	3	1·581	1·547	− ·034
	4	1·593	1·547	− ·046
	6	1·614	1·567	− ·057
Same sphere weighing in water 4204 grains	1	1·547	1·375	− ·172
	2	1·566	1·456	− ·110
	3	1·581	1·525	− ·056
	4	1·593	1·557	− ·036
	6	1·614	1·549	− ·065
Same sphere weighing in water 9216 grains	1	1·547	1·57	+ ·023
	2	1·566	1·553	− ·013
	3	1·581	1·59	+ ·009
	4	1·593	1·583	− ·010

	τ	n calc.	obs.	diff.
Another sphere of wood Diameter 6⅜ inches Weight in water 2102 grains	3	1·549	1·27	− ·279
	4	1·557	1·394	− ·163
	6	1·570	1·487	− ·083
	9	1·585	1·566	− ·019
	12	1·599	1·569	− ·030
	18	1·621	1·565	− ·056
Same sphere weighing in water 3204 grains	10·85	1·594	1·634	+ ·040
Same sphere weighing in water 4204 grains	3	1·549	1·651	+ ·102
	4	1·557	1·627	+ ·070
	6	1·570	1·654	+ ·084
	9	1·585	1·664	+ ·079
	12	1·599	1·674	+ ·075

70. If we strike out the experiments with the large sphere, which cannot well be compared with theory for a reason which will be explained further on, it will be observed that in seven out of the eight groups of experiments left, the signs in the last column are regularly *minus*. The preponderance of negative errors could be destroyed by using a much smaller value of $\sqrt{\mu'}$ in the reduction. We have seen, however, that the value of $\sqrt{\mu'}$ deduced from Coulomb's experiments on the decrement of the arc of oscillation of disks satisfied almost exactly Bessel's observations of the time of oscillation of a sphere about two inches in diameter oscillating in water. The very small errors which remained in this case had both the sign +, whereas in Dubuat's experiments on the 1-inch and 2½ inch spheres, the errors, which are far larger, have all the sign −. Since the experiments of Dubuat and Bessel, though made under similar circumstances, do not lead to the same result, it is of course impossible for any theory to satisfy them both. The numbers in the last column of the preceding table are, however, far too regular to be attributable to mere fortuitous errors of observation. If we suppose Bessel's results to have been nearly exact, there must have been something in the mode either of making or of reducing Dubuat's experiments which caused a tendency to error in one direction.

With respect to the reduction of the experiments it may be observed that the length l was measured from the centre of oscillation, whereas in the formula (148) it is supposed that the mass of which the weight is kP or $(\mathfrak{n}-1)P$ is collected at the centre of the sphere. If h be the distance of the centre of the sphere from the axis of suspension, the observed value of $\mathfrak{n}-1$ ought in strictness to be increased in the ratio of h^2 to l^2, or the calculated value diminished in the ratio of l^2 to h^2, before comparing the results of theory and experiment. In the case of the loaded spheres especially, the theoretical value of $\mathfrak{n}$ would thus be a little diminished; but except in a very few cases, in which either l or $a-l$ is small, the diminution is hardly worth considering. After having been for a good while at a loss to account for the regular occurrence of rather large negative errors, the following occurred to me as the probable solution of the difficulty.

When a pendulum oscillates in water, the arc of oscillation rapidly decreases; this rapid diminution forms in fact the grand difficulty in experiments of this kind. In Dubuat's experiments, it will be remembered, the suspending thread was lengthened or shortened till the time of oscillation was an exact number of seconds, or occasionally half a second. Now, it is probable that the observer occasionally gave the suspending thread a slight push as the pendulum was commencing its return, in order to keep the oscillations going for a sufficient time to allow of tolerable precision in rendering the time of oscillation equal to what it ought to be. If so, these pushes would slightly accelerate the oscillations, and therefore cause the length of thread fixed on by observation to be a little too great, which would make the effect of the water in retarding the oscillations appear a little too small. On inspecting the table of differences, it may be observed that sometimes when the same sphere differently loaded is swung in the same time as before, the numbers in the table of differences are altered more than appears to be attributable to merely fortuitous errors of observation. This accords very well with the conjecture just mentioned, and seems difficult to account for in any other way, inasmuch as everything relating to the fluid must have been almost exactly the same in the two cases.

The occurrences of positive differences in the case of the large wooden sphere may be accounted for by the limitation of the fluid mass by the sides and bottom of the vessel, and by the free surface, which, except in the case of very short oscillations, would

have much the same effect as a rigid plane, inasmuch as it would
be preserved almost exactly horizontal by the action of gravity.
The vessel which contained the water was 51 inches long and
17 broad, the water was 14 inches deep, and the spheres were
plunged to about 3 inches below the surface, so that the effect
of the confinement of the fluid mass would have been quite
sensible in the case of such large spheres. If it be objected that
the same sphere gave negative differences in the case of the first
group of experiments, it must be observed, that when the apparent
weight of so large a sphere was only 2102 French grains, the
resistance would quickly have caused the oscillations to subside if
an extraneous force had not frequently been applied.

71. In Dubuat's experiments on spheres oscillating in air, the
lightness of the fluid was compensated by the extreme lightness of
the spheres, which were composed, the first two of paper, and the
third of goldbeater's skin. In the following table the diameter
$2a$ of the sphere is expressed in French inches. The value of $\sqrt{\mu'}$
employed in the reduction is the same as was before used in the
reduction of observations made in air, namely 0·116 inch English,
or 0·1088 inch French.

Dubuat's experiments on light spheres oscillating in air.

No.	$2a$	τ	n calc.	n obs.	Diff.
337	4·0416	1·51	1·61	1·51	−0·10
338	6·625	1·84	1·57	1·63	+0·06
339	17·25	3·625	1·53	1·54	+0·01

The differences certainly appear very small when the delicacy
of the experiments and the simplicity of the apparatus employed
are considered.

72. The only comparison yet made in this section between
theory and observation in the case of pendulum experiments,
consists in comparing the observed times of vibration with the
results calculated with an assumed value of $\sqrt{\mu'}$. But according
to theory we ought to be able, without assigning a particular value
to any new disposable constant, to calculate the rate of decrease
of the arc of vibration. I have not met with any experiments

made with a view of investigating the decrease in the arc of vibration in the case of extremely small vibrations, such as those employed in pendulum experiments. The experiments of Newton and others, in which the arc of vibration was so large that the resistance depended mainly on the square of the velocity, would be quite useless for my purpose. The pendulum experiments of Bessel and Baily contain however the requisite information, or at least some portion of it, for the arcs are registered for the sake of giving the data for calculating the small reduction to indefinitely small vibrations.

In Bessel's experiments the arc is registered for the end of equal intervals of time during the motion. The number of such registrations in one experiment amounts in some cases to eleven, and is never less than three. So far the observations are just what are wanted; but there are other causes which prevent an exact comparison between theory and experiment. In the first place the spheres were swung so close to the back of the frame that the increase of resistance due to the confinement of the air must have been very sensible. In the second place the effect of the wire must have been very sensible, especially in the case of the long pendulum. For the table of Section III. Part I., shews that for the wire (for which m is very small) the value of k' is much larger than that of k, whereas for spheres of the size of those employed, when the time of oscillation is only one or two seconds, k' is a good deal smaller than k. Hence, if the formulæ of that section applied to such fine wires, the effect of the wire on the arc of vibration would be much greater than its effect on the time of vibration, and therefore would be quite sensible. But it has been shewn in Section IV., that the effect of the wire in diminishing the arc of vibration is probably greater than would be given by the formula, and therefore the uncertainty depending on the wire is likely to amount to a very sensible fraction of the whole amount. Again, since Bessel's experiments were all made in air, no data are afforded whereby to eliminate the portion of the observed result which was due to friction at the point of support, imperfect elasticity of the wire, or gradual dissipation of *vis viva* by communication of motion to the supporting frame. Moreover in the case of the long pendulum the observations were made with rather too large arcs, for the law of the decrease of the arc of vibration deviated sensibly from that of a geometric progression. In Baily's experiments, only the initial and final arcs are registered, and not even those in the case of the "additional experiments." Hence these experiments do not enable us to make out whether it would be sufficiently exact to suppose the decrease to take place in geometric progression. Moreover, the final arc was generally so small, that a small error committed in the measurement of it would cause a very sensible error in the rate of decrease concluded from the experiment. For these reasons it would be unreasonable to expect a near accordance between the formulæ and the results of the experiments of Bessel and Baily. Still, the formulæ might be expected to give a result in defect, and yet not so much in defect as not to form a large portion of the result given by observation. On this account it will not be altogether useless to compare theory and observation with reference to the decrement of the arc of vibration.

73. Let us first consider the case of a sphere suspended by a fine wire. Let the notation be the same as was used in investigating the expression for the effect of the air on the time of vibration, except that the factors k', k_1' come in place of k, k_1. Considering only that part of the resistance which affects the arc of vibration, we have for the portions due respectively to the sphere and to the element of the wire whose length is ds, and distance from the axis of suspension s,

$$k'M'n(l+a)\frac{d\theta}{dt}, \qquad k_1'\frac{M_1'}{l}ds \cdot ns\frac{d\theta}{dt},$$

and if we take the moment of the resistance, and divide by twice the moment of inertia, the coefficient of $\frac{d\theta}{dt}$ in the result, taken negatively, and multiplied by t, will be the index of ϵ in the expression for the arc. Hence if α_0 be the initial arc of vibration, and α_t the arc at the end of the time t

$$\log_e \alpha_0 - \log_e \alpha_t = \frac{k'M'(l+a)^2 + \frac{1}{3}k_1'M_1'l^2}{M(l+a)^2 + \frac{1}{3}M_1l^2} \cdot \frac{\pi t}{2\tau} \quad \dots(168),$$

$M'(l+a)^2$ being as before taken for the moment of inertia of the sphere, which will be abundantly accurate enough. If then we put l for the Napierian logarithm of the ratio of the arc at the beginning to the arc at the end of an oscillation, we must put $t = \tau$ in (168), whence, neglecting the effect of the wire, we obtain

$$I = \frac{\pi k'}{2} \cdot \frac{\sigma}{S} \quad \ldots\ldots\ldots\ldots \ldots\ldots\ldots (169).$$

If now $\Delta k'$ be the correction to be applied to k' in this formula on account of the wire, since k', k_1' are combined together in the expression for the arc just as k, k_1 in the expression for the time, we get

$$\Delta k' = \frac{k_1'}{k_1} \, \Delta k \quad \ldots\ldots\ldots\ldots\ldots (170),$$

and the approximate formulæ (115) give

$$\Delta k' = - \frac{4L}{\pi} \, \Delta k \quad \ldots\ldots\ldots\ldots\ldots (171),$$

whence the numerical value of $\Delta k'$ is easily deduced from that of Δk, which has been already calculated. We get also from (52)

$$k' = k - \tfrac{1}{2} + \tfrac{4}{9}(k - \tfrac{1}{2})^2 \quad \ldots\ldots\ldots\ldots\ldots (172),$$

whence k' may be readily deduced from k, which has been already calculated.

74. Before comparing these formulæ with Bessel's experiments, it will be proper to enquire how far the latter are satisfied by supposing the arcs of oscillation to decrease in geometric progression. In Bessel's tables the arc is registered in the column headed μ. This letter denotes the number of French lines read off on a scale placed behind the wire, and a little above the sphere, and is reckoned from the position of instantaneous rest of the wire on one side of the vertical to the corresponding position on the other side. The distance of the scale from the axis of suspension being given, as well as the correction to be applied to μ on account of parallax, the arc of oscillation may be readily deduced. However, for our present purpose, any quantity to which the arc is proportional will do as well as the arc itself, and μ, though strictly proportional to the tangent of the arc, may be regarded as proportional to the arc itself, inasmuch as the initial arc usually amounted to only about 50′ on each side of the vertical.

Now we may form a very good judgment as to the degree of accuracy of the geometric formula by comparing the arc observed in the middle of an experiment with the geometric mean of the initial and final arcs. I have treated in this way Bessel's experiments, Nos. 1, 2, 3, 4, and 5. Each of these is in fact a group of six experiments, four with the long pendulum and two with the short, so that the whole consists of 20 experiments with the long pendulum, and 10 with the short. In the case of the long pendulum, the observed value of μ regularly fell short of the calculated value, and that by a tolerably constant quantity. The mean difference amounted to 0·688 line, and the mean error in this quantity to 0·109. This mean error was not due entirely to errors of observation, or variations in the state of the air, &c., but partly also to slight variations in the initial arc, larger differences usually accompanying larger initial arcs. The initial arc usually corresponded to $\mu = 39$ or 40 lines, and the final to $\mu = 15$ or 16 lines. In the case of the short pendulum, the differences in 8 cases out of 10 had the same sign as before. The mean difference was 0·025, and the mean error 0·043. The arcs of oscillation were nearly the same as before; but inasmuch as the axis of suspension was nearer to the scale than before, the initial value of μ was only about 12 or 13 lines, and the final value about 7 lines. When the results of some of the experiments were laid down on paper, by abscissæ taken proportional to the times and ordinates to the logarithms of μ, it was found that in the case of the long pendulum the line so drawn was decidedly curved, the concavity being turned toward the side of the positive ordinates. The curvature of the line belonging to the short pendulum could hardly be made out, or at least separated from the effects of errors of observation. The experiments 9, 10, 11, having been treated numerically in the same way as the experiments 1—5, led to much the same result. In the 16 experiments with the ivory sphere and short pendulum contained in the experiments Nos. 12, 13, 14, and 15, the excess of the calculated over the observed value of μ was more apparent, the mean excess amounting to 0·129. The reason of this probably was, that the observations with the ivory sphere were made through a somewhat wider range of arc than those with the brass sphere.

It appears then that at least in the case of the long pendulum a correction is necessary, in order to clear the observed decrease in the arc of oscillation from the effect of that part of the resistance which increases with the arc more rapidly than if it varied as the first power of the velocity, and so to reduce the observed rate of decrease to what would have been observed in the case of indefinitely small oscillations.

75. In Coulomb's experiments it appeared that the resistance was composed of two terms, one involving the first power, and the other the square of the velocity. If we suppose the same law to hold good in the present case, and denote the amplitude of oscillation at the end of the time t, measured as an angle, by α, we shall obtain

$$\frac{d\alpha}{dt} = - A\alpha - B\alpha^2 \dots\dots\dots\dots\dots(173),$$

where A and B are certain constants. We must now endeavour to obtain A from the results of observation. Since the substitution for α of a quantity proportional to α will only change the constant B in (173), and the numerical value of this constant is not required for comparison with theory, we may substitute for α the number of lines read off on the scale as entered in Bessel's tables in the columns headed μ.

I have employed four different methods to obtain A from the observed results. The one I am about to give is the shortest of the four, and is sufficiently accurate for the purpose.

The equation (173) gives after dividing by α

$$\frac{d\log\alpha}{dt} = - A - B\alpha\dots\dots\dots\dots\dots(174).$$

Now, as has been already observed, the arcs of vibration decrease nearly in geometric progression. If this law were strictly true, we should have

$$\alpha = \alpha_0 \left(\frac{\alpha_2}{\alpha_0}\right)^{\frac{t}{T}} \dots\dots\dots\dots\dots(175),$$

where α_0 denotes the initial and α_2 the final arc, and T denotes the whole time of observation. We may, without committing any material error, substitute this value of α in the last term of (174). The magnitude of the error we thus commit is not to be judged of merely by the smallness of B. The approximate expression (175) is rather to be regarded as a well-chosen formula of interpolation, and in fact $T^{-1} \log_\epsilon (\alpha_0 \alpha_2^{-1})$ differs very sensibly from A. Making now this substitution in (174), integrating, and after integration restoring α in the last term by means of (175), we get

$$\log \alpha = - At - \frac{BT\alpha}{\log \alpha_2 - \log \alpha_0} + C \ldots\ldots\ldots(176),$$

C being an arbitrary constant. To determine the three constants A, B, C, let α_1 be the arc observed at the middle of the experiment, apply the last equation to the arcs α_0, α_1, α_2, and take the first and second differences of each member of the equation. Let Δ_1 denote the sum of the two first differences, so that $\Delta_1 t$ is the same thing as T. Then we may take for the two equations to determine A and B

$$\Delta_1 \log \alpha_0 = - A \Delta_1 t - \frac{B \Delta_1 t \cdot \Delta_1 \alpha_0}{\Delta_1 \log \alpha_0} ; \qquad \Delta^2 \log \alpha_0 = - \frac{B \Delta_1 t \cdot \Delta^2 \alpha_0}{\Delta^2 \log \alpha_0}.$$

Eliminating B, and passing from Napierian to common logarithms, which will be denoted by Log., we get

$$A = \frac{- \Delta_1 \operatorname{Log} \alpha_0}{\operatorname{Log} \epsilon \cdot \Delta_1 t} \left\{ 1 - \frac{\Delta^2 \operatorname{Log} \alpha_0 \cdot \Delta_1 \alpha_0}{\Delta_1 \operatorname{Log} \alpha_0 \cdot \Delta^2 \alpha_0} \right\} \ldots\ldots(177).$$

If we suppose the part of $- \dfrac{d\alpha}{dt}$ which does not vary as the first power of α to be $\alpha^2 \phi'(\alpha)$ instead of $B\alpha^2$, we shall get in the same way

$$A = \frac{- \Delta_1 \operatorname{Log} \alpha_0}{\operatorname{Log} \epsilon \cdot \Delta_1 t} \left\{ 1 - \frac{\Delta^2 \operatorname{Log} \alpha_0 \cdot \Delta_1 \phi(\alpha_0)}{\Delta_1 \operatorname{Log} \alpha_0 \cdot \Delta^2 \phi(\alpha_0)} \right\} \ldots\ldots(178).$$

76. I have not attempted to deduce evidence for or against the truth of equation (173) from Bessel's experiments. The approximate formula (175) so nearly satisfied the observations, that almost any reasonable formula of interpolation which introduced one new disposable constant would represent the experiments within the limits of errors of observation. It may be observed, that the factor outside the brackets in equations (177) and (178) is the first approximate value of A got by using only the initial and final arcs, and supposing the arcs to decrease in geometric progression. In the case of the long pendulum, the value of A, corrected in accordance with the formula (178), would be very sensibly different according as we supposed $\phi(\alpha)$ to be equal to $B\alpha$, in which case (178) would reduce itself to (177), or equal to

Ba^2. In the case of the long pendulum with the brass sphere, the corrected value of A, deduced from the formula (177), was equal to about 0·77 of the first approximate value.

I have not considered it necessary to go through all Bessel's experiments, as it was not to be expected that the formula should account for the whole observed decrement. I have only taken four experiments for each kind of pendulum, namely, I. a, b, e, and f for the long pendulum with the brass sphere; I. c and d and II. c and d for the short pendulum with the brass sphere; XII. a, b, c, and d for the long pendulum with the ivory sphere, and XII. a', b', c', and d' for the short pendulum with the ivory sphere. The formula (177) gave the following results.

First case,
   Log $\epsilon . \tau A = $ ·0000759 ; mean error = ·0000020.
Second case,
   Log $\epsilon . \tau A = $ ·0000504 ; mean error = ·0000075.
Third case,
   Log $\epsilon . A = $ ·000631 ; mean error = ·000046.
Fourth case,
   Log $\epsilon . A = $ ·000167 ; mean error = ·000074.

Now $l = \tau A$, and therefore, to get the values of l deduced from experiment, it will be sufficient to divide the numbers above given by the modulus of the common system of logarithms. The theoretical value of l will be got from (169), if we add to k' the correction $\Delta k'$ depending upon the wire. The following are the results :

	long p. brass s.	short p. brass s.	long p. ivory s.	short p. ivory s.
1000000 l for sphere alone in an unlimited mass of fluid, by theory	67	50	298	222
additional for wire	27	9	114	39
	94	59	412	261
1000000 l by experiment ...	175	116	1453	384

It appears then that the calculated rate of decrease of the arc amounts on the average to about half the rate deduced from observation. This is about what we might have expected, considering the various circumstances, all tending materially to augment the rate of decrease, which were not taken into account in the calculation.

77. Of Baily's pendulums I have compared the following with theory in regard to the decrement of the arc of vibration. No. 1 (the $1\frac{1}{2}$-inch platina sphere), experiments 1 to 8 ; No. 3 (the brass $1\frac{1}{2}$-inch sphere), experiments 9 to 16 ; No. 6 (the 2-inch brass sphere), experiments 33 to 40 ; No. 21 (the 0·410 inch long copper cylindrical rod), experiments 109 to 112 ; and Nos. 35—38 (the $1\frac{1}{2}$-inch long brass tube), experiments 167 to 174. I have not thought it worth while to compute the results obtained with the other $1\frac{1}{2}$-inch and 2-inch spheres, inasmuch as they were of the same size as the brass spheres, and moreover the observation of the decrement of the arc was not the object Baily had in view in making the experiments. The 3-inch sphere, and all the other cylindrical rods and combinations of cylindrical rods and spheres, belong to the "additional experiments" for which the arcs are not given.

The mode of performing the calculation will best be explained by an example. Take, for instance, the pair of experiments Nos. 1 and 2. In No. 1 the total interval was 4·22 hours, the initial arc was 0°·77, the final arc 0°·29, the mean height of the barometer 30·24 inches, and the temperature about $38\frac{1}{2}$° F. The difference of the common logarithms of the initial and final arcs is 0·424, and this divided by the total interval gives 0·1005 for the difference of logarithms for one hour. The second experiment, treated in a similar way, gives 0·0352, which expresses the effect of friction at the point of support, communication of motion to the support itself, &c., together with the resistance of highly rarefied air at a pressure of only 0·97 inch of mercury. Since we have reason to believe that μ' is independent of the density, we may get the effect of air at a pressure of 30·24 − 0·97 or 29·27 inches of mercury by subtracting 0·0352 from 0·1005, which gives 0·0653. Reducing to 29 inches of mercury for convenience of comparison, we get 0·0649. Each pair of experiments is to be treated in the same way. Since the temperature was nearly the same in the experiments made with the same pendulum, we may suppose it constant, and equal to the mean of the temperatures in the experiments made under the full atmospheric pressure. The experiments reduced consist of four pair for each pendulum, except No. 21, for which only two pair were performed. The following are the results. For the 1½-inch platina sphere 0·0644, mean error 0·0044. For the 1½-inch brass sphere 0·180, mean error 0·024. For the 2-inch brass sphere 0·094, mean error 0·013. For the copper rod 0·486, mean error 0·113. For the brass tube the results were 0·145, 0·363, 0·338, 0·305. Rejecting the first result as anomalous, and taking the mean of the others, we get 0·335, mean error 0·030. To obtain I from the mean results above given we have only to divide by 3600 times the modulus, and multiply by τ, and for the experiments with spheres we may suppose $\tau = 1$.

The mode of calculating I from theory in the case of a sphere suspended by a fine wire has already been explained. For the sake of exhibiting separately the effect of the wire, I will give one intermediate step in the calculation.

	1·44 inch sphere.	1·46 inch sphere.	2·06 inch sphere.
k', for sphere alone	0·326	0·320	0·220
$\Delta k'$, the correction for the wire	0·130	0·130	0·045
Total, to be substituted in (169)	0·456	0·450	0·265

The formula (168), which applies to a sphere suspended by a wire, will be applicable to a long cylindrical rod if we suppose $M = 0$. Hence the same formula (169) that has been used for a sphere may be applied to a cylindrical rod if we suppose k' to refer to the rod. For the copper rod $k' = 1\cdot107$, and for the tube $k' = 0\cdot2561$. The following are the results for the three spheres and two cylinders.

	No. 1.	No. 3.	No. 6.	No. 21.	Nos. 35—38.
1000000 I, from experiment	41	115	60	315	206
.... from theory......	39	106	60	237	156
Difference.........	+2	+9	0	+78	+50

It appears that the experiments with spheres are satisfied almost exactly. The differences between the results of theory and observation are much larger in the case of the long cylinders. Large as these differences appear, they are hardly beyond the limits of errors of observation, though they would probably be far beyond the limits of errors of observation in a set of experiments performed on purpose to investigate the decrement of the arc of vibration. It was to be expected beforehand that the results of calculation would fall short of those of observation, inasmuch as only two arcs were registered in each experiment, so that no data were afforded for eliminating the effect of that part of the resistance which did not vary as the first power of the velocity.

78. I have now finished the comparison between theory and experiment, but before concluding this Section I will make a few general remarks.

When a new theory is started, it is proper to enquire how far the theory does violence to the notions previously entertained on the subject. The present theory can hardly be called new, because the partial differential equations of motion were given nearly thirty years ago by Navier, and have since been obtained, on different principles, by other mathematicians ; but the application of the theory to actual experiment, except in some doubtful cases relating to the discharge of liquids through capillary tubes, and the determination of the numerical value of the constant μ', are, I believe, altogether new. Let us then, in the first instance, examine the magnitude of the tangential pressure which we are obliged by theory to suppose capable of existing in air or water.

For the sake of clear ideas, conceive a mass of air or water to be moving in horizontal layers, in such a manner that each layer moves uniformly in a given horizontal direction, while the velocity increases, in going upwards, at the rate of one inch per second for each inch of ascent. Then the sliding in the direction of a horizontal plane is equal to unity, and therefore the tangential pressure referred to a unit of surface is equal to μ or $\mu'\rho$. The absolute magnitude of this unit sliding evidently depends only on the arbitrary unit of time, which is here supposed to be a second. In the case supposed, it will be easily seen that the particles situated at one instant in a vertical line are situated at the expiration of one second in a straight line inclined at an angle of 45° to the horizon. Equating the tangential pressure $\mu'\rho$ to the normal pressure due to a height h of the fluid, we get $h = g^{-1}\mu'$, g being the force of gravity. Putting now $g = 386$, $\mu' = (0.116)^2$ for air, $\mu' = (0.0564)^2$ for water, we get $h = 0.00003486$ inch for air, and $h = 0.000008241$ inch for water, or about the one thirty-thousandth part of an inch for air, and less than the one hundred-thousandth part of an inch for water. If we enquire what must be the side of a square in order that the total tangential pressure on a horizontal surface equal to that square may amount to one grain, supposing the density of air to be to that of water as 1 to 836, and the weight of a cubic inch of water to be 252·6 grains, we get 25 feet 8 inches for air, and 1 foot 10 inches for water. It is plain that the effect of such small forces may well be insignificant in most cases.

79.   In a former paper I investigated the effect of internal friction on the propagation of sound, taking the simple case of an indefinite succession of plane waves*. It appeared that the effect consisted partly in a gradual subsidence of the motion, and partly in a diminution of the velocity of propagation, both effects being greater for short waves than for long. The second effect, as I there remarked, would be contrary to the result of an experiment of M. Biot's, unless we supposed the term expressing this effect to be so small that it might be disregarded. I am now prepared to calculate the numerical value of the term in question, and so decide whether the theory is or is not at variance with the result of M. Biot's experiment.

According to the expression given in the paper just mentioned, we have for the proportionate diminution in the velocity of propagation

$$\frac{8\pi^2\mu'^2}{9\lambda^2 V^2},$$

λ being the length of a wave, and V the velocity of sound. To take a case as disadvantageous as possible, suppose λ only equal to one inch, which would correspond to a note too shrill to be audible to human ears. Taking the velocity of sound in air at 1000 feet per second, there results for the common logarithm of the expression above written $\overline{11}\cdot0428$, so that a wave would have to travel near 100000000000 inches, or about 1578000 miles, before the retardation due to friction amounted to one foot. It is plain that the introduction of internal friction leaves the theory of sound just as it was, so far as the velocity of propagation is concerned, at least if the sound be propagated in free air.

----

* *Camb. Phil. Trans.* Vol. VIII. p. 302.   [*Ante*, Vol. I. p. 101.]

The effect of friction on the intensity of sound depends on the first power of μ'. In the case of an indefinite succession of plane waves, it appears that during the time t the amplitude of vibration is diminished in the ratio of 1 to ϵ^{-ct}, and therefore the intensity in the ratio of 1 to ϵ^{-2ct}, where

$$c = \frac{8\pi^2\mu'}{3\lambda^2}.$$

Putting $\lambda = 1$ and $t = 1$ we get 1 to 0·4923, or 2 to 1 nearly, for the ratio in which the intensity is altered during one second in the case of a series of waves an inch long. The rate of diminution decreases very rapidly as the length of wave increases, so that in the case of a series of waves one foot long the intensity is altered in one second in the ratio of 1 to 0·995095, or 201 to 200 nearly. It appears then that in all ordinary cases the diminution of intensity due to friction may be neglected in comparison with the diminution due to divergence. If we had any accurate mode of measuring the intensity of sound it might perhaps be just possible, in the case of shrill sounds, to detect the effect of internal friction in causing a more rapid diminution of intensity than would correspond to the increase of distance from the centre of divergence.

Section II.

Suggestions with reference to future experiments.

80. I am well aware that the mere proposal of experiments does not generally form a subject fit to be brought before the notice of a scientific society. Nevertheless, as it frequently happens in the division of labour that one person attends more to the theoretical, another to the experimental investigation of some branch of science, it is not always useless for the theorist to point out the nature of the information which it would be most important to obtain from experiment. I hope, therefore, that I may be permitted to offer a few hints with reference to experiments in which the theory of the internal friction of fluids is concerned. I shall omit all details, since they would properly come in connexion with the experiments.

Experiments with which the theory of internal friction in fluids has more or less to do may be performed for either of the following objects: first, to test still further the truth of the theory; secondly, to determine the index of friction of various gases, liquids, or solutions; to investigate the dependance of the index of friction of a gas on its state of pressure, temperature, and moisture; or to endeavour to make out the law according to which the index of friction of a mixture of gases depends upon the indices of friction of the separate gases; thirdly, to measure the length of the seconds' pendulum, or its variation from one part of the earth's surface to another.

81. *First object.* The theory has been already put to a pretty severe test by means of the experiments of Baily and others. Nevertheless there are some uncertainties in the comparison of theory and experiment arising from the influence of modifying causes of which the effect could only be estimated from theory, and yet was not so small as to be merged in errors of observation. Moreover, experiments on the decrement of the arc of vibration are almost wholly wanting. The following system of pendulums, meant to be swung in air and in vacuum, would afford a very good test of the theory.

No. 1. A 2-inch or $1\frac{1}{2}$-inch sphere swung with a fine wire.

No. 2. A very small sphere swung with the same kind of wire.

No. 3. A long cylindrical rod, a few tenths of an inch in diameter.

No. 4. A cylinder only three or four inches long, of the same diameter as No. 3, swung with the same kind of wire as No. 1.

The vacuum tube ought to be of sufficient size to render the estimated correction for confined space less than, or at most comparable with, errors of observation. The vacuum apparatus used by Col. Sabine would do very well. If the vacuum tube be not of sufficient size, it ought to admit of removal, and to be removed when the pendulums are swung in air.

In all the experiments the arc of oscillation ought to be carefully observed several times during the motion, the observation of the arc being quite as important for the purposes of theory as the observation of the time. Indeed, if it should be inconvenient to observe the time, the observation merely of the arc would be very valuable as a test of theory. In that case an approximate value of the time of oscillation in air would be required.

In the system proposed, Nos. 1 and 3 are the principal pendulums, Nos. 2 and 4 are introduced for the sake of making certain small corrections to the results of Nos. 1 and 3. No. 2 is meant for the elimination from No. 1 of the effect of the wire, and No. 4 for the elimination from No. 3 of the effect of the resistance experienced by a small portion of the rod near its end. The times of vibration of the four pendulums ought to be nearly the same, although for that purpose slightly different lengths of wire would be required in Nos. 1, 2, and 4.

It follows from theory that for a given pendulum the factor n is a function of the time of vibration. This is a result which seems to have been hardly so much as suspected by those who were engaged in pendulum experiments, or at most to have been mentioned as a mere possibility*, and therefore it might be thought advisable to verify it by direct experiment. For my own part I regard it as so intimately connected with the fundamental principles of the theory, that if the theory be confirmed in other respects I think this result may be accepted on the strength of theory alone. The direct comparison with experiment would be inconvenient, because it would require a clock which kept excellent time, and yet admitted of being adjusted so as to make widely different numbers of vibrations in a day. The result could, however, be confirmed indirectly by observing the arc of vibration, an observation which is as easy with one time of vibration as with another.

82. *Second object.* According to theory, the index of friction may be deduced from experiments either on the arc or on the time of vibration. It must be left to observation to decide which give the more consistent results. Should the results obtained from the arc appear as trustworthy as those obtained from the time, it would apparently be much the easiest way of determining μ' for an elastic fluid to observe the arc, because no particular accuracy would then be required in the observation of time. As to the

* It should be observed however that in a subsequent memoir (*Astronomische Nachrichten*, No. 223, p. 106) Bessel deduced from other experiments that the value of k was larger for the long than for the short pendulum.

form of the pendulum, a cylindrical rod would apparently be the best if only a single pendulum were employed. The observation of the arc seems the only practicable way of determining the influence of temperature on the index of friction, unless the pendulum be extremely light, or unless the observer be content with the limited range of temperature which may be procured by making observations at different times of year. For in an apparatus artificially heated or cooled, it would be difficult to prevent small unknown variations of temperature, which would cause variations in the rate of vibration, in consequence of the expansion and contraction of the pendulum; and these variations would vitiate the result of the experiment, so far as the time of vibration is concerned, because the effect of the gas on the time of vibration is deduced from the small difference between two large quantities which are directly observed. But the effect of the gas on the arc of vibration produces by far the greater part of the whole diminution observed, and therefore small fluctuations of temperature would not be of much consequence, except so far as they might occasion gentle currents; and even then would not be very important, because the forces thence arising would not be periodic, and dependent upon the phase of vibration of the pendulum.

The grand difficulty which besets the observation of the time of vibration of a pendulum oscillating in a liquid consists in the rapidity with which the oscillations subside. The best form of a pendulum to oscillate in a liquid would be a sphere suspended by a fine wire. The vessel containing the liquid and the sphere immersed in it ought to be so large as to render the correction for confined space insensible. But the index of friction of a liquid would probably be better determined by experiments more of the nature of those of Coulomb, or perhaps by the slow discharge of liquids through narrow tubes.

Among the gases for which μ' ought to be determined experimentally should be mentioned coal-gas, on account of the practical application which it appears possible to make of the result to the laying down of gas-pipes. The calculation of the resistance in a circular pipe is very simple, and is given in Art. 9 of my former paper. According to the equations of condition assumed in the present paper we must put $U = 0$, U denoting in that article the velocity close to the surface. It appears that the pressure spent in overcoming friction varies as the mean velocity divided by the square of the diameter of the pipe, or as the rate of supply divided by the fourth power of the diameter. This goes on the supposition that the motion is sufficiently slow to allow of our neglecting the pressure which may be spent in producing eddies, in comparison with that spent in overcoming what really constitutes internal friction.

83. *Third object.* With respect to experiments for determining the length of the seconds' pendulum, the theory of internal friction rather enables us to calculate for certain forms of pendulum the correction due to the inertia of the air than points out any particular mode of performing the experiments. Even the ordinary theory of hydrodynamics points out the importance of removing all obstacles to the free motion of the air in the neighbourhood of the pendulum if we would calculate from theory the whole correction for reduction to a vacuum.

Since the theoretical solution has been obtained in the case of a long cylindrical rod, or of such a rod combined with a sphere, we may regard a pendulum formed in this manner, and which is convertible in air, as also convertible in vacuum, for it is of small consequence whether the pendulum be or be not really convertible in vacuum, provided that if it be not we know the correction to be applied in consequence.

NOTE A, *Article* 65.

Let us apply the general equations (2), (3) to the fluid surrounding a solid of revolution which turns about its axis, with either a uniform or a variable motion, supposing the fluid to have been initially either at rest, or moving in annuli about the axis of symmetry.

In the first place we may observe, that the fluid will always move in annuli about the axis of symmetry. For let P be any point of space, and L any line passing through P, and lying in a plane drawn through P and through the axis of symmetry; and at the end of the time t let u' be the velocity at P resolved along L. Now consider a second case of motion, differing from the first in having the angular velocity of the solid and the initial velocity of the fluid reversed, everything else being the same as before. It follows from symmetry, that at the end of the time t the velocity at P resolved along L will be equal to u', since the motion of the solid and the initial motion of the fluid, which form the data of the one problem, differ from the corresponding quantities in the other problem only as regards the distinction between one way round and the other way round, which has no relation to the distinction between to and fro in the direction of a line lying in a plane passing through the axis of rotation. But since all our equations are linear as regards the velocity, it follows that in the second problem the velocity will be the same as in the first, with a contrary sign, and therefore the velocity at P in the direction of the line L will be equal to $-u'$. Hence $u' = -u'$, and therefore $u' = 0$, and therefore the whole motion takes place in annuli about the axis of rotation.

Let the axis of rotation be taken for the axis of z; let ω be the angle which a plane passing through this axis and through the point P makes with the plane of xy, and let v' be the velocity at P. Then

$$u = -v' \sin \omega, \qquad v = v' \cos \omega, \qquad w = 0,$$

and all the unknown quantities of the problem are functions of t, z, and ϖ, where $\varpi = \sqrt{(x^2 + y^2)}$. Substituting in equations (2) the above values of u, v, and w, and after differentiation putting $\omega = 0$, as we are at liberty to do, we get

$$\frac{dp}{d\varpi} = 0, \qquad \frac{dp}{dz} = 0,$$

$$\mu \left(\frac{d^2v'}{dz^2} + \frac{d^2v'}{d\varpi^2} + \frac{1}{\varpi}\frac{dv'}{d\varpi} - \frac{v'}{\varpi^2} \right) = \rho \frac{dv'}{dt} \quad \dots\dots\dots(179).$$

The first two of these equations give $p = $ a constant, or rather $p = $ a function of t, which for the same reason as in Art. 7 we have a right to suppose to be equal to zero. The third equation combined with the equations of condition serves to determine v'.

Now in the particular case of an oscillating disk, the equation (179) becomes according to the mode of approximation adopted in Art. 8

$$\mu \frac{d^2v'}{dz^2} = \rho \frac{dv'}{dt} \quad \dots\dots\dots\dots\dots(180),$$

which in fact is the same as the second of the equations (8). The solution thus obtained is as we have seen

$$v' = \varpi f(z, t) \dots\dots\dots\dots\dots(181),$$

f denoting a function the form of which there is no need to write down, which satisfies (180) when written for v'. Now it will be seen at once that the expression (181) satisfies the exact equation (179), and therefore the approximate solution obtained by the method of Art. 8 is in fact exact, except so far as regards the termination of the disk at its edge, which is what it was required to prove.

Passing from semi-polar to polar co-ordinates, by putting $z = r \cos \theta$, $\varpi = r \sin \theta$, we get from (179), after writing $\mu' \rho$ for μ,

$$\frac{d^2 v'}{dr^2} + \frac{2}{r} \frac{dv'}{dr} + \frac{1}{r^2 \sin \theta} \frac{d}{d\theta} \left(\sin \theta \frac{dv'}{d\theta} \right) - \frac{v'}{r^2 \sin^2 \theta} = \frac{1}{\mu'} \frac{dv'}{dt} \dots(182).$$

Suppose now the solid to be a sphere, having its centre at the origin. Let a be its radius, γ its angular velocity, and suppose the fluid initially at rest. Then v' is to be determined from the general equation (182) and the equations of condition

$$v' = 0 \text{ when } t = 0, \quad v' = a \gamma \sin \theta \text{ when } r = a, \quad v' = 0 \text{ when } r = \infty.$$

All these equations are satisfied by supposing

$$v' = v'' \sin \theta,$$

v'' being a function of r and t only. We get from (182)

$$\frac{d^2 v''}{dr^2} + \frac{2}{r} \frac{dv''}{dr} - \frac{2v''}{r^2} = \frac{1}{\mu'} \frac{dv''}{dt} \quad \dots\dots\dots(183).$$

If we suppose γ constant, v'' will tend indefinitely to become constant as t increases indefinitely, and in the limit $\frac{dv''}{dt} = 0$, whence we get from (183) and the equations of condition $v'' = a \gamma$ when $r = a$, $v'' = 0$ when $r = \infty$,

$$v'' = \frac{\gamma a^3}{r^2}, \qquad v' = \frac{\gamma a^3}{r^2} \sin \theta.$$

This is the solution alluded to in Art. 8 of my paper *On the Theories of the Internal Friction of Fluids in motion, &c.*

NOTE B, *Article* 65.

Let us resume the problem of Art. 7, but instead of the motion of the plane being periodic, let us suppose that the plane and fluid are initially at rest, and that the plane is then moved with a constant velocity V, and let the notation be the same as in Art. 7.

The general equations (8) remain the same as before, but the equations of condition become in this case

$$v = 0 \quad \text{when} \quad t = 0 \text{ from } x = 0 \text{ to } x = \infty,$$
$$v = V \quad \text{when} \quad x = 0 \text{ from } t = 0 \text{ to } t = \infty.$$

By Fourier's theorem and another theorem of the same kind, v may be expanded between the limits 0 and ∞ of x in the following form:

$$v = \frac{2}{\pi} \int_0^\infty \int_0^\infty \cos \alpha x \cos \alpha x' \, \phi(x', t) \, dx' d\alpha$$
$$+ \frac{2}{\pi} \int_0^\infty \int_0^\infty \sin \alpha x \sin \alpha x' \psi(x', t) \, dx' d\alpha \ldots\ldots(184).$$

In fact, v could be expanded by means of either of these expressions separately, and of course can be expanded in an infinite number of ways by the sum of the two. If however v had been expanded by means of the first expression alone, its derivatives with respect to x could not have been obtained by differentiating under the integral signs, inasmuch as the derivatives of an odd order do not vanish when $x = 0$, but would have been given by certain formulæ which I have investigated in a former paper*. A similar remark applies to the second expansion, in consequence of the circumstance that v itself and its derivatives of an even order do not vanish with x. But by combining the two expansions we may obtain the derivatives of v, up to any order i that we please to fix on, by merely differentiating under the integral signs. For we may evidently express the finite function v, and that in an infinite number of ways, as the sum of two finite functions $\phi(x, t)$, $\psi(x, t)$ which like v vanish when $x = \infty$, and which are such that the odd derivatives of the first, and the even derivatives of the

---

* On the critical values of the sums of periodic series. *Camb. Phil. Trans.*, Vol. VIII. p. 533. [*Ante*, Vol. I. p. 287.]

second, up to the order i, as well as $\psi(x, t)$ itself, vanish when $x = 0$. Substituting now in the second equation (8) the expression for v given by (184), we see that the equation is satisfied provided

$$\frac{d\phi}{dt} + \mu'\alpha^2\phi = 0, \qquad \frac{d\psi}{dt} + \mu'\alpha^2\psi = 0.$$

These equations give

$$\phi(x', t) = \chi(x') \epsilon^{-\mu'\alpha^2 t}, \qquad \psi(x', t) = \sigma(x') \epsilon^{-\mu'\alpha^2 t},$$

where χ, σ denote two new arbitrary functions. Substituting in (184), and then passing to the first of the equations of condition, we get

$$0 = \chi(x) + \sigma(x),$$

whence $\sigma(x) = -\chi(x)$ and

$$v = \frac{2}{\pi} \int_0^\infty \int_0^\infty \cos \alpha (x + x') \epsilon^{-\mu'\alpha^2 t} \chi(x')\, dx'\, d\alpha$$

$$= \frac{1}{\sqrt{\pi\mu't}} \int_0^\infty \epsilon^{-\frac{(x'+x)^2}{4\mu't}} \chi(x')\, dx' \quad \ldots\ldots\ldots(185).$$

The second of the equations of condition requires that

$$V = \frac{1}{\sqrt{\pi\mu't}} \int_0^\infty \epsilon^{-\frac{x'^2}{4\mu't}} \chi(x')\, dx' = \frac{2}{\sqrt{\pi}} \int_0^\infty \epsilon^{-s^2} \chi(2s\sqrt{\mu't})\, ds.$$

Since the second member of this equation must be independent of t, we get $\chi(x') =$ a constant, and this constant must be equal to V, since

$$\frac{2}{\sqrt{\pi}} \int_0^\infty \epsilon^{-s^2}\, ds = 1.$$

Substituting in (185) we get

$$v = \frac{V}{\sqrt{\pi\mu't}} \int_0^\infty \epsilon^{-\frac{(x+x')^2}{4\mu't}}\, dx' \ldots\ldots\ldots\ldots(186).$$

For the object of the present investigation nothing is required but the value of $\dfrac{dv}{dx}$ for $x = 0$, which we may denote by $\left(\dfrac{dv}{dx}\right)_0$. We get from (186)

$$\left(\frac{dv}{dx}\right)_0 = -\frac{V}{\sqrt{\pi\mu't}} \quad\ldots\ldots\ldots\ldots(187).$$

Now suppose the plane to be moved in any manner, so that its velocity at the end of the time t is equal to $f(t)$. We may evidently obtain the result for this case by writing $f'(t')\, dt'$ for V, and $t - t'$ for t in (187), and integrating with respect to t'. We thus get

$$\left(\frac{dv}{dx}\right)_0 = -\frac{1}{\sqrt{\pi\mu'}}\int_{-\infty}^{t} f'(t')\,\frac{dt'}{\sqrt{t-t'}} = -\frac{1}{\sqrt{\pi\mu'}}\int_{0}^{\infty} f'(t-t_1)\,\frac{dt_1}{\sqrt{t_1}} \quad\ldots(188).$$

To apply this result to the case of an oscillating disk, let $r\,\dfrac{d\theta}{dt} = rF(t)$ be the velocity of any annulus, and G the moment of the whole force of the fluid on the disk. Then

$$G = 4\pi\mu'\rho \int_0^a r^3 \left(\frac{dv_0}{dx}\right)_0 dr\,;$$

and $\left(\dfrac{dv}{dx}\right)_0$ will be got from (188) by substituting $rF(t)$ for $f(t)$. We find thus

$$G = -\sqrt{\pi\mu'}\,.\,\rho a^4 \int_0^{\infty} F'(t-t_1)\,\frac{dt_1}{\sqrt{t_1}}\ldots\ldots\ldots(189).$$

If we suppose the angular velocity of the disk to be expressed by $A\sin nt$, where A is constant, we must put $F(t) = A\sin nt$ in (189), and we should then get after integration the same expression for G as was obtained in Art. 8 by a much simpler process. Suppose, however, that previously to the epoch from which t is measured the disk was at rest, and that the subsequent angular velocity is expressed by $A_t \sin nt$, where A_t is a slowly varying function of t. Then

$$F(t) = 0 \text{ when } t < 0, \qquad F(t) = A_t \sin nt \text{ when } t > 0.$$

On substituting in (189) we get

$$G = -\sqrt{\pi\mu'}\,.\,\rho a^4 n \int_0^t A_{t-t_1} \cos n(t-t_1)\,\frac{dt_1}{\sqrt{t_1}} \quad\ldots\ldots(190).$$

Now treating A_t as a slowly varying parameter, we get from a formula given by Mr Airy, and obtained by the method of the variation of parameters,

$$\frac{dA_t}{dt} = \frac{G}{I}\sin nt \quad\ldots\ldots\ldots\ldots\ldots(191),$$

where I denotes the moment of inertia. In the expression for G we may replace A_{t-t_1} under the integral sign by A_t outside it, because A_t is supposed to vary so slowly that A_{t-t_1} does not much differ from A_t while t_1 is small enough to render the integral of importance. Making this simplification and substituting in (191) we get

$$\frac{dA_t}{A_t dt} = -c \sin nt \int_0^t \cos n(t-t_1) \frac{dt_1}{\sqrt{t_1}} \quad \ldots\ldots\ldots(192),$$

where $c = \sqrt{(\pi\mu')} \cdot \rho a^4 n I^{-1}$. If then A_0 be the initial and A the final value of A_t, we get from (192)

$$\log \frac{A_0}{A} = c \int_0^t \left\{ \sin nt \int_0^t \cos n(t-t_1) \frac{dt_1}{\sqrt{t_1}} \right\} dt \quad \ldots(193).$$

Let now $A_0 + \Delta A_0$ be what A_0 would become if, while the final arc A and the whole time t remained the same, the motion had been going on for an indefinite time before the epoch from which t is measured, in which case the superior limit in the integral involved in the expression for G would have been ∞ in place of t. Then

$$\log \frac{A_0 + \Delta A_0}{A} = c \int_0^t \left\{ \sin nt \int_0^\infty \cos n(t-t_1) \frac{dt_1}{\sqrt{t_1}} \right\} dt \ldots(194),$$

whence by subtracting, member from member, equation (193) from equation (194), we get

$$\log \frac{A_0 + \Delta A_0}{A_0} = c \int_0^t \left\{ \sin nt \int_t^\infty \cos n(t-t_1) \frac{dt_1}{\sqrt{t_1}} \right\} dt,$$

which becomes after integration by parts

$$\log \frac{A_0 + \Delta A_0}{A_0} = \frac{c}{4n} \left\{ \sqrt{\frac{\pi}{2n}} - 2\sqrt{t} \cdot \cos nt - \cos 2nt \int_t^\infty \cos nt \frac{dt}{\sqrt{t}} \right.$$
$$\left. + (2nt - \sin 2nt) \int_t^\infty \sin nt \frac{dt}{\sqrt{t}} \right\} \ldots\ldots(195).$$

Now t is supposed to be very large: in Coulomb's experiments in fact 10 oscillations were observed, so that $nt = 10\pi$. But when t is at all large the two integrals

$$\int_t^\infty \cos nt \frac{dt}{\sqrt{t}}, \qquad \int_t^\infty \sin nt \frac{dt}{\sqrt{t}}$$

can be expressed under the forms

$$-P \sin nt + Q \cos nt, \qquad P \cos nt + Q \sin nt,$$

where

$$P = n^{-1} t^{-\frac{1}{2}} - 1 \cdot 3 \cdot 2^{-2} n^{-3} t^{-\frac{5}{2}} + \ldots,$$
$$Q = 1 \cdot 2^{-1} n^{-2} t^{-\frac{3}{2}} - 1 \cdot 3 \cdot 5 \cdot 2^{-3} n^{-4} t^{-\frac{7}{2}} + \ldots,$$

series which are at first rapidly convergent, and which enable us to calculate the numerical values of the integrals with extreme facility. These expressions were first given by M. Cauchy, in the case of Fresnel's integrals, to which the integrals just written are equivalent. They may readily be obtained by integration by parts, though it is not thus that they were demonstrated by M. Cauchy. If now the above expressions be substituted for the integrals in (195) the terms containing $t^{\frac{1}{2}}$ destroy each other, and for general values of t the most important term after the first contains $t^{-\frac{1}{2}}$. Since however t is supposed to correspond to the end of an oscillation, so that nt is a multiple of π, the coefficient of this term vanishes, and the most important term that actually remains contains only $t^{-\frac{3}{2}}$. Hence neglecting insensible quantities we get from (195)

$$\log \frac{A_0 + \Delta A_0}{A_0} = \frac{c}{4n} \sqrt{\frac{\pi}{2n}} \quad \ldots\ldots\ldots\ldots(196).$$

We get from (194) by performing the integrations

$$\log \frac{A_0 + \Delta A_0}{A} = c \sqrt{\frac{\pi}{2n}} \int_0^t \sin nt \, (\cos nt + \sin nt) \, dt$$

$$= \frac{c}{4n} \sqrt{\frac{\pi}{2n}} \{2nt + 1 - \cos 2nt - \sin 2nt\},$$

which becomes since nt is a multiple of π

$$\log \frac{A_0 + \Delta A_0}{A} = \frac{c}{4n} \sqrt{\frac{\pi}{2n}} \cdot 2nt \ldots\ldots\ldots\ldots(197).$$

We get from (196) and (197)

$$2nt \log \frac{A_0 + \Delta A_0}{A_0} = \log \frac{A_0 + \Delta A_0}{A} = \log \frac{A_0 + \Delta A_0}{A_0} + \log \frac{A_0}{A},$$

whence

$$\log \frac{A_0 + \Delta A_0}{A_0} = (2nt - 1)^{-1} \log \frac{A_0}{A} \quad \ldots\ldots\ldots(198),$$

and the same relation exists between the common logarithms of the arcs, which are proportional to the Napierian logarithms. Now $\text{Log} \, A_0 - \text{Log} \, A$ is the quantity immediately deduced from experiment, and $\text{Log} \, (A_0 + \Delta A_0) - \text{Log} \, A_0$ is the correction to be applied, in consequence of the circumstance that the motion began from rest. Instead of applying the proportionate correction

$+(2nt-1)^{-1}$ to the difference of the logarithms, we may apply it to the deduced value of $\sqrt{\mu'}$, which is proportional to the difference of the logarithms. In Coulomb's experiments 10 oscillations were observed, and therefore $2nt = 20\pi$, and $(2nt-1)^{-1} = 0\cdot01617$, and the uncorrected value of $\sqrt{\mu'}$ being $0\cdot0555$, we get $0\cdot0009$ for the correction, giving $\sqrt{\mu'} = 0\cdot0564$.

NOTE C. *Article* 50.

The results mentioned in this article were originally given without demonstration; but as the mode in which they were obtained is short, and by no means obvious, I have thought it advisable to add the demonstrations.

In order that the right-hand members of equations (138) may be perfect differentials, we must have

$$\frac{d\delta}{dy} + \frac{d\omega'''}{dx} = 0, \qquad \frac{d\delta}{dz} + \frac{d\omega'}{dy} = 0, \qquad \frac{d\delta}{dx} + \frac{d\omega''}{dz} = 0 \ldots(a),$$

$$\frac{d\delta}{dz} - \frac{d\omega''}{dx} = 0, \qquad \frac{d\delta}{dx} - \frac{d\omega'''}{dy} = 0, \qquad \frac{d\delta}{dy} - \frac{d\omega'}{dz} = 0 \ldots(b),$$

$$\frac{d\omega''}{dy} + \frac{d\omega'''}{dz} = 0, \qquad \frac{d\omega'''}{dz} + \frac{d\omega'}{dx} = 0, \qquad \frac{d\omega'}{dx} + \frac{d\omega''}{dy} = 0 \ldots(c).$$

The equations (c) give

$$\frac{d\omega'}{dx} = 0, \qquad \frac{d\omega''}{dy} = 0, \qquad \frac{d\omega'''}{dz} = 0 \ldots\ldots\ldots\ldots(d).$$

In the particular case in which $\delta = 0$, the equations (a), (b), and (d) give

$$d\omega' = 0, \qquad d\omega'' = 0, \qquad d\omega''' = 0,$$

and therefore ω', ω'', and ω''' are constant as stated in Art. 50. In the general case the equations (a), (b), and (d) give for the differentials of ω', ω'', and ω''' the following expressions:

$$\left.\begin{aligned} d\omega' &= -\frac{d\delta}{dz}\, dy + \frac{d\delta}{dy}\, dz, \\[2mm] d\omega'' &= -\frac{d\delta}{dx}\, dz + \frac{d\delta}{dz}\, dx, \\[2mm] d\omega''' &= -\frac{d\delta}{dy}\, dx + \frac{d\delta}{dz}\, dy \end{aligned}\right\} \quad \ldots\ldots\ldots\ldots(e).$$

In order that the right-hand members of these equations may be perfect differentials, we must have

$$\frac{d^2\delta}{dy\,dz} = 0, \qquad \frac{d^2\delta}{dz\,dx} = 0, \qquad \frac{d^2\delta}{dx\,dy} = 0 \ \ldots\ldots(f),$$

$$\frac{d^2\delta}{dy^2} + \frac{d^2\delta}{dz^2} = 0, \qquad \frac{d^2\delta}{dz^2} + \frac{d^2\delta}{dx^2} = 0, \qquad \frac{d^2\delta}{dx^2} + \frac{d^2\delta}{dy^2} = 0,$$

and therefore

$$\frac{d^2\delta}{dx^2} = 0, \qquad \frac{d^2\delta}{dy^2} = 0, \qquad \frac{d^2\delta}{dz^2} = 0 \ \ldots\ldots\ldots\ldots(g).$$

The equations (f), (g) give

$$d\,\frac{d\delta}{dx} = 0, \qquad d\,\frac{d\delta}{dy} = 0, \qquad d\,\frac{d\delta}{dz} = 0,$$

so that $\dfrac{d\delta}{dx}$, $\dfrac{d\delta}{dy}$, and $\dfrac{d\delta}{dz}$ are constant. Substituting in (e) and integrating, and then substituting in (138) the resulting expressions for ω', ω'', ω''', and integrating again, we shall obtain the results given in Art. 50.

[The possibility of a more general kind of motion than that of a solid taking place in an elastic fluid without consumption of energy by internal friction, that is, without its being converted into the kinetic energy of heat, depends on the coefficient in the last term of (1) being $\mu/3$, or rather not greater than $\mu/3$; and that again on the assumption made in a former paper (Vol. I. p. 87) that in any elementary portion of the fluid a velocity of dilatation alike in all directions does not affect the hydrostatical relation between the pressure and density. Although I have shown (Vol. I., p. 119) that on the admission of a supposition which Poisson would probably have allowed the two constants in his equations of motion are reduced to one, and the equations take the form (1), and although Maxwell obtained the same equations from his kinetic theory of gases (*Philosophical Transactions* for 1867, p. 81) I have always felt that the correctness of the value $\mu/3$ for the coefficient of the last term in (1) does not rest on as firm a basis as the correctness of the equations of motion of an incompressible fluid, for which the last term does not come in at all. If the supposition made above be not admitted, we must replace the coefficient $\mu/3$ by a different coefficient, which may be written $\mu/3 + \varpi$, and ϖ must be positive, as otherwise the mere alternate expansion and contraction, alike in all directions, of a fluid, instead of demanding the exertion of work upon it, would cause it to give out work. But if the positive constant exists, the coefficient of the squared velocity of dilatation in the transformed expression for DV in p. 69, instead of being $-\frac{2}{3}\mu$, will be $-\frac{2}{3}\mu + \varpi$, and in order that the quantity under the sign of triple integration may vanish, we must have in addition to the equations (137) on p. 70 the further equation $\delta = 0$, and the conclusion is the same as in the case of an incompressible fluid.]

ADDITIONAL NOTE. (*See foot-note at p.* 77.)

[The fact that notwithstanding the great variety in the forms of Baily's pendulums, even when restricted to those to which the theory of the present paper is applicable, the results of his experiments manifest such a remarkable agreement with theory, in spite of the adoption in the calculation of a law as to the relation of μ to ρ which we now know to be wrong, paradoxical as that fact appears at first sight, admits of being easily explained by combining two considerations, relating the one to the way in which the experiments were conducted, the other to the character of the formulæ.

In the case of the 41 pendulums mentioned in Baily's paper as originally presented to the Royal Society, the high pressure under which each pendulum was swung was always the atmospheric pressure, and the low pressure did not much differ from that of 1 inch of mercury. There can be little doubt that the same practice was followed as regards the 45 additional pendulums for which the reduced results only, not the details, are given in the appendix. Hence the two pressures used would be nearly the same throughout, and nearly those measured by 30 inches and 1 inch of mercury. The ratio of the densities would be very nearly the same.

The expression (52) shows that for a sphere k is of the form

$$k = A + B\sqrt{\mu'} \quad \dots\dots\dots\dots\dots\dots(a),$$

where A is an abstract number, and B depends on the diameter and time of vibration of the sphere, but is constant when only the

nature and the density of the gas are changed.   In the case of a
long cylinder, the expression for k is complicated, and involves
infinite series.   Nevertheless the numerical table on p. 52 shows
that if we except the very early part of the table, which corresponds to very slender rods, or (according to Maxwell's law) to
very low pressures, the value of k nearly fits an expression of the
form (a), A being in this case 1, instead of $\frac{1}{2}$, which was its value
for a sphere, and B depending, as it did before, on the diameter
and time of vibration.   And if the pendulum be made up of a
sphere and a cylindrical rod, as was the case with many of Baily's,
we shall still not go far wrong if we take k to be expressed in the
same form (a), in which the constants A and B admit of being
obtained by calculation.   Now by the definition of k the effect of
the air on the time of vibration of a given pendulum varies as
$(1 + k)\rho$.   Baily's n, or $1 + k$, was got by dividing the observed
difference in the time of vibration at the high and low pressure,
corrected for everything but the effect of the air, by the calculated
difference for buoyancy alone.

Hence $n = 1 + \Delta k\rho/\Delta\rho$, where Δ denotes the difference at the
high and low pressures.   If we assume k to be given nearly
enough by the formula (a), we have

$$n = 1 + A + B\,\frac{\Delta\rho\sqrt{\mu'}}{\Delta\rho}.$$

If we put BC for $n - 1 - A$, we have according to the assumed
law $\sqrt{\mu'} = C$, and therefore if we denote the higher and lower
densities by ρ_1, ρ_0, we have at the atmospheric density

$$\sqrt{\mu} = C\sqrt{\rho_1},$$

which gives what μ is supposed to be.   But we ought to have
taken

$$C = \sqrt{\mu}\,.\,\frac{\Delta\sqrt{\rho}}{\Delta\rho},$$

so that if we denote the apparent coefficient by $'\mu$, reserving μ for
the true coefficient, we have

$$\sqrt{\frac{\mu}{'\mu}} = \frac{\Delta\rho}{\sqrt{\rho_1}\Delta\sqrt{\rho}} = \frac{\rho_1 - \rho_0}{\sqrt{\rho_1}(\sqrt{\rho_1} - \sqrt{\rho_0})} = 1 + \sqrt{\frac{\rho_0}{\rho_1}}.$$

Now in the same experiment the swings at high and low
pressure were taken at temperatures that did not much differ;

and indeed Baily arranged the order of the swings in such a manner as to eliminate the effect of a small progressive change. We may therefore take the ratio of the densities as being that of the pressures, which, as already stated, was nearly that of 30 to 1. Hence the values of μ which would be deduced independently from the different experiments on the supposition that μ' and not μ was independent of the density would all be wrong in nearly the same ratio, which would be nearly that of $(\sqrt{30} + 1)^2$ to 30. This accounts for the remarkable agreement between theory and experiment as regards the time of vibration, notwithstanding the employment of an erroneous law as to the relation between μ and ρ in making the correction for the residual air at the low pressure. Moreover, in order to arrive at the value of μ which would have been deduced from the experiments on the time of vibration had they been reduced according to Maxwell's law, we have merely to increase the value as obtained in this paper in a ratio which is nearly that given above, or 1 to 1·398. The mean high and low pressures for the four $1\frac{1}{2}$-inch spheres were 30·062 and 1·177, numbers which would give for the factor 1·558. Of the three pendulums in the table on p. 79, from which the adopted value of $\sqrt{\mu'}$ was deduced, the first is the only one for which the pressures are recorded in Baily's paper, and the calculated factor for it is 1·437.

The results of the most recent and trustworthy experiments for the determination of μ for air are brought together by Mr H. Tomlinson in a paper published in the *Philosophical Transactions* for 1886, p. 767. From the numbers given by him on p. 768 (first line in the table), and on pp. 784, 785, it appears that the true factor should be about 1·700. It is $\sqrt{\mu}$ that enters into the expression for the time of vibration, and the difference between the square roots of 1·7 and 1·4 is only ·0925 of the former; and it is only a portion of the correction for inertia in which the viscosity is involved at all. Thus in the case of the $1\frac{1}{2}$-inch spheres (see p. 86), that part of the correction for the air in which alone the viscosity is involved is little more than the one-seventh of the whole; and a fraction of this again which is barely one-tenth would amount to little more than one per cent. of the whole effect of the air. Considering the uncertainties as to some small corrections, such as that for the effect of the confinement of the space in which the pendulums were swung, the resistance to the wire in the case of the spheres, the deviation of the motion of the air near the bottom of long cylindrical rods from a motion in two dimensions, in a horizontal plane, such small discrepancies as that just noticed can hardly be affirmed to be real, that is, such as would emerge from mere casual errors of observation if the above corrections were made perfectly. It is however possible that with amplitudes of vibration as large as those actually used, amounting to about 1° to start with, there may have been a very slight pro-duction of eddies, the effect of which on the time of vibration may not have been wholly insensible.

But it is not only with regard to the time of vibration that the results of Baily's experiments manifest such a remarkable agree-ment with theory notwithstanding the adoption of an erroneous assumption as to the relation of μ to ρ; the observed reduction of the arc of vibration was also found not greatly to differ from that given by calculation.

As regards the spheres, the equation (172), p. 114, gives

$$k'/(k - \tfrac{1}{2}) = 1 + \tfrac{4}{9}(k - \tfrac{1}{2}).$$

Now even in the case of the 1½-inch spheres the right-hand member of this equation exceeds unity by only 0·127, so that the supposition that, in order to rectify the law connecting μ with ρ, k' requires to be altered in the same ratio as $k - \tfrac{1}{2}$ does not much differ from the truth. Accordingly the calculated numbers for the spheres (p. 120) differ but little from the numbers given by obser-vation. Calculation by aid of the table on p. 52 shows as regards the rod No. 21 that if the calculated number 237 represented the number given by observation, the assumed value (0·116)² of μ' for atmospheric density would have to be increased in the ratio of 1 to 1·400, bringing it considerably nearer to the true value as given by modern determinations. As observation gave for the logarithmic decrement 315, μ' would have to be still further increased, making it somewhat too high. Similar remarks apply to the tube Nos. 35—38, which however according to Baily's figure differs much from a plain cylinder, and which moreover requires a comparatively large correction for confined space. On the whole then the pretty close agreement between theory and observation as to the decrement of arc, notwithstanding the assumption of a wrong law in the reduction, may be considered to have been accounted for.

The very accurate experiments of Bessel are unfortunately not available for more than a very rude comparison, for the reason mentioned in the foot-note at p. 97.

As regards the correction on account of the air to the time of vibration of a pendulum, we have seen that in the case of a sphere, and very approximately in the case of a long cylindrical rod which is not extremely narrow, it is of the form

$$Gp + H\sqrt{(\mu\rho)},$$

where G and H depend on the form of the pendulum, but not upon the pressure, nor indeed on the nature, of the surrounding gas, which might be other than air. There can be little doubt that the same would apply as a very near approximation to any of the ordinary forms of pendulum, though in that case the constants G, H cannot in general be obtained by calculation. The first term depends partly on buoyancy, partly on the inertia of the gas regarded as a perfect fluid. As the latter part cannot be calculated, there is no need to calculate the former, since the two have to be determined as a whole by observation. As the value of μ for air is now well known, the constants G and H may be determined from the differences in the times of vibration at three suitably distributed pressures. These constants are determined once for all for the same pendulum. They may even be applied without a fresh experimental determination to any other pendulum of which the external form is geometrically similar, even though the internal distribution of mass be different, of course with due regard to the dimensions of the terms with respect to the units of length and time. Moreover unless we want to combine observations with different gases, or else to take account of the variation of μ with temperature, we may write the above formula

$$Gp + H'\sqrt{\rho},$$

and as we must appeal to experiment for the determination of H', we do not even need to know the value of μ.]

[From the *Philosophical Magazine*, Vol. I., p. 305 (*April*, 1851).]

An Examination of the possible Effect of the Radiation of Heat on the Propagation of Sound*.

INASMUCH as Laplace's formula is a rigorous deduction from the physical hypotheses adopted, there is no way of escaping from his result but by calling in question the hypotheses themselves. Now the development of heat and cold by sudden condensation and rarefaction is not merely a hypothetical cause, the only evidence of whose existence is that it explains the phenomena, but is a well-known physical fact, proved by direct experiment. That in the case of small sudden condensations (positive or negative) the increase of temperature is ultimately proportional, *cœteris paribus*, to the condensation, will not, it is presumed, be called in question. The only way, then, of escaping from the conclusion that the velocity of sound is really increased by the cause assigned is, to suppose that the heat produced by condensation passes away so rapidly by radiation that the result is the same as though condensation and rarefaction were incapable of changing the temperature of air. The main object of the present communication is to examine the consequences of such a supposition, in order to make out whether it be tenable or not.

Let us take the case of an infinite mass of homogeneous elastic fluid, acted on by no external forces, and having throughout a uniform temperature, and consequently a uniform pressure, except

---

* [This examination was made in consequence of the publication in the *Philosophical Magazine* of some papers in which the correctness of Laplace's explanation of the excess of the observed velocity of sound over that calculated by Newton was called in question. In the reprint, a few passages which are merely controversial, and of ephemeral interest, are omitted.]

in so far as the pressure, and consequently the temperature, are affected by small vibratory movements. Let the fluid be referred to the rectangular axes of x, y, z; let u, v, w be the components of the velocity, t the time, p the pressure, ρ the density in equilibrium, $\rho(1 + s)$ the actual density, so that s is the condensation. The three ordinary equations of motion and the equation of continuity become in this case, on neglecting as usual the squares of small quantities,

$$\frac{dp}{dx} = -\rho \frac{du}{dt}, \quad \frac{dp}{dy} = -\rho \frac{dv}{dt}, \quad \frac{dp}{dz} = -\rho \frac{dw}{dt} \quad \ldots\ldots\ldots(1),$$

$$\frac{ds}{dt} + \frac{du}{dx} + \frac{dv}{dy} + \frac{dw}{dz} = 0 \quad \ldots\ldots\ldots\ldots\ldots\ldots\ldots(2).$$

Let θ_0 be the temperature in equilibrium, $\theta_0 + \theta$ the actual temperature. Then $p = k_0 \rho (1 + s)(1 + \alpha_0 \overline{\theta_0 + \theta})$. Putting k for $k_0 (1 + \alpha_0 \theta_0)$, α for $\alpha_0 (1 + \alpha_0 \theta_0)^{-1}$, and neglecting the product of s and θ, which are both small quantities of the first order, we get

$$p = k\rho (1 + s + \alpha\theta) \quad \ldots\ldots\ldots\ldots\ldots(3).$$

It remains to form the equation relating to the changes of temperature. Let βs be the elevation of temperature produced by a sudden small condensation s. The condensation which a given element of the fluid receives in the time dt is equal to $s'dt$, where

$$s' = \frac{ds}{dt} + u \frac{ds}{dx} + v \frac{ds}{dy} + w \frac{ds}{dz} = \frac{ds}{dt}, \text{ nearly };$$

and the elevation of temperature due to this condensation is equal to $\beta s'dt$. We know that heat radiates freely to great distances in air, and therefore, of the heat which radiates from the element considered, we may neglect the small portion which may be absorbed by the air in its neighbourhood, and consider only what goes to great distances. Hence the result will be sensibly the same as if the element radiated into a medium having the constant temperature θ_0, which is the mean temperature of the whole. The quantity, then, which escapes from the element during the time dt, will be proportional to the small excess θ of the temperature of the element over the mean temperature of the medium; and the consequent depression of temperature may be expressed by $q\theta dt$, where q is a constant which may be called the *velocity of*

cooling referred to a difference of temperature unity. We have, therefore,

$$\frac{d\theta}{dt} = \beta\frac{ds}{dt} - q\theta \quad\dots\dots\dots\dots(4).$$

The six general equations (1), (2), (3), (4) serve, along with the equations of condition relating to any particular problem, to make known the six unknown quantities u, v, w, p, s, θ.

To simplify the question as much as possible, I shall take the case of plane waves. Taking the axis of x perpendicular to the planes of the waves, we have $v = 0$, $w = 0$, and u, p, s, θ will be functions of only x and t. The equations (1) and (2) become

$$\frac{dp}{dx} = -\rho\frac{du}{dt}, \quad \frac{ds}{dt} + \frac{du}{dx} = 0 \quad\dots\dots\dots(5);$$

and eliminating p and u from these equations and (3), we get

$$\frac{d^2s}{dt^2} = k\left(\frac{d^2s}{dx^2} + \alpha\frac{d^2\theta}{dx^2}\right) \quad\dots\dots\dots\dots(6).$$

Eliminating θ between (4) and (6), we get

$$\left(\frac{d}{dt} + q\right)\frac{d^2s}{dt^2} = k\left\{(1 + \alpha\beta)\frac{d}{dt} + q\right\}\frac{d^2s}{dx^2} \quad\dots\dots(7).$$

This equation is satisfied by

$$s = A'e^{m'x+n't}\dots\dots\dots\dots\dots(8).$$

where A' is an arbitrary constant, real or imaginary, and m', n' are two real or imaginary constants connected by the equation

$$m'^2 = \frac{(n' + q)n'^2}{k\{(1 + \alpha\beta)n' + q\}} \quad\dots\dots\dots\dots(9).$$

If we suppose m' wholly imaginary, the formulæ will refer to an infinite series of waves, the expressions for s, &c., involving x under the circular functions *sine* and *cosine*. In this case our formulæ would make known the manner in which the motion alters with the time. If we suppose n' wholly imaginary, the motion will be periodic as regards the time. In this case we must not suppose the fluid unlimited, but bounded in one direction by a vibrating plane which keeps up the motion. I shall select the second case for consideration, inasmuch as it is analogous to that of the vibrations propagated along a long tube from a sonorous body at one end of it, and accordingly will bear on the

experiments by which M. Biot proved that the velocity of propagation of sound in air is independent of the pitch.

Let the origin be situated at the vibrating plane, and let us
consider the motion of the fluid situated at the side of x positive.
Let m be what m' becomes when n' is replaced by $\sqrt{-1}n$. The
equation (9) furnishes two values of m, corresponding to two
series of waves, which travel, one in the positive, and the other
in the negative direction. Of course we are only concerned with
the former. We get from (9)

$$m^2 = -\mu^2(\cos 2\psi - \sqrt{-1}\sin 2\psi)\dots\dots(10),$$

where

$$\mu = \frac{n}{\sqrt{k}}\left(\frac{n^2+q^2}{(1+\alpha\beta)^2 n^2 + q^2}\right)^{\frac{1}{4}}\dots\dots(11)$$

$$2\psi = \tan^{-1}\frac{(1+\alpha\beta)n}{q} - \tan^{-1}\frac{n}{q} = \tan^{-1}\frac{\alpha\beta nq}{(1+\alpha\beta)n^2+q^2}\dots(12).$$

Choosing that root of m^2 which corresponds to waves travelling
in the positive direction, we get from (10)

$$m = -\sqrt{-1}\mu(\cos\psi - \sqrt{-1}\sin\psi).$$

Substituting in (8), introducing another function got by changing
the sign of $\sqrt{-1}$ and taking a new arbitrary constant, changing
the arbitrary constants so as to get rid of the imaginary quantities, and altering the origin of the time so as to get rid of one of
the circular functions, we get

$$s = Ae^{-\mu\sin\psi\cdot x}\cos(nt - \mu\cos\psi\cdot x)\dots\dots(13).$$

It will be easily seen that the expressions for θ, u, and p are
of the same form, that is to say, that they involve the same exponential multiplied by a sine or cosine of the same angle. Had
the actual expressions been required, it would have been shorter
to defer the substitution of real for imaginary quantities until
after the imaginary expressions for θ, u, and p had been obtained.

Now the formula (13) shows, that unless $\sin\psi$ be insensible;
sound cannot be propagated to a distance, but must be stifled in
the neighbourhood of the vibrating body by which it is excited.
Since we know very well that this is not the case, we are taught
that $\sin\psi$ is insensible, and therefore ψ itself, since ψ denotes an
angle lying between 0 and $\pi/4$. The formula (13) shows, that
if V be the velocity of propagation, $V = n\mu^{-1}\sec\psi$, which, when

ψ is insensible, reduces itself to $n\mu^{-1}$. Referring to (12), we see that, in order that ψ may be insensible, it is necessary to suppose, either that q is incomparably greater than n, or that n is incomparably greater than q. On the former supposition the formula (11) gives $V = \sqrt{k}$, which is equivalent to Newton's result. On the latter supposition we get $V = \sqrt{k(1 + \alpha\beta)}$, which is equivalent to Laplace's result.

The reason why sound would be so rapidly stifled were q and n comparable with each other, may be easily seen on taking a common-sense view of the subject. Conceive a mass of air contained in a cylinder in which an air-tight piston fits, which is capable of moving without friction, and which has its outer face exposed to a constant atmospheric pressure; and suppose the air alternately compressed and rarefied by the motion of the piston. If the motion take place with extreme slowness, there will be no sensible change of temperature, and therefore the work done on the air during compression will be given out again by the air during expansion, inasmuch as the pressure on the piston will be the same when the piston is at the same point of the cylinder, whether it be moving forwards or backwards. Similarly, the work done in rarefying the air will be given out again by the atmosphere as the piston returns towards its position of equilibrium, so that the motion would go on without any permanent consumption of labouring force. Next, suppose the motion of the piston somewhat quicker, so that there is a sensible change of temperature produced by condensation and rarefaction. As the piston moves forward in condensing the air, the temperature rises, and therefore the piston has to work against a pressure greater than if there had been no variation of temperature. By the time the piston returns, a good portion of the heat developed by compression has passed off, and therefore the piston is not helped as much in its backward motion by the pressure of the air in the cylinder as it had been opposed in its forward motion. Similarly, as the piston continues its backward motion, rarefying the air, the temperature falls, the pressure of the air in the cylinder is diminished more than corresponds merely to the change of density, and therefore the piston is less helped in opposing the atmospheric pressure than it would have been had the temperature remained constant. But by the time the piston is returning towards its position of equilibrium, the cold has diminished in consequence of the supply of heat from the sides of the cylinder, and therefore the force urging the piston forward, arising, as it does, from the excess of the external over the internal pressure, is less than that which opposed the piston in moving from its position of equilibrium. Hence in this case the motion of the piston could not be kept up without a continual supply of labouring force. Lastly, suppose the piston to oscillate with great rapidity, so that there is not time for any sensible quantity of heat to pass and repass between the air and the sides of the cylinder. In this case the pressures would be equal when the piston was at a given point of the cylinder, whether it were going or returning, and consequently there would be no permanent consumption of labouring force. I do not speak of the disturbance of the external air, because I am not now taking into account the inertia of the air either within or without the cylinder. The third case, then, is similar to the first, so far as regards the permanence of the motion; but there is this difference; that, in consequence of the heat produced by compression and the cold produced by rarefaction, the force urging the piston towards its position of equilibrium, on whichever side of that position the piston may happen to be, is greater than it would have been had the temperature remained unaltered.

Now the first case is analogous to that of the sonorous vibrations of air when the heat and cold produced by sudden condensation and rarefaction are supposed to pass away with great rapidity. For we are evidently concerned only with the *relative* rates at which the phase of vibration changes, and the heat causing the excess of temperature θ passes away, so that it is perfectly immaterial whether we suppose the change of motion to be very slow, or the cooling of heated air to be very rapid. The second case is analogous to that of sound, when we suppose the constants q and n comparable with each other; and we thus see how it is, that, on such a supposition, labouring force would be so rapidly consumed, and the sound so rapidly stifled. The third case is analogous to that of sound when we make the usual supposition, that the alternations of condensation and rarefaction take place with too great rapidity to allow a given portion of air to acquire or lose any sensible portion of heat by radiation. The increase in the force of restitution of the piston, arising from the alternate elevation and depression of temperature, is analogous to the increase in the forces of restitution of the particles of air arising from the same cause, to which corresponds an increase in the velocity of propagation of sound.

Another consequence follows from the formula (13), which deserves to be noticed. We have already seen that this formula gives $n\mu^{-1} \sec \psi$ for the value of V, the velocity of propagation. Putting for shortness

$$1 + \alpha\beta = K \dots\dots\dots\dots(14),$$

we get from (11) and (12),

$$V^2 = \frac{2k\,(K^2 n^2 + q^2)}{Kn^2 + q^2 + \sqrt{\{(K^2 n^2 + q^2)\,(n^2 + q^2)\}}} \quad \dots\dots(15).$$

Hence if q be comparable with n, V, which is a function of the ratio of q to n, will change with n, and therefore the velocity of propagation will depend upon the pitch, which is contrary to observation. But if q be either incomparably greater or incomparably smaller than n, V will assume one or other of its limiting values $\sqrt{k}$, $\sqrt{kK}$; and the velocity of propagation will be independent of the pitch, as observation shows it to be. We are thus led, by considering the velocity of propagation, to the same conclusion as was deduced from the circumstance that sound is capable of travelling to a distance.

Since, then, we are driven to one or other of the alternatives above mentioned, it only remains to decide which we must choose. But before entering on this subject, it will be proper to consider whether the formula (13) is of sufficient generality.

In the first place we may observe, that the formula (13) is only a particular integral of (7). It is adapted to the case in which the motion is kept up by a vibrating plane, which agrees most nearly with the circumstances of ordinary experiments; but a particular law of disturbance as regards the time is assumed, namely, that expressed by a single circular function. Now we know that any periodic function of the time, having τ for its period, may be expressed by the sum of a finite or infinite number of circular functions having for their periods τ and its submultiples; and even a non-periodic function may be expressed by a definite integral, of which each element denotes a circular function. So far, therefore, the formula (13) is of sufficient generality.

In the next place, the formula (13) applies to motion in one dimension only. But had we employed the general equations (1), (2), which relate to motion in three dimensions, we should have obtained the same partial differential equation as (7), with the exception that the last term outside the brackets would have been replaced by

$$\frac{d^2 s}{dx^2} + \frac{d^2 s}{dy^2} + \frac{d^2 s}{dz^2}.$$

If now we take the case next in order of simplicity, in which the motion is symmetrical about a centre, and put r for the distance of any point from the centre, we shall get for the determination of rs the same partial differential equation as (7), with the exception that x will be replaced by r. To obtain, therefore, the integral corresponding to (13), it will be sufficient to replace x by r and divide the second member by r. This integral would apply to the case of the disturbance produced by a vibrating spherical body, in which the motion is supposed to be symmetrical with respect to the centre. And in the more general case of a vibrating body of irregular form, or a musical instrument, or any other source of sound, the conclusions would doubtless be the same as to their leading features.

There remains a more important point to be considered before we apply the formula (13) to the vibrations of air within a long tube. At first sight it might seem that the radiation of heat within a tube must take place in a manner altogether different from that in which it would take place in free air. But a little consideration will show, I think, that such is not the case. Of the heat radiating from any particle of air which has been slightly heated by condensation, any particular ray is incident on the side of the tube, where it is partly absorbed, partly reflected, and, it may be, partly scattered. The reflected ray, or any one of the scattered rays, is again incident on the side of the tube, where a good portion is absorbed, and so on. The small quantity of radiant heat which remains after three or four reflexions may be regarded as insensible. Now since radiant heat travels with a velocity equal to, or at any rate comparable with, that of light, we may neglect as altogether insensible the time which any portion of heat, once become radiant, takes to be absorbed. Moreover, we may neglect the small portion of heat reabsorbed by the air itself, because a ray of heat has only to traverse a length of air comparable with three or four diameters of the tube before it is absorbed by the tube. Hence we may conceive a small periodic flux of heat as taking place across the inner surface of the tube. Now it follows from the mathematical theory of heat, that when a periodic flux of heat takes place at the surface of a solid, the corresponding variation in the temperature of the solid near the surface is very small if the period be very small. If we suppose the flux expressed by the sine or cosine of an angle proportional to the time, the expression for the fluctuation of temperature will involve in its coefficient the square root of the period. In the present case, the period with which we have to deal is that of a sonorous vibration, a time which must be regarded as extremely small in questions relating to the conduction of heat. Hence, if τ be the period of vibration, the fluctuation of temperature of the tube will be a small quantity of the order $\sqrt{\tau}$ compared with the flux of heat. Now if H, h be the interior and exterior conductivities, ν a normal to the inner surface of the tube, drawn from the tube inwards, θ' the excess of temperature of the tube above the mean temperature θ_0; and if we suppose the surface to be plane, and to radiate into an infinitely extended medium at a temperature $\theta_0 + \theta$, where θ is supposed to be constant as regards space, but to be a periodic function of the time, we must have at the inner surface of the tube

$$H\frac{d\theta'}{d\nu} + h\left(\theta' - \theta\right) = 0.$$

Now, according to what has been already remarked, θ' is a small quantity of the order $\sqrt{\tau}$ compared with $d\theta'/d\nu$; and it follows from the above equation, that $d\theta'/d\nu$ is comparable with θ, and therefore θ' is a small quantity of the order $\sqrt{\tau}$ compared with θ. Hence, even in the case above supposed, the fluctuation of temperature of the tube at the surface would be very small. But in the actual case, the tube radiates, not into an infinite medium, but merely across the air contained within it, beyond which is situated the opposite face of the tube, at a temperature equal to the first face; and therefore the fluctuation of temperature of the inner surface of the tube will be far smaller than in the case supposed above, so as to be altogether insensible. Hence the air radiates within an envelope at a temperature θ_0, so that the radiation takes place as if the air heated by compression radiated into an infinite medium at a temperature θ_0. Of course the same reasoning will apply to the apparent radiation of cold. Hence the formula (13) may be applied without change to the vibration of the air within a long tube, and accordingly may be employed in considering the experiments of M. Biot above alluded to.

The preceding view of the effect of radiation within a tube is very different from that taken by M. Poisson in his *Traité de Mécanique* (vol. ii. art. 665). The latter, however, is contained in a mere passing remark offered by way of conjecture, and probably written without much consideration, and therefore ought hardly to be regarded as supported by Poisson's authority.

Let us now pass to numerical values, in order to make out, independently of any assumption respecting the true explanation of the velocity of sound, whether q must be regarded as very great or very small compared with n. It follows from (13) that the decrease of intensity in going one wave's length in the direction of propagation is a maximum when $\tan\psi$, and therefore ψ, is a maximum. Now (12) shows that ψ is a maximum when

$$\frac{q}{n} = \sqrt{K} \quad \dots\dots\dots\dots\dots\dots (16),$$

K being the quantity defined by (14). For the above value of q we get from (11) and (12),

$$\mu = nk^{-\frac{1}{2}}K^{-\frac{1}{4}}, \quad 2\psi = \tan^{-1}K^{\frac{1}{4}} - \tan^{-1}K^{-\frac{1}{4}} \dots\dots(17).$$

The velocity of propagation, which is equal to $n\mu^{-1}\sec\psi$, does not much differ from $n\mu^{-1}$, since ψ, as will immediately appear, is not very large. It may be observed, that the expression for $n\mu^{-1}$ given by the first of equations (17), is a geometric mean between the velocities of propagation resulting from the theories of Newton and Laplace.

The value of K, deduced from experiments in which the theory of sound is not assumed, is about $1\cdot36^*$, whence $2\psi = 8° 47'$. If

---

* Poisson, *Traité de Mécanique*, vol. ii. art. 637. The value deduced from the observed velocity of sound is somewhat larger, and is more likely to be correct. I have employed the value $1\cdot36$ in order to avoid arguing in a circle, because I am reasoning as if the received theory of sound were not established.

$-l$ be the index of e in (13) when x is equal to one wave's length, we have

$$\mu \sin \psi . x = l, \ \mu \cos \psi . x = 2\pi, \text{ whence } l = 2\pi \tan \psi, \ e^{-l} = 0.6172;$$

so that the intensity, supposed to vary as the square of the amplitude of vibration, would be diminished in the ratio of 2·625 to 1.   Supposing the period of vibration to be the $\frac{1}{300}$th part of a second, which would correspond to a note of moderate pitch, and taking the velocity of propagation at 1100 feet per second, we should have 44 inches for the length of one wave. Hence in travelling 20 yards, or 16·36 wave-lengths, the intensity would be diminished in the ratio of $(2.625)^{16.36}$ to 1, or about 7 millions to 1.   A decrease of intensity like this is utterly contrary to observation, and therefore we are really compelled to suppose that the ratio of q to n is either very much greater or very much less than what has just been determined.   Since in the case supposed $n = 2\pi\tau^{-1} = 600\pi$, we get from (16)

$$q = 2198 \quad\dots\dots\dots\dots\dots\dots\dots(18),$$

which, it is to be remembered, is referred to a second as the unit of time.

Let us now, adopting this value of q, examine a little at what rate a small portion of heated air, situated in other air which has not been heated, would cool by radiation.   If θ be the excess of the temperature of the heated air over that of the surrounding air, we should have, supposing θ to be sufficiently small to allow us to adopt Newton's law of cooling,

$$\frac{d\theta}{dt} = - q\theta;$$

from which it follows that the excess of temperature would be diminished during the time t in the ratio of e^{qt} to 1.   It would follow from the numerical value of q above given, that, even in so short a time as the hundredth part of a second, the temperature would be reduced in the ratio of about 3514 millions to 1.   Such rapidity of cooling as this is utterly contrary to observation.   Put a poker into the fire, and when it is hot look along it, and an ascending stream of heated air will be rendered visible by the distortion which it produces in objects seen through it, in consequence of the diminution of refractive power accompanying the rarefaction produced by heat.   But were the rate of cooling anything like what has just been determined, no such stream could exist. Yet we have seen that the observed fact, that sound is propagated to a distance, obliges us to suppose that the rate of cooling is either immensely greater or immensely less than corresponds to $q = 2198$. It is needless now to say which alternative we must choose. Accordingly, no doubt whatever exists as to the correctness of Laplace's explanation of the excess of the observed velocity of sound over that calculated by Newton.

Now that it has been decided which of the two ratios $n : q$ and $q : n$ we must regard as extremely small, we may simplify the formula (13) by retaining only the first power of the ratio in question, and we shall thus be the more readily enabled to see in what direction we must look for the first faint indications of the effect of radiation. Retaining only the first power of q, and putting $n = V\mu$, $\mu = 2\pi\lambda^{-1}$, where $V = \sqrt{(kK)}$, the velocity of propagation, we get from (11), (12) and (13),

$$ s = Ae^{-(1-K^{-1})\frac{qx}{2V}} \cos \frac{2\pi}{\lambda} (Vt - x) \dots\dots\dots\dots(19). $$

Hence it is to a diminution of intensity, rather than to an alteration of velocity corresponding to an alteration of pitch, that we are to look for the effect of radiation. Now that the objection raised against Laplace's explanation of the velocity of sound has been answered, we may take 1·414 for the value of k, this being the mean of the values quoted by Poisson in art. 664, which were deduced from the velocity of sound, and are probably nearer the truth than the somewhat smaller values determined by a different process. Putting $K = 1·414$, $V = 1100$, taking the square of the coefficient as a measure of the intensity, and putting $N : 1$ for the ratio in which the intensity is diminished while the sound travels, without divergence, over a length x, we get

$$ \log_{10} N = 0·0001156 qx \dots\dots\dots\dots(20), $$

the units of time and space to which q and x are respectively referred being a second and a foot.

From the account of M. Biot's experiments given by Sir John Herschel in art. 24 of his Treatise on Sound*, it would seem that the diminution of intensity which we can by any possibility refer to radiation must be very small, especially when we remember

---

* *Encyclopædia Metropolitana*, art. Sound.

that, in the case of these experiments, the intensity would be diminished by a sort of reflexion at the bendings of the tube, as well as by the friction of the air against the sides of the tube, and the internal friction of the air itself. That the cause last mentioned would produce a small but not utterly insensible effect in causing a diminution of intensity, I have shown in the course of a paper " On the Effect of the Internal Friction of Fluids on the Motion of Pendulums," recently read before the Cambridge Philosophical Society*. If we suppose, at a venture, that a diminution of intensity in the ratio of 2 to 1 is the utmost which we can attribute to radiation in the case of M. Biot's experiments, putting $N = 2$, and $x = 3120$, the length of the tube in feet, we get from (20) $q = 0.834$ for a superior limit of q. If we suppose $q = 0.834$, we get for the ratio in which the temperature of a small portion of slightly heated air would be diminished in the course of one second, 1 to e^{-q}, or 1 to 0.4343, or 7 to 3 nearly. It is curious that it should, theoretically speaking, be possible to assign a superior limit to the velocity of cooling of heated air by observations on sound; but I imagine that the real value of q is a good deal smaller than any limit which it would be practically possible to assign in this way.

I have supposed, as was already observed, that radiant heat is capable of traversing great lengths of air before any considerable portion of it is absorbed. This is especially the case with heat of such high refrangibility as to place it within the limits of the visible spectrum; whereas heat of low refrangibility, such as that which would emanate from slightly heated air, is absorbed more rapidly. Should the distance to which radiant heat can proceed in air before a given fraction of it, such as one-half, is absorbed, not be extremely great compared with the length of a wave of sound, it may be seen after a little reflection that the general conclusions arrived at will be unchanged, though the numerical details would be somewhat altered. I have not met with any experiments relating to the absorption of non-luminous heat by air which could be made a foundation for numerical calculation.

* [*Ante*, p. 123.]

[From the *Transactions of the Cambridge Philosophical Society*, Vol. IX. p. (147)].

ON THE COLOURS OF THICK PLATES.

[Read *May* 19, 1851.]

THE expression "colours of thick plates," has been appropriated to a class of phenomena discovered by Newton, and described by him in the fourth part of the second book of his Optics. In Newton's experiment, the sun's light was admitted into a darkened room through a hole in the window-shutter, and allowed to fall perpendicularly upon a concave mirror, formed of glass quick-silvered at the back. A white opaque card pierced with a small hole being then interposed, at the distance of the centre of curvature of the mirror, so that the regularly reflected light returned by the same small hole by which it entered, a set of coloured rings was seen depicted on the card encompassing the hole. The existence of these rings was attributed by Newton to the light scattered on entering the glass, and then regularly reflected and refracted; and he succeeded in deducing from his theory of fits the laws of the rings, both as regards the relation between the diameters of successive rings, the order of the colours, the variation of the diameter of a given ring corresponding to a variation either in the radius of curvature of the surfaces or in the thickness of the glass, and even the absolute magnitude of the system formed under given circumstances. The phenomena which present themselves when the mirror is inclined a little, so as to throw the image of the hole to one side, are very curious, and have been accurately described by Newton in his tenth and eleventh Observations.

In the course of a series of experimental researches on these rings, the Duke de Chaulnes* discovered accidentally that their brilliancy was greatly increased by breathing on the glass. Since the moisture soon evaporated, in order to procure a permanent tarnish, he spread over the surface a small quantity of a mixture of milk and water, which on drying left a degree of dimness very suitable to the experiments. By substituting for the glass mirror a metallic speculum, in front of which there was placed a plate of tarnished mica, it was easy to observe the variation in the diameter of the rings corresponding to a variation in the distance of the mica from the speculum. In this form of the experiment the glass plate was replaced by the plate of air comprised between the mica and the speculum. Rings were also produced when the tarnished mica was replaced by a screen of fine muslin. In this case, however, according to the Duke's statement, the rings were nearly square, though rounded off a little at the angles. A set of parallel wires gave merely a bright band intersected by short bands which were vividly coloured. Even the blade of a knife produced a similar appearance, weak indeed, but sufficient to establish the identity of the effect. It is unnecessary here to discuss the theoretical views of the Duke de Chaulnes, since the progress of optical science has since led to a complete explanation of the formation of the rings.

The colours of thick plates were first explained on the undulatory theory by Dr Young†, by whom they were attributed to the interference of two streams of light, of which one is scattered on entering the glass, and then regularly reflected and refracted, and the other regularly refracted and reflected, and then scattered on its return through the first surface. Dr Young's explanation is however excessively brief; and he has rather pointed out the application of the grand and newly-discovered principle of interference to the explanation of the phenomenon, than followed the subject into any of its details. At the same time, it appears evident, from an attentive perusal of what he has written, that at least the broad outlines of the complete explanation were clearly present to his mind.

In the course of a paper entitled "Experiments for investigating the cause of the coloured concentric rings, discovered by Sir Isaac

* *Mémoires de l'Académie*, 1755, p. 136.
† *On the Theory of Light and Colours. Philosophical Transactions* for 1802, p. 41.

Newton, between two object-glasses laid upon one another," Sir William Herschel mentions an experiment in which rings of the nature of those of thick plates were produced by merely strewing hair-powder in the air in front of a metallic speculum placed as the mirror in Newton's experiment*. The result of this experiment was justly regarded by Herschel as inexplicable on the theory of fits. It may here be remarked that it is in perfect accordance with the theory of undulations.

In the *Annales de Chimie et de Physique*†, will be found a report by Ampère and Poisson on a memoir by M. Pouillet, containing some experiments on the rings. The experiments were mostly the same as those of the Duke de Chaulnes, but accompanied by measures. M. Pouillet found that the rings were produced by placing in front of a metallic speculum an opaque screen containing an aperture of any form. In this case the rings were round, whatever might be the form of the aperture. The experiments are mentioned by M. Pouillet in his *Élémens de Physique*‡.

A complete explanation of the rings, according to the theory of undulations, has been given by Sir John Herschel in his treatise on Light§. The rings are supposed to be formed in Newton's manner with a glass mirror, the luminous point being situated in the axis. Having investigated the elementary system of rings which would be produced by the two streams scattered in passing and repassing at the point of the first surface where it is cut by the axis, Sir John Herschel shews that if the surfaces be supposed to be a pair of concentric spheres, having the luminous point in their centre, the elementary systems corresponding to the several elements of the first surface will be superposed, in such a manner that a distinct system would be thrown on a screen held at the distance of the luminous point. The laws of the rings resulting from theory are precisely those which had been discovered by Newton, and the calculated magnitudes were found to agree almost exactly with Newton's measures.

A set of coloured bands have since been observed by Dr Whewell, which are formed when the image of a candle held near

* *Philosophical Transactions* for 1807, p. 231.
† Tom. i. (1816) p. 87.   ‡ Tom. ii. p. 476.
§ *Encyclopædia Metropolitana*, Arts. 676, &c.

the eye is viewed by reflexion in a plane mirror of quicksilvered glass, placed at the distance of some feet. This observation was communicated to M. Quetelet, by whom it has been published*. In repeating the experiment together, Dr Whewell and M. Quetelet found that it was an essential condition of success that the surface should not be perfectly bright, and that to ensure the production of the bands it was sufficient to breathe gently on the surface of a cool mirror. Instead of vapour, which soon evaporates, M. Quetelet recommends a tarnish of grease †.

In closing this sketch of the history of the subject, I may be allowed to express my obligations to Dr Lloyd for his valuable Report on Physical Optics, which contains a brief account of all that was known about the subject, accompanied by references to the original papers.

My attention was called to the subject by the Master of Trinity College, who shewed me the bands above-mentioned, which he shortly afterwards brought before the notice of this Society‡. It seemed to me from the first that these bands were of the nature of the coloured rings of thick plates, so that the theory of the former only required to be worked out, that of the latter being known. Had I felt any doubt on the subject, it would soon have been dissipated when I came to make experiments; for by properly varying the experiments the two systems were seen to be incontestably of the same nature.

The mirrors, whether plane or curved, prepared in the following manner, which I can recommend to any one who wishes to repeat the experiments, as being both easy and efficacious. The mirror being held horizontally, a mixture consisting of three or four parts of water to one of milk was poured on it, and allowed to spread over the surface. The mirror was then held in a vertical position in front of a fire, when the greater part of the mixture ran off, and the remainder dried in two or three minutes, when the mirror was ready for use. To prevent disappointment on the part of any one who may be looking for Dr Whewell's bands, I will here mention that in order to see them properly the image of

---

* *Correspondance Mathématique et Physique*, Tom. v. (1829) p. 394.

† Tom. vi. p. 69.

‡ See the *Philosophical Magazine* for April, 1851, p. 336.

the flame must be seen distinctly, so that a short-sighted person requires an eye-glass or spectacles.

A concave mirror prepared with milk and water is well adapted for performing Newton's experiment in his manner, or rather by substituting, as in the Duke de Chaulnes's experiments, the image of the sun in the focus of a convex lens for the small hole employed by Newton. The experiment may however be varied in the following manner. Whatever appearance is presented on a screen may be seen without a screen by receiving the rays directly into the eye, and adapting it for distinct vision of an object at the distance of the screen. Accordingly, in order to see the rings which in Newton's experiment were thrown on a screen, it is sufficient to place a small flame in front of the mirror, in such a position as to coincide with its inverted image, when a remarkably beautiful system of rings is seen in air, surrounding the flame. Not the least striking circumstance connected with these rings is their apparent corporeity, since they seem to have a definite position in space like an actual object. The striking and beautiful phenomena so accurately described by Newton in his tenth and eleventh Observations may be seen in this manner by moving the flame sideways. By altering in various ways the positions of the flame and of the eye, both in this experiment and in that with a plane mirror, the rings or bands seen in the two cases may be perceived, independently of any theory, to be evidently of the same nature. It is unnecessary here to describe at length the various appearances presented, since they are noticed in the body of the paper, in connexion with the theory.

The first section contains the theory of the rings formed in Newton's manner. The investigation, though differing a little in the mode in which it is conducted, is the same in principle as that given by Sir John Herschel, but is somewhat more general, inasmuch as the curvatures of the two surfaces are supposed to be any whatsoever, and the luminous point is not supposed to be situated in the axis. The distance, too, of this point from the axis is at first supposed to be arbitrary, in order to investigate under what circumstances the rings can be formed most distinctly on a screen. The second section contains the theory of the bands and rings formed by a plane mirror. The expression for the retardation is deduced as a particular case from the formula investigated in the first section; but on account of the interest which attaches itself to these bands, and the simplicity of their theory, a separate investigation is likewise given. The next two sections are devoted to cases of more generality, and on the whole less interest: still, a few results of some interest are obtained. The last two sections contain a closer examination of the precise mode in which the phenomena are produced.

Although the present paper is a little long, the reader must not suppose that the theory of the rings and bands is anything but simple. The length arises partly from the detail in which the subject has been considered, partly from the generality of some of the investigations, partly from the description of experiments which accompanies the theoretical investigations.

SECTION I.

Rings thrown on a screen by a concave mirror consisting of a lens dimmed at the first surface, and quicksilvered at the back. Condition of distinctness when the rings are thrown on a screen, or of fixity when they are viewed in air. Investigation of the phenomena observed when the luminous point is moved in a direction perpendicular to the axis of the mirror.

1. Let a luminous point L be situated either in or not far out of the axis of a mirror such as that just described; and let it be required to investigate the illumination, at the point M of a screen, due to two streams of light, of which one is scattered at the first surface, and then regularly reflected and refracted, and the other is regularly refracted and reflected, and then scattered in coming out, the point M being supposed to be situated not far out of the axis. Let the mirror be referred to the rectangular axes of x, y, z, the axis of z being the axis of the mirror, and the origin being situated in the first or dimmed surface. Let r be the radius of the first surface, s that of the second, t the thickness of the glass, μ its index of refraction; and suppose r and s positive when the concavities of both surfaces are turned in the direction of z positive. Let a, b, c be the co-ordinates of L; a', b', c' those of M; and suppose a, b, a', and b' small compared with c, c', r, and s. Let x, y, z be the co-ordinates of any point P on the dimmed

surface, R the retardation of the stream which was scattered at P on emergence, relatively to that which was scattered at the same point on entrance. Let L_1, L_2, L_3 be the images of L after refraction, reflexion, and second refraction, respectively; M_1, M_2, M_3, the images of M; and let a, b, c, or a', b', c', with the suffixes 1, 2, 3, denote the co-ordinates of L_1, L_2, L_3, or M_1, M_2, M_3. In approximating to the value of R, let the squares of the small quantities, a, b, x, y, &c. be retained, so that the terms neglected are of the fourth order, since all the terms are of even orders, as will be immediately seen when the approximation is commenced.

2. The rays diverging from L may after refraction be supposed to diverge from L_1, notwithstanding the spherical aberration of direct pencils, and the astigmatism of oblique pencils. For, first, let L be in the axis. The supposition that the rays diverge from L_1 is equivalent to supposing that the front of a wave is a sphere having L_1 for centre, whereas it is really a surface of revolution such that L_1 is the centre of curvature of a section made by a plane through the axis. This plane cuts the sphere above mentioned in a circle, which, being a circle of curvature, cannot have with the curve a contact lower than one of the second order. But the contact is actually of the third order, since the curve and circle touch without cutting. Hence the error produced in the calculation of R by supposing the front of a wave to be a sphere, instead of that surface which it actually is, is only a small quantity of the fourth order, and quantities of this order are supposed to be neglected.

Next, consider an oblique pencil. Let L' and L'' be two points in the axis of the pencil which are the centres of curvature of its principal sections. If the distance of L' and L'' from each other, and from L_1, were not small, the front of the wave would have a contact of the first order with a sphere described round L_1, with such a radius as to pass through the point where the front is cut by the axis of the pencil; and in that case the error committed by taking the sphere for the actual front would be of the second order. But L' and L'' are situated at distances from L_1 which are small quantities of the second order, whence it will readily be seen that the actual error is only of the fourth order.

3. Let the expression (L to L_3) denote the retardation of a wave proceeding from L to L_3, or rather, in case L_3 be a virtual

focus, the difference of retardations of two waves starting from L and L_3 and reaching the same given point. Then

$$R = (L \text{ to } L_3) + PM - PL_3 - \{(M \text{ to } M_3) + PL - PM_3\} = K + V,$$

where

$$K = (L \text{ to } L_3) - (M \text{ to } M_3),$$
$$V = PM - PL_3 - (PL - PM_3) = V' - V'', \text{ suppose.}$$

According to the explanation given in the preceding article, when the position of P changes K remains constant, to the degree of approximation which it is proposed to employ, but the value of V depends upon the position of P. We have

$$PM = \sqrt{\{(a' - x)^2 + (b' - y)^2 + (c' - z)^2\}},$$
$$PL_3 = \sqrt{\{(a_3 - x)^2 + (b_3 - y)^2 + (c_3 - z)^2\}},$$

and $z = \dfrac{1}{2r}(x^2 + y^2)$, nearly. Expanding, we get

$$V' = c' - \frac{1}{2r}(x^2 + y^2) + \frac{1}{2c'}\{(a' - x)^2 + (b' - y)^2\} - c_3$$
$$+ \frac{1}{2r}(x^2 + y^2) - \frac{1}{2c_3}\{(a_3 - x)^2 + (b_3 - y)^2\}.$$

Let $V' = c' - c_3 + V_x' + V_y'$, where V_x' is the sum of the terms containing a and x, and V_y' the sum of the terms containing b and y. Then

$$V_x' = \frac{(a' - x)^2}{2c'} - \frac{(a_3 - x)^2}{2c_3}.$$

There is no occasion to write down V_y', since it may be deduced from V_x' by writing b, y for a, x. Taking V_x to denote for V what V_x' denotes for V', we get by interchanging a and a', a_3 and a_3', and subtracting,

$$V_x = \frac{(a' - x)^2}{2c'} - \frac{(a_3 - x)^2}{2c_3} - \frac{(a - x)^2}{2c} + \frac{(a_3' - x)^2}{2c_3'} \quad \ldots\ldots(1).$$

In order that the rings may be formed on a screen with perfect distinctness, it is necessary that the difference of phase of the several pairs of streams which come from the several points of the dimmed surface should be the same; in other words, that the retardation R should be independent of x and y. Deferring for the present the investigation of the conditions of distinctness, we may observe that when these conditions are satisfied the expression for R must be the same as if x and y were each equal to zero. We have therefore

$$R = [L, \ M] - [M, \ L],$$

where

$$[L, \ M] = (L \text{ to } L_3) + c' - c_3 + \frac{e'^2}{2c'} - \frac{e_3^2}{2c_3},$$

and $[M, \ L]$ is formed from $[L, \ M]$ by interchanging the co-ordinates of L and M. In the above expression e^2 has been written for shortness, in place of $a^2 + b^2$. Now supposing c, c_1, c_2, c_3, to be all positive, and denoting by A, B the points in which the first and second surfaces respectively are cut by the axis of the mirror, we have

$$(L \text{ to } L_3) = AL - \mu AL_1 + \mu BL_1 + \mu BL_2 - \mu AL_2 + AL_3; \ ...(2)$$

which gives, on expanding,

$$(L \text{ to } L_3) = c + \frac{e^2}{2c} - \mu c_1 - \frac{\mu e_1^2}{2c_1} + \mu (c_1 + t) + \frac{\mu e_1^2}{2(c_1 + t)} + \mu (c_2 + t)$$

$$+ \frac{\mu e_2^2}{2(c_2 + t)} - \mu c_2 - \frac{\mu e_2^2}{2c_2} + c_3 + \frac{e_3^2}{2c_3}$$

$$= c + c_3 + 2\mu t + \frac{e^2}{2c} + \frac{e_3^2}{2c_3} - \frac{\mu t e_1^2}{2c_1 (c_1 + t)} - \frac{\mu t e_2^2}{2c_2 (c_2 + t)}.$$

We have therefore

$$R = \frac{\mu t}{2} \left\{ \frac{e_1'^2}{c_1' (c_1' + t)} + \frac{e_2'^2}{c_2' (c_2' + t)} - \frac{e_1^2}{c_1 (c_1 + t)} - \frac{e_2^2}{c_2 (c_2 + t)} \right\} ..(3).$$

Although this formula was obtained on the supposition that the points $L, L_1, L_2, L_3, M, M_1, M_2, M_3$, lay on the positive side of the plane of xy, it is true independently of that restriction. For when one of the foci L, L_1, L_2, L_3, from having been real becomes virtual, or from having been virtual becomes real, the corresponding ordinate $c, c_1, c_2 + t$, or c_3 changes sign. At the same time, in the expression for the retardation *distance passed over* is converted into *distance saved*, and *vice versâ*. Hence in any such expression as (2) the sign of one or more of the lines is changed. But in the expansion of the radical by which the length of such line is expressed, the sign of $c, c_1, c_2 + t$, or c_3 must be changed at the same time, and therefore no change is required in the expanded expressions.

To eliminate e_1 and e_2 from (3), we may observe that we have very nearly

$$\frac{e_1}{e} = \frac{1}{\mu}\frac{c_1}{c}, \quad \frac{e_2}{e_1} = -\frac{c_2 + t}{c_1 + t},$$

and similar expressions hold good for e', &c.   Hence

$$R = \frac{e'^2 t}{2\mu c'^2}\frac{c_1'}{c_1' + t}\left(1 + \frac{c_1'}{c_1' + t}\frac{c_2' + t}{c_2'}\right)$$

$$- \frac{e^2 t}{2\mu c^2}\frac{c_1}{c_1 + t}\left(1 + \frac{c_1}{c_1 + t}\frac{c_2 + t}{c_2}\right)\ldots\ldots\ldots(4).$$

We might, if required, express c_1 and c_2 in terms of c by the formulæ of common optics, without making any supposition as to the magnitude of t.   In practice, however, t is usually small compared with c, c_1, &c., so that we may simplify the above expression by retaining only the first power of t.   We thus get

$$R = \frac{t}{\mu}\left(\frac{e'^2}{c'^2} - \frac{e^2}{c^2}\right) = \frac{t}{\mu}\left(\frac{a'^2 + b'^2}{c'^2} - \frac{a^2 + b^2}{c^2}\right)\ldots\ldots\ldots(5).$$

4.   Before proceeding to apply this expression, let us investigate the conditions of distinctness.   Denoting by $\Delta_x R$, $\Delta_y R$ the additions to R on account of the terms involving x, y, we get from (1)

$$\Delta_x R = \tfrac{1}{2}\left(\frac{1}{c'} + \frac{1}{c_3'}\right)x^2 - \left(\frac{a'}{c'} + \frac{a_3'}{c_3'}\right)x$$

$$- \left\{\frac{1}{2}\left(\frac{1}{c} + \frac{1}{c_3}\right)x^2 - \left(\frac{a}{c} + \frac{a_3}{c_3}\right)x\right\};\ \ldots\ldots\ldots(6),$$

and $\Delta_y R$ may be obtained by interchanging a and b, x and y.

We have by the formulæ of common optics

$$\frac{\mu}{c_1} = \frac{\mu - 1}{r} + \frac{1}{c}, \quad \frac{1}{c_2 + t} = \frac{2}{s} - \frac{1}{c_1 + t}, \quad \frac{\mu}{c_2} = \frac{\mu - 1}{r} + \frac{1}{c_3};\ldots(7).$$

whence, supposing t small, expanding as far as the first power of t, and putting for shortness

$$\frac{\mu}{s} - \frac{\mu - 1}{r} = \frac{1}{s} - (\mu - 1)\left(\frac{1}{r} - \frac{1}{s}\right) = \frac{1}{\rho},\ldots\ldots\ldots(8),$$

so that ρ is the radius of a speculum having the same focal length as the mirror, we obtain

$$\frac{1}{c} + \frac{1}{c_3} = \frac{2}{\rho} + \frac{2t}{\mu}\left\{\frac{2}{\rho^2} + \frac{2(\mu - 1)}{\rho r} + \frac{(\mu - 1)^2}{r^2} - \frac{2}{\rho c} + \frac{1}{c^2}\right\};$$

and therefore

$$\frac{1}{2}\left(\frac{1}{c'}+\frac{1}{c_3'}-\frac{1}{c}-\frac{1}{c_3}\right)=\frac{t}{\mu}\left(\frac{1}{c'}-\frac{1}{c}\right)\left(\frac{1}{c'}+\frac{1}{c}-\frac{2}{\rho}\right) \quad \ldots\ldots (9).$$

Again

$$a_1=\frac{a}{\mu}\frac{c_1}{c}, \quad a_2=-a_1\frac{c_2+t}{c_1+t}, \quad a_3=\mu a_2\frac{c_3}{c_2};$$

whence

$$\frac{a}{c}+\frac{a_3}{c_3}=\frac{a}{c}\left\{1-\frac{c_1(c_2+t)}{(c_1+t)c_2}\right\}=\frac{at}{c}\left(\frac{1}{c_1}-\frac{1}{c_2}\right), \text{ nearly.}$$

Neglecting t altogether in the formulæ (7), we get

$$\frac{1}{c_1}-\frac{1}{c_2}=\frac{2}{\mu}\left(\frac{1}{c}-\frac{1}{\rho}\right);$$

and therefore

$$\frac{a'}{c'}+\frac{a_3'}{c_3'}-\frac{a}{c}-\frac{a_3}{c_3}=\frac{2t}{\mu}\left\{\frac{a'}{c'}\left(\frac{1}{c'}-\frac{1}{\rho}\right)-\frac{a}{c}\left(\frac{1}{c}-\frac{1}{\rho}\right)\right\}\ldots\ldots (10).$$

Let $\Delta_x R+\Delta_y R=\Delta R$. The formulæ (6), (9), (10), and the corresponding formulæ relating to $\Delta_y R$, give

$$\Delta R=\frac{t}{\mu}\left(\frac{1}{c'}-\frac{1}{c}\right)\left(\frac{1}{c'}+\frac{1}{c}-\frac{2}{\rho}\right)(x^2+y^2)$$

$$-\frac{2t}{\mu}\left\{\frac{1}{c'}\left(\frac{1}{c'}-\frac{1}{\rho}\right)(a'x+b'y)-\frac{1}{c}\left(\frac{1}{c}-\frac{1}{\rho}\right)(ax+by)\right\}\ldots(11).$$

The condition of distinctness, as has been already observed, is that ΔR shall vanish independently of x and y, in which case the elementary systems of rings corresponding to the several elements of the dimmed surface will be superimposed on the screen. The coefficient of x^2+y^2 will vanish when either of the following equations is satisfied:

$$c'=c, \text{ or } \frac{1}{c'}+\frac{1}{c}=\frac{2}{\rho} \quad \ldots\ldots\ldots\ldots\ldots(12).$$

In order that the coefficients of x and y may vanish independently of particular values of a' and b' we must have

$$c'=c=\rho \quad \ldots\ldots\ldots\ldots\ldots\ldots(13),$$

which equations satisfy at the same time both of the equations (12), of which it would have been enough that one should have been satisfied. Hence the rings are formed most distinctly when the luminous point and the screen are both at a distance from the mirror equal to that of coincident conjugate foci. This agrees with observation*.

Whatever be the position of the luminous point, if a', b', c' be the co-ordinates of its image, we have

$$\frac{a'}{c'} = -\frac{a}{c}, \quad \frac{b'}{c'} = -\frac{b}{c};$$

and the second line in the expression for ΔR becomes

$$\frac{2t}{\mu c}\left(\frac{1}{c'}+\frac{1}{c}-\frac{2}{\rho}\right)(ax+by),$$

which vanishes since c' satisfies the second of equations (12). If a screen be held in a direction perpendicular to the axis of the mirror, at such a distance as to receive a distinct image of the luminous point, and if a', b', c' be now taken to denote the co-ordinates, not of the image itself, but of a point of the screen very near the image, the part of ΔR which involves the squares of x and y will continue to vanish, inasmuch as c' remains the same as before, and the part which contains first powers, though not absolutely evanescent, will be very small; and therefore a portion of the system of rings in the neighbourhood of the image will be formed distinctly.

5. This agrees with observation. In repeating Newton's experiment in his way, except that a lens of short focus was employed instead of a small hole, and that the surface of the glass was purposely dimmed with milk and water, I found that when the mirror was placed at a distance from the luminous point widely different from its radius of curvature, and inclined a little, so as to allow of receiving the image on a sheet of paper without stopping the incident light, and when the paper was held at such a distance from the mirror as to receive a distinct image of the luminous point, the image was accompanied by very distinct arcs of rings.

---

* See Newton's *Optics*, Book II., Part IV., Obs. 1, for the case in which the curvatures of the two surfaces are alike, and an experiment by the Duke de Chaulnes (*Mém. de l'Académie*, 1755, p. 141) for the case in which they are unlike. In this experiment a plano-convex lens was employed. Each face in succession, after having been tarnished, was turned towards the incident light. It appears from a passage at the end of Newton's twelfth Observation that he had himself made experiments of a similar nature, the results of which however are not described.

Whatever appearance is presented on a screen may be seen without a screen, by placing the eye in such a position as to receive the rays, and adapting it to distinct vision of an object at the distance of the screen in its former position. It is found universally that when the image of the luminous point is seen distinctly it is accompanied by a portion, more or less extensive, of a system of coloured rings or bands. In this way the rings may be seen when the image is virtual, in which case they cannot, of course, be thrown on a screen.

In the experiment described in the introduction, in which a small flame is placed in such a position as to coincide with its inverted image, and viewed directly, the rings seen are remarkable for their fixity, appearing like a bodily object surrounding the flame, and having a definite parallax, whether judged of by the motion of the head, or by the convergence of the axes of the two eyes. The same is true of the system of rings formed when the flame is moved sideways out of the position above mentioned. The reason of this fixity is, that inasmuch as the retardation is independent of x and y, a given point of an imaginary plane drawn through the flame perpendicular to the axis of the mirror belongs to a ring of the same order, whatever be the point of the mirror against which it is seen projected.

6. Having investigated the conditions of distinctness, let us now proceed to consider the magnitude and character of the rings, supposing the luminous point to be situated at a distance ρ from the mirror, and the rings to be thrown on a screen at the same distance, or else viewed in air. In this case $c = c' = \rho$; and if the luminous point be in the axis $e = 0$, which reduces (5) to

$$R = \frac{te'^2}{\mu c^2} \quad \dots\dots\dots\dots\dots\dots\dots\dots (14).$$

It readily follows from this expression that a system of rings is formed similar to the transmitted rings of the system to which Newton's name is especially attached. The rings in the present case, however, especially when viewed in air, are far more brilliant, and in this respect more resemble the reflected system. If e_1 be the radius of the first bright ring, for which $R = \lambda$, the length of a wave of light,

$$e_1 = \sqrt{\frac{\mu\lambda}{t}} \cdot c ;$$

and for the bright ring of the order n, $e' = \sqrt{n} \cdot e_1$. The formula
(14) has already been discussed by Sir John Herschel, and compared with Newton's measures, with which it manifests a very
close agreement.

With white light, only a moderate number of rings can be
seen, on account of the variation of the scale of the system
depending on a variation in the refrangibility of the component
parts of which white light is made up. When the rings were
formed in air, and the source of light was the flame of an oil-lamp
with a small wick, I have counted seven or eight surrounding the
central bright spot. But when the system is viewed through a
prism, or when the flame of a spirit-lamp is used, an immense
number of rings may be seen.

7. Next, suppose the luminous point out of the axis. Referring to the formula (5), we see that the retardation is not now
equal to zero at the axis, but throughout a circle whose radius e'
is equal to e. Hence the *achromatic line** of the system, which
was formerly reduced to a point, is now a circle having its centre
in the axis, and passing through the luminous point and its image,
which are situated at the opposite extremities of a diameter. The
fringes of the first order will be a pair of circles having their
centre in the axis, and lying, one outside, and the other inside the
central fringe: the fringes of the second order will be another
pair of circles lying, one outside the larger, and the other inside
the smaller fringe of the first order, and so on. It is to be
remarked, however, that only a finite number of fringes are formed
inside the central white fringe. If the value of R when $e' = 0$ be
denoted by $- n_0 \lambda$, n_0 will be a numerical quantity, a function of λ,
which determines the number of fringes and the fraction of a
fringe, belonging to the light of which the wave-length is λ,
which are formed inside the central white fringe. The value of n_0
may be got from equation (5) on putting $e' = 0$, which gives

$$ n_0 = \frac{te^2}{\mu \lambda c^2}. $$

If white light be used, and if n_0 exceed 8 or thereabouts for rays of

---

* I use this term to denote the locus of the points for which the retardation is
equal to zero, which forms a curve on either side of which the colours are arranged
in descending order.

mean refrangibility, all the fringes which the overlapping of the different colours allows to be visible are formed inside as well as outside the central white fringe; and if the luminous point be moved still further from the axis, a portion of the field of view around the axis will appear free from rings. If the radius of the central white fringe, or, which is the same, the distance of the luminous point from the axis, be denoted by $\sqrt{n_0}$, the radii of the bright fringes of the first, second, ... orders will be denoted, on the same scale, by $\sqrt{(n_0 \pm 1)}$, $\sqrt{(n_0 \pm 2)}$ ... and those of the dark rings by $\sqrt{(n_0 \pm \frac{1}{2})}$, $\sqrt{(n_0 \pm \frac{3}{2})}$ ...

The manner in which the rings open out from the centre as the luminous point is moved sideways out of the axis is very striking, and has been accurately described by Newton. The explanation of it is obvious. It may be remarked that the system of rings, regarded as indefinite, is formed on the same scale whatever be the distance of the luminous point from the axis, but the portion of the indefinite system which alone is visible, in consequence of the coincidence or approximate coincidence of the maxima and minima of intensity corresponding to the several colours, depends altogether upon that distance. Since in passing from the interior to the exterior boundary of a given fringe the square of the radius receives a given increase, it follows that the area of the fringe is constant, that is, independent of the perpendicular distance of the luminous point from the axis. Hence the breadth of the fringe continually decreases as the diameter of the circle which forms either boundary increases. When a small flame is used for the source of light, and is moved sideways from the axis, the fringes soon become confused, because a flame which does very well for forming the broad fringes of comparatively small radius seen near the axis, will not answer for the fine fringes of large radius which are formed at a distance from the axis. But on using for the source of light the image of the sun in the focus of a small concave mirror belonging to a microscope apparatus, I found that the fringes were formed quite distinctly even when their diameters became very large and consequently their breadths very small.

SECTION II.

Bands formed by a plane mirror, and viewed directly by the eye.

8. In the case of a plane mirror $\rho = \infty$; and if R be the retardation of the stream scattered at emergence relatively to the stream scattered at entrance, R will be obtained by adding together the second members of equations (5) and (11). Hence we have

$$R = \frac{t}{\mu} \left[\frac{1}{c'^2} \{(x - a')^2 + (y - b')^2\} - \frac{1}{c^2} \{(x - a)^2 + (y - b)^2\} \right] (15).$$

It is to be remembered that in this formula a, b, c denote the co-ordinates of the luminous point; x, y those of any point in the dimmed surface; a', b', c' those of any point M of space towards which the eye is directed, and for distinct vision of which it is adapted; and that the formula is only approximate, the approximation depending both upon the smallness of the obliquity, and upon the smallness of the thickness t of the glass in comparison with the distances of the luminous point and the point M from the mirror.

As regards the illumination at a given point M, we are evidently concerned with so much only of the dimmed surface as lies within a cone having M for vertex and the pupil of the eye for base; and the bands will be seen distinctly if R do not alter by more than a small fraction of λ when x, y alter from one point to another of the portion of the dimmed surface which lies within this cone. Now we have seen already that the bands are in all cases seen distinctly in the neighbourhood of the image when the image itself is seen distinctly, so that when the image is real the bands may even be thrown on a screen, in which case a comparatively large portion of the dimmed surface is concerned in their formation. We may conclude that in the present case the bands will be seen with sufficient distinctness throughout, provided the image of the luminous point be seen distinctly.

9. In considering the distinctness or indistinctness of the bands, we are concerned with the finite size of the pupil of the eye; but in investigating only their form and magnitude we may

suppose the pupil a point, and reduce each pencil entering it to a single ray, which forms the axis of the pencil. Let E be the eye, or rather the centre of the pupil, h its perpendicular distance from the mirror, and suppose the axis of z to pass through E. The ray by which a portion of a band is seen as if at M cuts the mirror in a point whose co-ordinates x, y are equal to a', b' altered in the ratio of h to $h - c'$, so that

$$a' = \left(1 - \frac{c'}{h}\right)x, \qquad b' = \left(1 - \frac{c'}{h}\right)y.$$

Substituting in (15), we get

$$R = \frac{t}{\mu}\left[\frac{1}{h^2}(x^2 + y^2) - \frac{1}{c^2}\{(x - a)^2 + (y - b)^2\}\right]\ldots\ldots(16),$$

from which it may be observed that c' has disappeared, as evidently ought to be the case. The expression (16) might have been at once deduced from (15) by putting the co-ordinates of the eye in place of a', b', c'. The reason of this is evident, because the retardation is constant for the same ray, and a ray may be defined by the positions of any two points through which it passes. We may therefore employ the points E and P, instead of M and P, to define the ray, and may therefore at once substitute the co-ordinates of E for those of M in the expression for the retardation.

10. To determine the forms, &c. of the bands, nothing more will be requisite than to discuss the formula (16). As however this formula was obtained as a particular case from a very general, and consequently rather complicated investigation, in which the curvatures of the surfaces were supposed to have any values, and as the bands to which it relates are of great interest, the reader may be pleased to see a special investigation of the formula for the case of a plane mirror.

Retaining the same notation as before, except where the contrary is specified, let L_0, E_v be the feet of the perpendiculars let fall from L, E on the plane of the dimmed surface, and let $L_0P = s$, $E_0P = u$. Let R_1 be the retardation of a ray regularly refracted and reflected, scattered at emergence at P, and so reaching E; R_2 the retardation of a ray reaching E after having been scattered at P on entering into the glass, and let $R_1 - R_2 = R$. Let $LSTPE$ be the course of the first ray, which emanates from

L, is regularly refracted at S, regularly reflected at T, and scattered on emergence at P. The lines LS, ST, TP will evidently lie in the plane LL_0P. Let ϕ, ϕ' be the angles of incidence and refraction at S. We have

$$R_1 = LS + 2\mu ST + PE,$$
$$= c \sec \phi + 2\mu t \sec \phi' + \sqrt{(h^2 + u^2)} \ldots\ldots\ldots\ldots(17),$$

and

$$c \tan \phi + 2t \tan \phi' = s, \qquad \sin \phi = \mu \sin \phi' \ldots\ldots(18).$$

The obliquity being supposed small, we may expand, and retain only the squares of small quantities, the terms thus neglected involving only fourth and higher powers. We get in the first place from (17)

$$R_1 = c + 2\mu t + h + \tfrac{1}{2}\left(c\phi^2 + 2\mu t\phi'^2 + \frac{u^2}{h}\right)\ldots\ldots\ldots(19).$$

But equations (18) give

$$\phi = \mu\phi' = \frac{\mu s}{\mu c + 2t};$$

and substituting in (19) we get

$$R_1 = c + 2\mu t + h + \frac{\mu s^2}{2(\mu c + 2t)} + \frac{u^2}{2h}.$$

To obtain R_2, it will be sufficient to interchange c and h, s and u, since if we supposed the course of the ray reversed it would emanate from E, be regularly refracted and reflected, then scattered on emergence at P, and so would reach L. Interchanging, subtracting, and reducing, we obtain

$$R = t\left\{\frac{u^2}{h(\mu h + 2t)} - \frac{s^2}{c(\mu c + 2t)}\right\} \ldots\ldots\ldots(20).$$

This formula is more general than (16), since no approximation has yet been made depending on the magnitude of t. In practice, however, t is actually small compared with c and h, so that we may simplify the formula by retaining only the first power of t, which reduces (20) to (16), inasmuch as

$$u^2 = x^2 + y^2, \quad \text{and} \quad s^2 = (x-a)^2 + (y-b)^2.$$

11. Let us now proceed to apply the formula (16) to the explanation of the phenomena. In discussing this formula, it is to be remembered that x, y are the co-ordinates of the point of the mirror on which a fringe is seen projected. Since the direction

of the axis of y is disposable, we may make the plane of y, z pass through the luminous point, in which case $b = 0$, and

$$R = \frac{t}{\mu}\left\{\left(\frac{1}{h^2} - \frac{1}{c^2}\right)(x^2 + y^2) + \frac{2ax}{c^2} - \frac{a^2}{c^2}\right\} \dots\dots(21).$$

For a given fringe R is constant. Hence the fringes form a system of concentric circles, the centre of the system lying in the axis of x. If α be the abscissa of the centre

$$\alpha = \frac{ah^2}{h^2 - c^2} = \tfrac{1}{2}\left(\frac{ah}{h+c} + \frac{ah}{h-c}\right) \dots\dots\dots(22).$$

Now $ah(h-c)^{-1}$ and $ah(h+c)^{-1}$ are the abscissæ of the points in which the plane of the mirror is cut by two lines drawn through the eye, one to the luminous point, and the other to its image. Hence we have the following construction: join the eye with the luminous point and with its image, and produce the former line to meet the mirror; the middle point of the line joining the two points in which the mirror is cut by the two lines drawn from the eye will be the centre of the system.

Hence if the luminous point be placed to the right of the perpendicular let fall from the eye on the plane of the mirror, and between the mirror and the eye, the concavity of the fringes will be turned to the right. If the luminous point, lying still on the right, be now drawn backwards, so as to come beside the eye, and ultimately fall behind it, the curvature will decrease till the fringes become straight, after which it will increase in the contrary direction, the convexity being now turned towards the right. This agrees with observation.

12. The expression for R shews that the circle which forms the achromatic line of the system passes through the two points mentioned in the last paragraph but one. This is always observed to be true in experiment as far as regards the image, and is found to be true of the luminous point also when it is in front of the eye, so as to be seen along with the fringes, provided the fringes reach so far.

Denoting by $n_0\lambda$ the value of R at the centre of the system of circles, taken positively, we get from (21) and (22)

$$n_0 = \frac{a^2 t}{\mu\lambda(h^2 \sim c^2)} \dots\dots\dots(23).$$

The numerical quantity n_0 may conveniently be called the *central order*, since when it lies between $i - \frac{1}{2}$ and $i + \frac{1}{2}$, where i is any integer, the colour at the centre belongs to the bright ring of the ith order. If v be the radius of the central fringe, v will be equal to the semi-difference of the quantities $ah(h + c)^{-1}$ and $ah(h - c)^{-1}$, whence

$$v = \frac{ach}{h^2 \sim c^2} \dots\dots\dots\dots\dots\dots\dots\dots(24).$$

Having found the centre of the system of circles and the projection of the image, or the point where the line joining the eye and the image cuts the mirror, describe a circle passing through this projection. This will be the central line of the bright fringe of the order 0, and its radius will be equal to v. Now describe a pair of circles whose radii are to v as $\sqrt{(n_0 \pm 1)}$ to $\sqrt{n_0}$. These will be the central lines of the two bright fringes of the first order, for the particular colour to which the assumed value of λ relates. The central lines of the two bright fringes of the second order will be a pair of circles with radii proportional to $\sqrt{(n_0 \pm 2)}$, and so on. The fringes will be broader on the concave than on the convex side of the central white fringe. When the fringes become straight, n_0 becomes infinite, and the system becomes symmetrical with respect to the central fringe. This agrees with observation.

13.   When the luminous point is situated in a line drawn through the eye perpendicular to the mirror $a = 0$, and we have simply

$$R = \frac{(c^2 - h^2)\,t}{\mu c^2 h^2}\,(x^2 + y^2).$$

In this case the achromatic line of the system is reduced to a point, and the rings are analogous in every respect to the transmitted system of Newton's rings. For the bright ring of the first order $R = \pm \lambda$, and therefore the radius of the ring is equal to

$$\frac{\sqrt{\mu\lambda}\,.\,ch}{\sqrt{\{(c^2 \sim h^2)\,t\}}}\,,$$

which becomes infinite when $c = h$. Hence if the luminous point be at first situated in front of the eye, and be then conceived to move backwards through the eye till it passes behind it, the rings will expand indefinitely, and so disappear, and will reappear again when the luminous point has passed the eye.

14. All the preceding conclusions agree perfectly with experiment, so far as qualitative results are concerned; for I have not taken any measures. The change in the direction of curvature, which I had not noticed till it was pointed out by the formula, may be readily seen, when the flame of a candle is the source of light, by holding the candle at arm's length nearly in front of the eye, but a little to the right, then drawing it back beside the eye, and finally holding it at arm's length behind the head, and as nearly in a line drawn through the eye perpendicular to the mirror as the shadow of the head will allow.

When the candle is held near the eye, a portion only of the circles can be seen; the circles are in fact reduced to circular arcs, and these arcs may even become perfectly straight. But when the candle is placed at a good distance from the eye, suppose half-way between the eye and the mirror, and a small piece of card is placed as a screen in front of the flame to keep off the glare of the direct light, the circles, or at least several of them, may be seen complete, except that it must be left to the imagination to fill them up where they are hid by the screen. In this way the manner in which the rings open out from the centre of the circles may be observed, though not for the first ring or two, which open out while the centre is hid by the screen. Instead of a candle with a screen, it is better to use the image of the sun in the focus of a small convex or concave mirror.

15. The conclusion deduced from theory which was mentioned in Art. 13 cannot, of course, be compared with experiment directly. But the experiment may be successfully performed by substituting for either the luminous point or the eye a virtual image. Using for the luminous point the image of the sun in the focus of a small concave mirror, at the distance of some feet in front of a plane mirror of which the surface had been prepared with milk and water, I placed a piece of plate glass between the mirrors, inclined at an angle of about 45°. The greater part of the light coming from the image of the sun was transmitted through the plate of glass; and on returning from the large mirror a portion of this light was reflected sideways, so that the rings could be seen by reflexion in the plate of glass without obstructing the incident light. The system of rings thus seen was very beautiful, and there was no direct light glaring in the eye, and yet no screen to hide

any part of the system. It was easy to know when the image of the eye in the inclined plate lay in a line drawn through the luminous point perpendicular to the plane mirror, by observing when the image of the luminous point seen by reflexion, first at the plane mirror, and then at the plate of glass, lay in the centre of the system of rings. Supposing the image to have this position, on moving the head sideways the opening out of the rings could be traced from its very commencement. By moving back the head so as to keep the image of the luminous point in the centre of the system of rings, it was easy to try the experiment to which Art. 13 relates, the virtual image of the eye being thus kept in a line drawn through the luminous point perpendicular to the mirror, and the eye moving relatively to the luminous point, which is as good as if the luminous point had moved while the eye remained fixed. I found, in fact, that on moving back the head the rings expanded till the bright central patch surrounding the image filled the whole field of view, and on continuing to move back the head the rings appeared again. In the position in which the central patch filled the whole field of view, the least motion of the eye sideways was sufficient to bring into the field portions of excessively broad coloured bands.

Between the system of rings seen when the eye was respectively nearer to and further from the mirror than when in the position in which the rings became infinite, there was one difference which may here be mentioned. So long as the image occupied the centre of the systems, they were similar to each other; but when the head was moved sideways, the centre of the circles passed in the first case to the side of the image towards which the head was moved, and in the second case to the contrary side. This affords another way of comparing with experiment the result of theory already mentioned relating to the direction of curvature, and it will be readily seen that the result of experiment agrees with the prediction of theory. For, suppose the distance of the eye from the oblique plate less than that of the luminous point, so that the virtual image of the eye lies between the luminous point and the mirror, and let the eye move to the right. Then its virtual image moves to the left, and therefore, according to theory, the centre of curvature ought to fall to the left of the image of the luminous point, right and left being estimated with reference to an eye supposed to occupy the position of the virtual image of the actual eye. But on account of reflexion at the glass plate, there is reversion from right to left, and therefore to the eye in its actual position the centre of curvature falls to the right of the image of the luminous point, which agrees with observation. The experiments described in this article may be tried very well with the flame of a taper, but in examining what becomes of the rings when they expand it is more satisfactory to use sun-light.

16. In describing the disappearance of the rings, I said that the central spot expanded till it filled the whole field. In truth, when the rings had expanded a faint luminous central spot of finite size remained visible, which was surrounded by a dark ring, and then a faint luminous ring. It would have been more correct to speak of the *dark* ring as faint, since these rings consisted merely in slight alternations of intensity in a generally bright field. These rings, however, had evidently nothing to do with the former rings, which had disappeared; for they continued to have the image of the luminous point for their centre when the head was moved to one side. They were doubtless of the same nature as those which are seen when a luminous point, or the flame of a candle, is viewed through a piece of glass powdered with lycopodium seed, and arose from the interference of pairs of streams of light which passed on opposite sides of the globules of dried milk. I merely mention these rings lest any one in repeating the experiment should observe them, and mistake them for something relating to the colours of thick plates.

17. The formula (21) determines the breadths of the several fringes, which are unequal, except in the case in which the eye and the luminous point are at the same distance from the mirror. It will be convenient, however, to investigate a simple formula to express what may be regarded as a sort of mean breadth. Let the mean breadth be defined to be that which would be the breadth of one fringe if the rate of variation of the order of a fringe, for variation of position in a direction perpendicular to the length of a fringe, were constant, and equal to the rate in the neighbourhood of the projection of the image, and let β be this mean breadth.

Putting $y = 0$ in (21), differentiating on the supposition that

R and x vary together, and after differentiation putting $ah\,(h + c)^{-1}$ for x, we find

$$dR = \frac{2ta}{\mu ch}\,dx;$$

and since, according to the definition of β, $\lambda^{-1}dR = \beta^{-1}dx$, we have

$$\beta = \frac{\mu ch\lambda}{2ta} \quad \dotfill (25).$$

When $c = h$ the bands are straight, and of uniform breadth, that breadth being equal to β; and when the bands are not very much curved β may still be taken as a convenient measure of the scale of the system; but the formula (25) is not meant to be applied to cases in which the projection of the image of the luminous point falls at all near the centre of the circles.

Section III.

Rings formed by a curved mirror, and viewed directly by the eye, when the luminous point and its image are not in the same plane perpendicular to the axis.

18.   The rings and bands of which the theory has been considered in the two preceding sections may be regarded as forming the two extreme cases of the general system.   In the first case, the rings appear to have a definite position in space; in the second case, everything depends upon the position of the eye.   These are the cases of most interest, but there are some properties of the general system which deserve notice.

In order that rings may be thrown on a screen, it is necessary that the retardation of one of the interfering streams relatively to the other should be sensibly constant over the whole of the dimmed surface, or at least over a large portion of it.   But when the rings are viewed directly by the eye, we are concerned with so small a portion of the dimmed surface, in viewing a given point of a ring, that the rings may be seen very well in cases in which they could not be thrown on a screen.   Moreover, we have seen that even independently of the small size of the pupil, a portion at least of the system is seen distinctly when the image of the luminous point is seen distinctly. Omitting further consideration of the conditions of distinctness, let us regard the eye as a point, and investigate the form and character of the rings.

19. Let f, g, h be the co-ordinates of the eye. To find the retardation, it will be sufficient, as in Art. 9, to write f, g, h for a', b', c', and take x, y to denote the co-ordinates of that point of the mirror on which a given point of a ring is seen projected. The whole retardation is the sum of the expressions in (5) and (11); and making the above substitution we find

$$R = \frac{t}{\mu}\left(\frac{f^2 + g^2}{h^2} - \frac{a^2 + b^2}{c^2}\right) + \frac{t}{\mu}\left(\frac{1}{h} - \frac{1}{c}\right)\left(\frac{1}{h} + \frac{1}{c} - \frac{2}{\rho}\right)(x^2 + y^2)$$

$$- \frac{2t}{\mu}\left\{\frac{1}{h}\left(\frac{1}{h} - \frac{1}{\rho}\right)(fx + gy) - \frac{1}{c}\left(\frac{1}{c} - \frac{1}{\rho}\right)(ax + by)\right\} \ldots(26).$$

Hence the bands still form a system of concentric circles. If Y, Y be the co-ordinates of the centre of the system,

$$X = \frac{\dfrac{f}{h}\left(\dfrac{1}{h} - \dfrac{1}{\rho}\right) - \dfrac{a}{c}\left(\dfrac{1}{c} - \dfrac{1}{\rho}\right)}{\left(\dfrac{1}{h} - \dfrac{1}{c}\right)\left(\dfrac{1}{h} + \dfrac{1}{c} - \dfrac{2}{\rho}\right)} \quad \ldots\ldots\ldots\ldots(27),$$

and Y may be written down from symmetry.

The equations of a line joining the eye and the luminous point are

$$\frac{\xi - a}{f - a} = \frac{\eta - b}{g - b} = \frac{\zeta - c}{h - c}.$$

At the point in which this line cuts the mirror $\zeta = 0$, or at least is a very small quantity, which may be neglected. Hence we have

$$\xi = \frac{\dfrac{a}{c} - \dfrac{f}{h}}{\dfrac{1}{c} - \dfrac{1}{h}} \quad \ldots\ldots\ldots\ldots\ldots\ldots(28),$$

from whence η may be written down if required. If ξ_1, η_1 denote the co-ordinates of the point in which the line joining the eye and the image meets the mirror, ξ_1, η_1 may be obtained from ξ, η by writing a_1, b_1, c_1 for a, b, c, where a_1, b_1, c_1 denote the co-ordinates of the image. Observing that

$$\frac{a_1}{c_1} = -\frac{a}{c}, \qquad \frac{1}{c_1} = \frac{2}{\rho} - \frac{1}{c},$$

we find

$$\xi_1 = \frac{\dfrac{a}{c}+\dfrac{f}{h}}{\dfrac{1}{h}+\dfrac{1}{c}-\dfrac{2}{\rho}} \quad\dotfill(29).$$

The formulæ (27), (28), and (29) shew that X is equal to the semi-sum of ξ and ξ_1, and for the same reason Y is equal to the semi-sum of η and η_1. Hence the geometrical construction given in Art. 11 for finding the centre of the system in the case of a plane mirror applies equally to a curved mirror, even when the curvatures of the two surfaces are different. Since the retardation vanishes for the image itself, it follows that the achromatic line is a circle having the two points of intersection above mentioned for opposite extremities of a diameter.

20.   It follows from the expressions for X and Y, or from the geometrical construction to which they lead, that if the eye be not in the line joining the luminous point and its image, whenever it crosses either of two planes drawn perpendicular to the axis, and passing, one through the luminous point, and the other through its image, the centre of curvature of the bands moves off to an infinite distance, and the bands become straight, and then bend round the other way.

When the eye coincides with the luminous point, f, g, h become equal to a, b, c, respectively, and R vanishes independently of x and y. The same takes place when the eye coincides with the image, since in this case

$$\frac{f}{h}=-\frac{a}{c}, \quad \frac{g}{h}=-\frac{b}{c}, \quad \frac{1}{h}+\frac{1}{c}=\frac{2}{\rho}.$$

Hence, when the eye crosses either of the planes above mentioned, remaining in the line joining the luminous point and its image, instead of bands which become straight and then change curvature, we have rings which disappear by moving off to infinity, and then appear again.

I have verified these conclusions by experiment, substituting when necessary a virtual image of the eye for the eye itself, in the manner explained in Art. 15. The experiments embraced the following cases, in the description of which O will be used to denote the centre of curvature of the mirror, F its principal focus, L the luminous point, and L_3 its image.

Concave mirror: L beyond O.   Eye (1) beyond L; (2) passing L; (3) between L and L_3; (4) passing L_3; (5) between L_3 and the mirror.—Concave mirror: L between O and F.   Eye (1) beyond L_3; (2) passing L_3; (3) between L_3 and L; (4) passing L; (5) between L and the mirror.—Concave mirror: L between F and the mirror.   Eye (1) beyond L; (2) passing L; (3) between L and the mirror.—Convex mirror.   Eye (1) beyond L; (2) passing L; (3) between L and the mirror.

The mirrors employed were formed, as usual, with surfaces of equal curvature.   When the observation was made directly, there was no difficulty in determining at which side of the line joining the luminous point and its image the eye lay, and consequently in deciding whether the direction of curvature agreed with theory or not.   When the observation was made by reflexion in a plate of glass, the eye was placed so that its virtual image fell in the line LL_3 by moving the head till the image of the luminous point was seen in the centre of the system of rings.   The radii of the two surfaces of the mirror being the same, or only differing by a small quantity comparable with the thickness of the glass, the surfaces may be regarded as forming a pair of concentric spheres; and therefore, everything being symmetrical with respect to the line joining L and L_3, when the image of the eye is in this line the bands necessarily become rings, having the image for their centre.   Hence the theory of the rings or bands, which it is the object of the experiment to compare with observation, is not involved in the assumption that the image of the eye was in the line LL_3 when the system of rings appeared arranged symmetrically around the image of the luminous point.   By moving the head a little to one side, and observing whether the centre of the system of rings then lay to the right or left of the image, it was easy to compare theory with observation as to the direction of curvature.

There was no difficulty in telling when the virtual image of the eye coincided with the image of the luminous point, since in that case the latter image expanded indefinitely.   The phenomena observed offered no direct test of the coincidence of the virtual image of the eye with the luminous point, except what arose from the appearance of the bands themselves.   I did not think it worth while to take any measures, but contented myself with observing that when the eye was in the expected position, or thereabouts, the rings expanded indefinitely when the image was kept in the centre of the system, and the bands formed when the image was allowed to pass to one side changed curvature as the head moved backwards and forwards.

21.   The bands may be considered as completely characterized by the position and magnitude of the achromatic line, and by the value of the numerical quantity which has been already defined as the central order.   A simple geometrical construction has already been given for determining the achromatic circle.   Substituting X, Y for x, y in (26), and denoting the resulting value of R by $-n_0\lambda$, we find

$$n_0 = \frac{t}{\mu\lambda}\frac{\left\{\frac{f}{h}\left(\frac{1}{c}-\frac{1}{\rho}\right)-\frac{a}{c}\left(\frac{1}{h}-\frac{1}{\rho}\right)\right\}^2 + \left\{\frac{g}{h}\left(\frac{1}{c}-\frac{1}{\rho}\right)-\frac{b}{c}\left(\frac{1}{h}-\frac{1}{\rho}\right)\right\}}{\left(\frac{1}{h}-\frac{1}{c}\right)\left(\frac{1}{h}+\frac{1}{c}-\frac{2}{\rho}\right)}\ldots(30).$$

In the application of this formula n_0 is to be taken positively.

Denoting as before the radius of the achromatic circle by v, we find from (28), (29), and the formulæ thence derived which give η, η_1,

$$v^2 = \frac{\left\{\frac{f}{h}\left(\frac{1}{c}-\frac{1}{\rho}\right)-\frac{a}{c}\left(\frac{1}{h}-\frac{1}{\rho}\right)\right\}^2 + \left\{\frac{g}{h}\left(\frac{1}{c}-\frac{1}{\rho}\right)-\frac{b}{c}\left(\frac{1}{h}-\frac{1}{\rho}\right)\right\}^2}{\left(\frac{1}{h}-\frac{1}{c}\right)^2\left(\frac{1}{h}+\frac{1}{c}-\frac{2}{\rho}\right)^2}\ldots(31).$$

When the bands are nearly straight, instead of the central order it is more convenient to consider the mean breadth of a fringe.   According to the definition of β,

$$v + \beta dn_0 : v :: \sqrt{(n_0 + dn_0)} : \sqrt{n_0},$$

since the radii of the several rings are as the square roots of their orders.   We have therefore

$$\beta = \frac{v}{2n_0}\ldots\ldots\ldots(32).$$

22.   In the case of a concave mirror, if a small flame be so placed as to coincide with its image, and be then moved a little towards the mirror, or from it, it is possible to see a single

system of rings with both eyes at once, if the eyes be situated symmetrically with respect to the flame and its image. The rings so seen appear to be situated between the flame and its image. Let E be the right and E' the left eye, and suppose the head so placed that the line LL_3 bisects EE' at right angles. On account of the similarity of position of the two eyes, the system of rings seen with one eye must be exactly like the system seen with the other, and therefore, in order that a single system may be seen with both eyes at once, it is necessary and sufficient that the axes of the eyes be directed to the centres of the respective systems. It has been shewn already that the projected place on the mirror of the centre of the system seen with either eye, suppose the right eye, bisects the line joining the projected places of the flame and its image. On account of the supposed smallness of the obliquities, this is the same thing as saying that the centre of the system seen by the right eye appears in the direction of a line bisecting the angle LEL_3. Similarly, the centre of the system seen by the left eye appears in the direction of a line bisecting the angle $LE'L_3$. In order therefore that a single system only may be seen, the axes of the eyes must be made to converge to the point in which the bisecting lines intersect LL_3, and therefore the system of rings will appear to be situated between the flame and its image.

Since the angles LEL_3, $LE'L_3$, are bisected by the axes of the eyes when the system of rings is seen single, it follows that the flame and its inverted image are each seen double, in such a manner that the erect flame seen by either eye is superposed on the inverted flame seen by the other. This agrees with observation: in fact, I was led by experiment to the above rule for determining the apparent position of the rings before I had deduced it from theory. The observation was made when the flame was in front of its image, in which case the position of the rings in space appears more definite than when the image is in front of the flame.

Section IV.

Straight bands formed by a plane mirror at a considerable angle of incidence, and viewed by the eye, either directly, or through a telescope.

23.　As the angle of incidence increases the bands become finer and finer, and after they have become too fine to be distinguished by the naked eye they may still be seen through a telescope, provided the source of light be sufficiently small. When the source of light was the image of the sun in a lens of short focus, I saw traces of the bands when the angle of incidence was about $24° 50'$, but they were not at all well formed beyond an angle of about $10° 40'$, after which they began to be confounded with rays which shot in all directions from the image of the luminous point. With a mirror made of thinner glass they would probably have been visible at a still larger angle of incidence. The theory of Section II. sufficiently explains their origin and general character; but inasmuch as the obliquity was supposed small in investigating the formulæ of that section, it may be desirable to obtain an expression for their breadth, in which no approximation shall be made depending on the smallness of the obliquity, in order to meet the case of any future measures which may be taken at a large angle of incidence.

24.　The notation being the same as in Art. 10, we have merely to employ the equations (17) and (18), without making any approximation depending on the smallness of ϕ and ϕ'. The thickness t may still be supposed small compared with c and h. Neglecting t for a first approximation, and then substituting in the small terms the values of $\tan \phi'$ and $\sec \phi'$ got from the first approximation, we find

$$R_1 = \sqrt{c^2 + s^2} + \sqrt{h^2 + u^2} + 2t \sqrt{\frac{(\mu^2 - 1)s^2 + \mu^2 c^2}{s^2 + c^2}}.$$

Interchanging c and h, s and u, and subtracting, we get finally

$$R = 2t \left(\sqrt{\mu^2 - \frac{s^2}{s^2 + c^2}} - \sqrt{\mu^2 - \frac{u^2}{u^2 + h^2}} \right) \ldots\ldots(33).$$

25. For the achromatic line $R = 0$, and therefore $s : c :: u : h$. Hence s is to u in the constant ratio of c to h, and therefore, by a well-known geometrical theorem, the achromatic curve is a circle, having its centre in the line $L_0 E_0$ produced, and cutting this line in the two points in which it is divided internally and externally in the given ratio. The latter of these points may be formed by producing LE to meet $L_0 E_0$ produced, and the former by producing LL_0 till the produced part is equal to the line itself, and then joining the extremity of the produced part with E. Hence the construction given in Art. 11 for determining the bright band of the order zero continues to hold good whatever be the angle of incidence.

26. In the neighbourhood of the image the bands are sensibly straight, being arcs of circles of very large radius. To find the mean breadth of a band, it will be sufficient to suppose the point P to lie in the line $L_0 E_0$, to differentiate equation (33) making R, s, and u vary together, while $s + u$ remains constant, replace $\dfrac{dR}{du}$ by $\dfrac{\lambda}{\beta}$, and after differentiation take u and s to refer to the small pencil, regarded as a ray, by which the image is seen. If i be the angle of incidence, we may put after differentiation $s = c \tan i$, $u = h \tan i$. We thus find

$$\beta = \frac{\lambda c h}{2t(c+h)} \frac{\sqrt{\mu^2 - \sin^2 i}}{\sin i \cos^3 i} \dots\dots\dots\dots(34).$$

On account of the largeness of the angle of incidence, the breadth of the bands is sensibly uniform, and therefore β may be regarded as the breadth of any one band. It is to be remembered that β denotes the linear breadth of a band as seen projected on the mirror. If we denote the angular breadth by ϖ, we have on account of the smallness of ϖ.

$$\varpi = \frac{\beta \cos i}{h \sec i} = \frac{\lambda c}{t(c+h)} \frac{\sqrt{\mu^2 - \sin^2 i}}{\sin 2i} \dots\dots\dots(35).$$

SECTION V.

On the nature of the deflection of the two interfering streams from the course of the regularly refracted light.

27. It was suggested to me by a friend, to whom I was shewing some of the experiments described in this paper, that in order to see bands or rings of the same nature it would probably be sufficient to dim the two faces of a plate of glass, and view a luminous point through it. But having prepared the two faces of a piece of plate glass with milk and water in the same manner as for mirrors, taking care to treat the two faces as nearly as possible alike, on viewing a luminous point through the plate I found not the slightest trace of the rings or bands, whatever were the distance of the eye from the plate. There were indeed one or two indistinct rings surrounding the luminous point; but these were of a totally different nature, being analogous to the rings seen with lycopodium seed, and arising from the interference of pairs of streams which passed on opposite sides of the milk globules. There was no difficulty in distinguishing them from the system sought for, since they continued to have the luminous point for their centre when the plate was inclined to the line joining that point with the eye. The absence of rings or bands indicates therefore that the streams scattered at the opposite sides of a plate are incapable of interfering.

The rays scattered so as to make infinitely small angles with the regularly refracted rays belong to a point in the bright band of the order zero, and are therefore brought to a focus on the retina when the luminous point is seen distinctly. The same must be at least very approximately true for neighbouring points of the system of rings, did any such exist, and therefore a portion at least of the system would be seen distinctly when the luminous point was seen distinctly. The distances of the luminous point from the glass plate, and of the glass plate from the eye, were comparable with the corresponding distances in the experiment with a plane mirror, and the thickness of glass was comparable likewise; and with a mirror the bands are seen with the utmost facility within wide limits of the thickness of the glass, and of the distances of the luminous point and of the eye from the mirror.

But to prevent any doubt as to whether the bands might not have been too small to be seen when formed by transmission, I have calculated the retardation in the same manner as in Art. 10. The result is

$$R = -\frac{t}{2\mu}\left(\frac{1}{c}+\frac{1}{h}\right)^2 (x^2 + y^2),$$

where R is the retardation of the stream scattered at emergence relatively to that scattered at entrance, c is the distance of the luminous point from the plate, h that of the plate from the eye, t the thickness, and μ the refractive index of the plate, and x, y are the co-ordinates of the point in which the plate is cut by any small pencil (regarded as a ray) which enters the eye, and are measured from the point in which the plate is cut by a line joining the luminous point and the eye, a line to which the plane of the plate is supposed to be perpendicular. On substituting numerical values in the above formulæ, it appeared that the dimensions of the rings were such that they could not possibly have escaped notice had they really been formed. The non-appearance of the rings leads to the following law.

In order that two streams of scattered light may be capable of interfering, it is necessary that they should be scattered, in passing and repassing, by the same set of particles. Two streams which have been scattered by two different sets of particles, although they may have come originally from the same source, behave with respect to each other like two streams coming from different sources.

According to this law, in all calculations relating to the colours of thick plates, we must consider the elementary system of rings or bands corresponding to each element of the dimmed surface of the mirror, and then conceive these elementary systems superposed. We must not compound the vibrations corresponding to streams which have been scattered by different elements, and then find the resulting illumination.

28. The reason of this law will be apparent if it be considered that particles of dust, &c. small as they may be, are usually large in comparison with waves of light, so that the light scattered at entrance, taken as a whole, is most irregular; and the only reason why regular interference is possible at all is, that each particle of dust acts twice in a similar manner, once when the general wave is going, and again when it is returning.

To examine more particularly the mode of action, let P be any particle of dust, and consider a wave of light which emanates from any particular element of the flame or source of light whatever it be. When this wave reaches P and proceeds along it, a portion is reflected externally in all directions, and with this we have nothing more to do. When the wave has just passed P, we may conceive it as having in a certain sense *a hole* in its front, corresponding in size to P, that is to say, there will be a certain portion of the surface forming the general front of the wave where the ether is quiescent. As the wave proceeds, the disturbance diverges from the neighbourhood of this hole by regular diffraction, and when the disturbance reaches the quicksilvered surface the general wave suffers reflection, as well as the secondary waves, which, having diverged from the neighbourhood of P, do not constitute a wave with an unruffled front, in consequence of the absence of secondary waves diverging from the hole, which would be necessary to complete a wave with a front similar to that of the original wave. If we consider any particular diffracted ray, the chances are that on its return it will get out by regular refraction, since the dust is supposed to cover a moderate portion only of the first surface of the mirror. A portion of the original wave which entered the glass by regular refraction at a certain distance from P, after regular reflexion is incident on P from within. The chances are that the portion thus incident on P does not correspond to a spot where the front of the wave is materially ruffled by diffraction at entrance, so that in considering the wave incident on P we may neglect the previous diffraction. The wave, then, just after refraction, is incident on P, by which a portion is reflected back again in various directions, with which we have nothing to do, a portion, it may be, is refracted or absorbed by P, and the remainder passes on. The wave so passing on diverges from the neighbourhood of P by ordinary diffraction, and the two diffracted streams, having been diffracted in a similar manner by the same particle, are in a condition to interfere. The similarity of the two diffractions will be considered in more detail in the next section.

Now while the light is still in the glass conceive the particles of dust removed, and then replaced at random. The chances are that no particle will now occupy the position formerly occupied by

P. Let *P'* be the particle nearest to the former position of *P*; and, to make a supposition as favourable as possible to interference, let *P'* be the very particle *P* moved a little along the surface without rotation. Although the interval of retardation *R* of the two streams diffracted by *P* in its first position, and reaching a given point of space, is sensibly the same as the interval of retardation of the two streams diffracted by *P* in its second position, and reaching the same point, yet this interval would be changed altogether were the transference of position to take place during the interval of time which elapses between the departure of the wave from *P* and its return after reflexion, as may very readily be seen. The amount, too, by which the interval of retardation would be changed would vary in an irregular manner from one particle to another, and therefore no regular interference would take place. Now the purely ideal case just considered is precisely analogous to the case of actual experiment when a luminous point is viewed through a plate of glass with both faces dimmed, since the particles on one face have no relation to those on the other. We ought not therefore in such a case as this to expect to see rings or bands.

29. According to the formula (35), the angular breadth of one of the bands formed by a plane mirror becomes considerable when *i* becomes nearly equal to 90°, so that, apparently, bands ought to be visible at a large angle of incidence. But if the courses of the two streams scattered by the same set of particles be traced, it will be found that they are so widely separated that, for various reasons, no regular interference can be expected to take place. Accordingly, the bands are not seen at a large angle of incidence.

30. In the preceding sections I have spoken of the light by which the rings are formed as having been *scattered* at the dimmed surface. And so it really is, if by that term we merely understand deflected from the course it would have followed according to the regular law of refraction. But according to the explanation given in the preceding article the light is not scattered, in the strict sense of the term, but regularly diffracted. Scattered light is, strictly speaking, such as that by which objects are commonly seen, or again, such as that which is transmitted through white paper and similar substances. The preceding view of the nature

of the light by which the rings are formed is confirmed by the results of several experiments.

In the experiments of Sir William Herschel and M. Pouillet mentioned in the introduction, as well as in some of those of the Duke de Chaulnes, rings of the same nature as those formed by a tarnished mirror of quicksilvered glass were produced in cases in which the deflection of the light from its regular course was incontestably of the nature of diffraction. From the similarity of effect we have a right to infer a probable similarity of cause, unless such a supposition should entail some peculiar difficulty, which does not seem to be the case in the present instance, but quite the contrary.

31. Having cleaned the surface of a concave mirror which had been employed in forming the bands, I rubbed a little tallow on it, and then wiped the mirror in one direction with a handkerchief, so as to have a finely striated tarnish on it. The tarnish was not sufficient to cause much obscurity; but the image of a candle seen in the mirror was accompanied, as is usual in such cases, by two tails of light, which ran out in a direction perpendicular to the striæ. Having placed a small flame in the centre of curvature of the mirror, I found that the rings were formed with great brilliancy where they were intersected by the tails of light, but elsewhere they were almost wholly invisible.

Now the tails of light are known to be a phenomenon of diffraction: the striated tarnish may in fact be regarded as a sort of irregular grating, and the tails of light are of the nature of Fraunhofer's spectra. If a tarnish in general were capable of producing rings independently of diffraction, there appears no reason why a tarnish of tallow should not be capable; for the particles of tallow are as fine as those of most other kinds of tarnish. But if in the case of a tarnish of tallow the deflection of the light from its regular course be not a phenomenon of diffraction, there appears no reason why the rings should be confined to the tails of light in the experiment described above.

32. The phenomena of polarization seem however to lead to a crucial experiment for deciding whether the deflection of the light from its regular course, which enables the rings to be formed, be a phenomenon of diffraction, or of scattering in the strict sense of

the term.   When polarized light is scattered, as for example when it is reflected from or transmitted through white paper, it loses its polarization, but when polarized light suffers regular diffraction it retains its polarization.

Having placed a small flame near the centre of curvature of a concave mirror, of which the surface had been prepared with milk and water, I placed a Nicol's prism close to the flame, so as to polarize the light incident on the mirror.   On examining the rings with another Nicol's prism, they proved to be perfectly polarized.

33.   It may not be considered out of place here to point out what appears to be the cause of a phenomenon observed by M. Pouillet.   In an experiment in which rings were occasioned simply by the straight edge of an opaque body held in front of a metallic speculum, it was found that they were formed distinctly in only one half of their circumference.   The reason of this appears to be simply as follows.   As the waves of light pass the diffracting edge in their progress towards the mirror, those rays which are diffracted inwards, so as to enter the geometrical shadow, after being regularly reflected at the mirror fall upon the opaque body, by which they are stopped.   As these rays are required for the formation of that half of the system which lies on the same side as the opaque body, the other half only is well formed.   The first half may be formed obscurely by a few rays which are diffracted in the required direction at such a distance from the edge that on their return they pass clear of the edge, and so proceed to interfere with other rays diffracted by the edge on the return of the general wave.

Section VI.

Investigation of the angles of diffraction.

34.   Something yet remains to be done in order to complete the theory of these rings and bands, namely, to compare the two diffractions which a wave of light experiences at its entrance into the glass and on its return, respectively.   For the phase and intensity of a ray diffracted in a given direction depend altogether on the circumstances under which the diffraction takes place; and

were these circumstances materially different in the case of the two diffractions above mentioned, the rings might be modified, or might even disappear altogether.

Let us consider first the case of a concave mirror when the luminous point and its image are in the same plane perpendicular to the axis. In this case, if we consider any point P on the dimmed surface, and any point M in the plane of the rings, the angle of diffraction for the ray diffracted at emergence will be L_3PM^*. For the ray diffracted at entrance, the angle of diffraction *measured in air* will be LPM_3, that is to say, M_3P is the course of a ray in air which by regular refraction into glass would be brought into the direction of the ray diffracted at P. If C be the intersection of the axis and the plane of the rings, C will be the centre of the system, and the middle point of both the lines LL_3 and MM_3, and therefore LM_3 will be equal and parallel to ML_3. Hence, on account of the smallness of the obliquities, the angles of diffraction LPM_3, L_3PM are sensibly equal, and their planes sensibly coincident, but the deviations take place in opposite directions. But between the two diffractions the light undergoes reflexion; and since the mutual inclination of two rays is reversed by reflexion, we must conceive the direction of deviation reversed in the first diffraction, in order to compare the circumstances of the two diffractions. Allowing for this reversion, we see that not only are the angles of diffraction sensibly coincident, but the directions of deviation are the same.

Accordingly, the interference connected with diffraction, and the interference which gives rise to the colours of thick plates, take place independently of each other. For, let I, I' denote the vibrations at M due to two streams of light diffracted by any particle of dust P on entering the glass, and passing on opposite sides of P; let J, J' denote the vibrations due to two streams diffracted at emergence, and passing on the same sides of P as I, I', respectively; and let $I + I'$ denote the resultant of I and I', and similarly in other cases. Let χ be the difference of phase corresponding to the retardation R, and ω the difference of phase of I, I', and therefore also of J, J', on account of the similarity of

---

* In speaking of angles of diffraction, such as L_3PM, I shall distinguish between L_3PM and MPL_3, using the former notation to denote that the deviation takes place from PL_3 to PM, and the latter to denote that it takes place from PM to PL_3.

the two diffractions. We may represent the phases of the four vibrations by $\theta + \chi + \omega$, $\theta + \chi$, $\theta + \omega$, θ, respectively. Writing down for greater clearness the phases along with the symbols of the vibrations, we may express the resultant of the whole four vibrations by

$$(I_{\theta + \chi + \omega} + I'_{\theta + \chi}) + (J_{\theta + \omega} + J_{\theta}).$$

Moreover, on account of the similarity of the two diffractions, the coefficients of the two vibrations I, J may be supposed equal to each other, and likewise those of the vibrations I', J'. It is true that the diffractions take place at different distances from the source of light, on account of the finite thickness of the glass, but the difference of distance compared with either of the absolute distances is a small quantity of the order t, which may be neglected. Hence the two resultants $I + I'$, $J + J'$ belong to a diffraction ring of the same kind, and in fact differ in nothing but in phase; the phase of the former exceeding that of the latter by χ. Hence the two kinds of interference go on independently of each other. It is true that in the preceding reasoning we have considered only two interfering streams I, I', and that in calculations of diffraction we have to consider the resultant of an infinite number of streams. But the same reasoning would evidently hold good whatever were the number of streams I, I', I''... with their correspondents J, J', J'' ...

35. When an irregular powder, or anything of the kind, is used to scatter the light, no diffraction rings are visible, because a given point M in the plane of the rings would belong to a diffraction ring of one kind so far as one particle of dust was concerned, and to a diffraction ring of another kind so far as another such particle was concerned; and therefore nothing is seen but the interference rings belonging to thick plates. But when lycopodium seed is used the lycopodium rings and the interference rings are seen together. The former are always arranged symmetrically around the image, as ought to be the case, since they depend only on the angle of diffraction, which is the same for all points of a circle described round L_3 as a centre. By this circumstance they are at once distinguished from the latter, the centre of which falls half way between the luminous point and its image. On scattering some lycopodium seed on a concave mirror, and placing a small flame near the centre of curvature, at such a distance laterally

that the two systems of rings intersected each other, I found in fact that whatever colour appeared in that part of a lycopodium ring which lay outside the interference system was predominant in the latter system throughout the remaining part of a circle described round the image. When the flame was placed in the axis, an abnormal inequality in the brilliancy of the rings of the interference system became very apparent. This inequality was easily seen to correspond to the alternations of intensity in the lycopodium system.

36. Let us now turn to the general case, in which the luminous point and the eye are supposed to have any positions, either in the axis or not far out of it.

The equations of the lines PL_3, PE are

$$\frac{\xi - a_3}{a_3 - x} = \frac{\eta - b_3}{b_3 - y} = \frac{\zeta - c_3}{c_3},$$

$$\frac{\xi - f}{f - x} = \frac{\eta - g}{g - y} = \frac{\zeta - h}{h}.$$

Let the small angle L_3PE be projected on the planes of zx and zy, and let α, β be the projections, measured positively towards x, y, and from PL_3 towards PE. The preceding equations give

$$\alpha = \frac{x - a_3}{c_3} - \frac{x - f}{h},$$

which becomes, when a_3 and c_3 are expressed in terms of a and c,

$$\alpha = \left(\frac{2}{\rho} - \frac{1}{c} - \frac{1}{h}\right) x + \frac{a}{c} + \frac{f}{h} \ldots\ldots\ldots\ldots(36).$$

If α', β' be the projections of the angle of diffraction LPE_3, (where E_3 denotes the image of the eye,) we may find α', β' from α, β by interchanging a, b, c, and f, g, h, and changing the sign. If now we change the signs of the resulting expressions, in order to allow for reflexion at the back, and so compare the circumstances of the two diffractions, we shall obtain the very same expressions as at first, since (36) and the corresponding expression which gives β remain unchanged when a, b, c and f, g, h are interchanged. Hence in the general case, as well as in the particular case first considered, the two diffractions take place under the same circumstances, and therefore the interference rings are not affected by any irregularities which may attend the mode of diffraction. Furthermore, should the diffraction take place with a certain degree of regularity, as in the case of lycopodium seed, so as to exhibit rings or fringes in the aggregate effect of all the particles which send light into the eye in such a direction as to be brought to a given point on the retina, the diffraction rings and the interference rings are seen independent of each other*.

37. If δ be the small angle of diffraction, $\delta^2 = \alpha^2 + \beta^2$, whence from (36) and the other equation which may be written down from symmetry,

$$\delta^2 = \left\{ \left(\frac{2}{\rho} - \frac{1}{c} - \frac{1}{h} \right) x + \frac{a}{c} + \frac{f}{h} \right\}^2 + \left\{ \left(\frac{2}{\rho} - \frac{1}{c} - \frac{1}{h} \right) y + \frac{b}{c} + \frac{g}{h} \right\}^2 \quad (37).$$

Hence the loci of the points for which the angle of diffraction has given values form a system of concentric circles. Referring to (29), we see that the co-ordinates of the centre of the system are ξ_1, η_1, so that the centre is situated in the point in which the mirror is cut by the line joining the eye and the image of the luminous point. This result might have been foreseen, since δ vanishes only for the regularly refracted light, and this enters the eye only in the direction of the line joining the eye and the image. By introducing the co-ordinates ξ_1, η_1, the equation (37) may be put under the form

$$\delta^2 = \left(\frac{1}{h} + \frac{1}{c} - \frac{2}{\rho} \right)^2 \left\{ (x - \xi_1)^2 + (y - \eta_1)^2 \right\} \quad \ldots\ldots\ldots (38).$$

Since the diffraction becomes very sparing when the angle of diffraction becomes at all considerable, it follows that the interference rings are but weak at a considerable angular distance from the image of the luminous point. This agrees with observation. In the experiment in which a flame is placed in the centre of curvature of a concave mirror, and is then moved to one side, although the rings are symmetrical with respect to the flame and

* From some rough experiments which I have myself made with gauze stretched in front of a concave glass mirror, of which the surface was clean, I am inclined to think that the squarish rings observed by the Duke de Chaulnes in the experiment with muslin, already mentioned, were due to a combination of the coloured rings of thick plates, and of the appearance produced by a cross-bar grating. If so, the independence of the two systems would have been rendered evident by slightly inclining the mirror, when the latter system would have had the image for its centre, whereas the former would have had for its centre a point situated midway between the luminous point and the image.

its image, so far as regards their forms and colours, they are not symmetrical so far as regards their intensities, but are decidedly more brilliant on the side of the image than on the side of the flame itself. That this is not due merely to the glare of the direct light, may be proved by holding a small object in front of the flame, so as to screen the eye from the direct light, when the rings, though better seen than before in the neighbourhood of the flame, are still much weaker than on the opposite side, if the distance of the flame from the axis is at all considerable. For the same reason, in the case of a plane mirror, when the luminous point is placed a good distance in front of the eye, so that the rings do not run out of the field of view, they cannot be traced throughout the whole extent if the angular distance between the luminous point and its image be too great, but only throughout a portion, more or less considerable, on the side of the image.

38. In the case of a concave mirror when the luminous point is not far from the centre of curvature, and the rings are viewed by an eye placed at no great distance off, the first factor in the expression for δ^2 (equation 38) is not large, and the angle of diffraction does not increase rapidly in passing away from the image. In the case of a plane mirror $\rho = \infty$, and if we suppose c and h equal to what they were in the former case, or thereabouts, in order to make the two cases comparable in every respect except the curvature of the mirror, the factor in question, though larger than before, is still sufficiently small to prevent δ from increasing very rapidly on receding from the image. Accordingly, in both these cases, the rings and bands are seen with brilliancy at a considerable angular distance from the image. But in the case of a convex mirror of considerable curvature ρ is negative, and not large, so that the factor in the expression for δ^2 becomes considerable, and accordingly the angle of diffraction increases rapidly on receding from the image. I found, in fact, that such a mirror was peculiarly ill suited for producing rings or bands, inasmuch as only a comparatively small portion of the system usually seen was visible, namely, the portion which lay in the immediate neighbourhood of the image.

[From the *Report of the British Association* for 1851, Part II. p. 14.]

On a new Elliptic Analyser.

AFTER alluding to various methods which had been employed in investigating experimentally the nature of elliptically-polarized light, that is to say, the elements of the ellipse described, the author exhibited and described a new instrument which he had invented for the purpose. In its construction he had aimed at being in all important points independent of the instrument-maker, assuming nothing but the accuracy of the graduation.

The construction is as follows :—A brass rim, or thick annulus, is fixed on a stand, so as to have its plane vertical. A brass circle, graduated to degrees, turns round within the annulus, and the angle through which it is turned is read by verniers engraved on the face of the annulus. The brass circle is pierced at its centre, and carries on the side turned towards the incident light a plate of selenite, of such a thickness as to produce a difference of retardation in the oppositely polarized pencils amounting to about a quarter of an undulation for rays of mean refrangibility. On the side next the eye the brass circle carries a projecting collar, and round this collar there turns a moveable collar carrying verniers, and destined to receive a Nicol's prism.

The observation consists in extinguishing the light by a combination of the two movements. The retarding plate converts the elliptically-polarized light which has to be examined into plane-polarized, and this plane-polarized light is extinguished by the Nicol's prism. There are two distinct positions of the retarding plate and the Nicol's prism in which this takes place. In each of these principal positions the retarding plate and the Nicol's prism may be reversed (*i. e.* turned through 180°), and the means of the readings in these four subsidiary positions may be taken for greater accuracy. The readings of the fixed and moveable verniers in each of the two principal positions are four quantities given by observation, which determine four unknown quantities, namely, (1) the index error of the fixed verniers, or, which comes to the same, the azimuth of the major axis of the ellipse described by the particles of ether, measured from a plane fixed in the graduated circle; (2) the ratio of the axes of the ellipse; (3) the index error of the moveable verniers; (4) the retardation due to the retarding plate. The unknown are determined by the known quantities by certain simple formulæ given by the author.

Let these unknown quantities be denoted by I, tan ϖ, i, and ρ, respectively, the latter being reckoned as an angle, at the rate of 360° to an undulation. Let R, r be the readings of the fixed and moveable verniers respectively in one of the principal positions, R', r' the corresponding readings in the other; then

$$I = \frac{R' + R}{2}; \quad i = \frac{r' + r}{2};$$

$$\cos 2\varpi = \frac{\sin(r' - r)}{\sin(R' - R)}; \quad \cos \rho = \frac{\tan(r' - r)}{\tan(R' - R)}.$$

The author stated that he had made a good many observations with this instrument for the sake of testing its performance, and that he had found it very satisfactory. Inasmuch as light is not homogeneous, the illumination never vanishes, but only passes through a minimum, and in passing through the minimum the tint changes rapidly. This change of tint is at first somewhat perplexing; but after a little practice, the observer is able to point mainly by intensity, taking notice of the tint as an additional check against errors of observation. The accuracy of the observations is a little increased by the use of certain rather pale coloured glasses.

To give an idea of the degree of accuracy of which the instrument is susceptible, suppose the ratio of the axes of the ellipse described to be about 3 to 1. In this case the author found that the mean error of single observations amounted to about a quarter or the fifth part of a degree in the determination of the azimuth, three or four thousandths in the determination of the ratio of the minor to the major axis, and about the thousandth part of an undulation in the determination of the retardation.

On account of the accuracy with which the retardation is determined, and the largeness of the chromatic variations to which it is subject, the instrument may be considered as determining, not only the elements of the ellipse described, but also the refrangibility of the light employed, or its length of wave, which corresponds to the refrangibility. The author stated that the error of the thousandth part of an undulation, to which the determination of the retardation was subject, corresponded to an error of only the twentieth or thirtieth part of the interval between the fixed lines D and E of Fraunhofer.

[Apart from the details of construction of the instrument, the distinctive feature of the method of observation here described consists in regarding the retardation due to the plate as an unknown quantity, which is eliminated by the method of observation. The author was not aware when the paper was communicated to the British Association that he had been anticipated in this by Mac Cullagh. (See *Proceedings of the Royal Irish Academy*, Vol. II. p. 384, or *Collected works of Mac Cullagh*, p. 239.) Mac Cullagh's actual observations were made with a Fresnel's Rhomb, not a retarding plate, but of course the principle is the same in the two cases.

In the use of the instrument here described, the observer has got to extinguish the transmitted light, or in case the incident light used be not homogeneous, to reduce it to a minimum, by the alternate employment of the two angular motions of which such an instrument must be susceptible.

In designing the instrument, the choice of the two angular motions which shall be independent of each other is not altogether a matter of indifference. It might have been constructed either (a) so that the Nicol's prism and the retarding plate should rotate independently, or (b) so that the prism and plate should rotate together and the prism independently, or (c) so that the prism and plate should rotate together and the plate independently. It is evidently desirable that in adjusting one of the rotations so as to reduce the intensity of the transmitted light to a minimum the adjustment of the other should not be much put out.

Let θ be the azimuth of the plane of polarization of the light transmitted by the Nicol, measured from a fixed direction, ϕ that of one of the neutral axes of the plate when the incident light, supposed to be homogeneous and perfectly elliptically polarized, is wholly extinguished, $\theta + \delta\theta$, $\phi + \delta\phi$ the actual azimuths in the course of an observation. In the true positions of the plate and Nicol, the elliptically polarized light presented for observation is converted by the plate into plane-polarized, the plane of polarization having the azimuth θ. In consequence of the errors of azimuth $\delta\theta$, $\delta\phi$, the light falling on the Nicol has a component polarized perpendicularly to the azimuth $\theta + \delta\theta$. Let us determine the intensity (Q) of this light as a function of the errors of pointing.

Let O, E denote the neutral axes of the retarding plate, N the plane of polarization of the light transmitted by the Nicol, all in the true positions, O', E', N' the same in the actual positions, P' a plane perpendicular to N'. Let us take the intensity of the elliptically polarized light presented for observation as unity, which may also be taken as the intensity of the light falling on the Nicol, whatever be its azimuth, since the small loss by reflection at the surfaces of the plate is almost rigorously the same for the two components, and we are only concerned with the relative intensities. In the true positions, light of intensity 1 polarized in the plane N falls on the Nicol; and if we suppose the direction of this light reversed until it has passed through the plate, substituting in the plate acceleration for retardation, and then reverse the direction again, we shall obtain the light presented. This incident light is now to be supposed to fall on the plate and Nicol in their actual, not their true positions.

The light, whether reversed or direct, falling on the plate must be resolved into its components polarized along the neutral axes, and these again must be resolved so as to retain the components polarized in the plane P'. The four components with which we are concerned may conveniently be designated as $NOO'P'$, $NOE'P'$, $NEO'P'$, $NEE'P'$, from their successive planes of polarization. The magnitude of each component will be got by taking the product of the cosines of the successive differences of azimuth; thus for $NOO'P'$ it is

$$\cos(\theta - \phi) \cos \delta\phi \cos(\theta + \delta\theta + \tfrac{1}{2}\pi - \phi - \delta\phi).$$

The relative retardations of phase for the four are $0, -\rho, \rho, 0$. The intensity is found by the usual formula. For the object in view, $\delta\theta$ and $\delta\phi$ may be supposed very small. The result is

$$Q = (\delta\theta - \delta\phi + \cos\rho\,\delta\phi)^2 + (\sin\rho\,\cos 2\psi\,\delta\phi)^2,$$

where ψ is put for $\theta - \phi$, the azimuth of the Nicol relatively to the plate.

When the retardation given by the plate does not differ considerably from $90°$, the first term in Q does not differ greatly from $\delta\psi^2$. The expression for Q shews therefore that it is best that ψ and ϕ, not θ and ϕ, should be the angles that are altered independently; that is, that of the three constructions mentioned at the bottom of p. 199, that marked (b) should be the one chosen. This is the one described at the beginning, and is that with which the trials of the working of the instrument were made, the general result of which is mentioned above.

The employment of a plate giving a retardation of about a quarter of an undulation introduces considerable chromatic variations, which we might sometimes desire to avoid. Suppose for example that we were working with light only slightly differing from plane-polarized, and did not wish to reduce the intensity by the use of absorbing media or by selecting a portion of a spectrum; it might seem unreasonable to introduce such large chromatic variations merely to determine a small ellipticity. But in such a case we are not bound to use a retarding plate such as hitherto supposed; we may use a thin plate of mica giving a comparatively small retardation; all that is requisite being that the retardation should be large enough to command the ellipticity of the light that we have to observe, that is, as may readily be shewn, that it should not be less than 2ϖ, where $\tan\varpi$ denotes the ratio of the minor to the major axis of the ellipse. With the diminished retardation, the chromatic changes which the retarding plate introduces into the light observed are of course diminished. The expression for Q shews that in this case it would be best that the motions of the plate and prism should be mechanically independent; but as the instrument does not lend itself to that, we must take our choice of the two other arrangements, and it may be shewn from theory that the instrument ought to work well provided we give to the Nicol, not to the plate, the independent motion which we have at our disposal.

With a view to its possible employment in such case, the instrument as originally constructed was furnished with adapting pieces enabling the Nicol and the plate to exchange places, the whole instrument being of course in that case turned round, so that the graduation faces the incident light instead of facing the observer.

If on p. 198 I was called the azimuth of the *major* axis of the ellipse described, it was merely to facilitate the conception of the four unknown quantities that we had to determine. It is hardly necessary to observe that all we are concerned with in the formulæ is that it is the azimuth of a principal axis.]

[From the *Cambridge and Dublin Mathematical Journal*, Vol. VI., p. 215 (*November*, 1851).]

ON THE CONDUCTION OF HEAT IN CRYSTALS.

THE 21st, 22nd, and 23rd volumes of the *Annales de Chimie et de Physique* contain three very interesting papers by M. de Senarmont, describing a series of experimental researches on the conduction of heat in crystals, as well as in bodies subject to mechanical pressure in one direction. The mode of experimental examination employed consisted in cutting a plate from the crystal to be examined, drilling a small hole through it near the middle, covering the faces with a thin coating of wax, and then heating the crystal by a wire or fine tube inserted into the hole. The heat caused the wax to melt in the neighbourhood of the hole, and thus a certain isothermal line was rendered visible to the eye, namely, the line corresponding to the temperature of melting wax. The variation of conductivity in different directions was indicated by the elliptical, or at least oval form of the line bounding the melted wax. This line remained sufficiently visible after the plate had cooled, and thus the eccentricity of the ellipse and the azimuth of its major axis could be examined at leisure. On allowing for errors of observation, it was found that, for a plate cut in a given direction from a given crystal, the axes of the ellipse had a determinate ratio, and the major axis a determinate azimuth. Universally it was found that the thermic corresponded with the crystallographic symmetry, so that for example in crystals belonging to the cubical system, the propagation of heat took place as it would have done in a homogeneous uncrystallized medium; in crystals belonging to the rhombohedral system the axis was a direction of thermic symmetry, and similarly in other cases. When the plate was not perpendicular to an axis of thermic symmetry, the circumstance was indicated by the non-correspondence of the ovals formed on the opposite faces, the line joining the centres of the ovals being in that case oblique to the faces.

The subject of crystalline conduction had previously been investigated theoretically by M. Duhamel, in a memoir presented to the Academy of Sciences in 1828, and printed in the 21st *Cahier* of the *Journal de l'École Polytechnique*, p. 356. In this memoir the author deduces the general expressions for the flux of heat, and the equation of motion of heat, from the hypothesis of molecular radiation, and applies the general equations to the solution of a few simple problems, or at least problems which may very simply be reduced to the corresponding problems relating to uncrystallized media. After the publication of the researches of M. de Senarmont, M. Duhamel was induced to resume the subject, and in a memoir printed in the 32nd *Cahier* of the above-mentioned *Journal*, he has deduced from theory a number of general consequences which are directly applicable to the experiments of M. de Senarmont.

In the following paper, I propose to present the theory of crystalline conduction in a form independent of the hypothesis of molecular radiation—a hypothesis which for my own part I regard as very questionable. The subject will thus be considerably simplified, for in fact the results flow readily from certain very general assumed laws, which no doubt follow as consequences of the hypothesis of molecular radiation, but which are of such simplicity that they would seem to follow from almost any reasonable hypothesis relating to the manner in which the passage of heat takes place in the interior of a solid body. As regards the mathematical deduction of consequences from the general formulæ, I have introduced the consideration of what may be called an *auxiliary solid*, by which means problems relating to crystallized bodies are reduced to corresponding problems relating to ordinary media. All the principal results of M. Duhamel, of which one at least was obtained by him in a very artificial manner, are thus rendered almost self-evident, or else directly reduced to known results relating to ordinary media; and some results of still greater generality follow with equal facility.

1. Let P be any point of a solid body, homogeneous or hetero-geneous, crystallized or uncrystallized; suppose the temperature of the body to vary from point to point, and let dS be an elementary plane area drawn through P in a given direction. The quantity of heat which passes across the element dS in the elementary time dt will be ultimately proportional to $dSdt$, and may be expressed by $fdSdt$. This quantity f is the *flux* of heat referred to a unit of surface. Its value will depend upon the time, upon the position of the point P, and upon the direction of the elementary plane drawn through P. For the present, suppose the time and the position of the point P given, and consider only the variation of f in different directions about P.

If we suppose the values of f given in the direction of each of three planes, rectangular or not, passing through P, its value in the direction of any fourth plane follows. For make P the vertex of a triangular pyramid, of which the sides are in the direction of the first three planes, and the base is parallel to the fourth, and then conceive the base, remaining parallel to itself, to approach indefinitely to P. The quantity of heat gained by the pyramid during the time dt is equal to the quantity which enters by the faces, diminished by the quantity which escapes by the base. Now when the pyramid is indefinitely diminished, the gain of heat in a given indefinitely short time will vary ulti-mately as the volume, or as the cubes of homologous lines, whereas the quantity which passes across any one of the four faces of the pyramid will vary ultimately as the area of the face, and therefore as the squares of homologous lines. Hence in the limit the quantity of heat which escapes by the base will be equal to the sum of the quantities which enter by the sides, and consequently if the flux across each be given, the flux across the base is determinate.

In particular, if we suppose the medium referred to the rect-angular axes of x, y, z, and if f_x, f_y, f_z be the fluxes across three planes drawn through P in directions perpendicular to the axes of x, y, z; f the flux across a plane drawn in any other direction through P; l, m, n the cosines of the angles which the normal to this plane makes with the axes, we have

$$f = lf_x + mf_y + nf_z \dots\dots\dots\dots\dots\dots(1).$$

This equation shews that if we represent the fluxes across planes perpendicular to the axes of x, y, z, by three forces or three velocities, the flux across any other plane will be represented by the resolved part of the forces or velocities along the normal to this plane. Hence the flux across one particular plane passing through P is a maximum, and the flux across any other plane is equal to this maximum flux multiplied by the cosine of the angle between the two planes.

2.   Let u be the temperature at P at the end of the time t, and consider the portion of the solid which is contained in the elementary volume $dx\,dy\,dz$ adjacent to P. The quantity of heat which enters this element during the time dt by the first of the faces $dy\,dz$ is ultimately equal to $f_x\,dy\,dz\,dt$, and the quantity which escapes by the opposite face is ultimately equal to $(f_x + df_x/dx \,.\, dx)\,dy\,dz\,dt$. Subtracting the former from the latter, and treating in the same way each of the other two pairs of opposite faces, we find that the loss of heat in the element is ultimately equal to

$$\left(\frac{df_x}{dx} + \frac{df_y}{dy} + \frac{df_z}{dz}\right) dx\,dy\,dz\,dt.$$

But if ρ be the density, $\rho\,dx\,dy\,dz$ will be the mass; and if c be the specific heat, the loss of heat will also be equal to

$$-\rho\,dx\,dy\,dz\,.\,c\,\frac{du}{dt}\,dt$$

ultimately.   Equating the two results, and passing to the limit, we get

$$c\rho\,\frac{du}{dt} = -\left(\frac{df_x}{dx} + \frac{df_y}{dy} + \frac{df_z}{dz}\right) \dotfill (2).$$

3.   The formulæ (1) and (2) are general, but for the future I shall suppose the medium to be homogeneous, and the temperature to differ by only a small quantity from a certain fixed standard which we may suppose to be the origin from which u is measured.   Since the medium is homogeneous ρ is constant*, and c moreover will be constant, except so far as relates to a change of specific heat produced by a change of temperature. But since u is supposed to be small, the terms arising from the

---

* The expansion of the solid produced by heat is not here taken into account.

variation of c would be small quantities of the second order, since c only appears multiplied by du/dt, and therefore c may be regarded as constant.

It remains to form the expressions for f_x, f_y, f_z. By the conduction of heat we mean that sort of communication which takes place between the contiguous portions of bodies. In the case of bodies which are partially *diathermous*, that is to say, which behave with respect to heat, or at least heat of certain degrees of refrangibility, in the same way in which semi-opaque bodies behave with respect to light, or rather in which a green glass behaves with respect to red rays, heat may be communicated from one portion of the body to another situated at a sensible distance. But this is, properly speaking, internal radiation, and not conduction. Again, if the solid be perfectly diathermous to heat of certain degrees of refrangibility, a portion in the interior of the mass may by radiation send heat out of the solid altogether. For my own part I believe conduction to be quite distinct from internal radiation, although the theory which makes conduction to be nothing more than molecular radiation and absorption seems to be received by many philosophers with the most implicit reliance. No doubt internal radiation may, and I believe generally if not always does, accompany conduction; and when the distance which a ray of heat can travel before it is absorbed is insensible, we may include internal radiation in the mathematical theory of conduction, and even, if we please, in our definition of the word *conduction*. Of course the distance which we may regard as insensible will depend partly on the dimensions of the body, partly upon certain lengths relating to the state of temperature in the interior, and depending upon the problem with which we have to deal. As an example of a length of this sort, we may take the distance between consecutive maxima, if we are considering the internal temperature of a solid of which the surface has a temperature that is subject to periodic variations.

4. Let us now confine ourselves to conduction, using that term with the extensions and restrictions above explained. The temperature u is supposed to be sufficiently small to allow us to superpose different systems of temperature without mutual disturbance. If the temperature were the same at all points, there

would be equilibrium of temperature, and the flux at any point in any direction would be equal to zero. Let then a uniform temperature, equal and opposite to that of P, be superposed on the actual system. Then the temperature at P will be reduced to zero without any change being made in the fluxes f_x, f_y, f_z. Hence these quantities will depend, not upon the absolute temperature at P, but only on its variation in the neighbourhood of P. Since, by hypothesis, these fluxes have nothing to do with the temperatures at points situated at sensible distances from P, they may be assumed to depend only on the differential coefficients du/dx, du/dy, du/dz, which define the variation of temperature in the neighbourhood of P. Since moreover different systems of temperature may be superposed, it follows that f_x, f_y, f_z are linear functions of the three differential coefficients above written. Hence equation (2) may be put under the form

$$c\rho \frac{du}{dt} = A' \frac{d^2u}{dx'^2} + B' \frac{d^2u}{dy'^2} + C' \frac{d^2u}{dz'^2}$$

$$+ 2D' \frac{d^2u}{dy'dz'} + 2E' \frac{d^2u}{dz'dx'} + 2F' \frac{d^2u}{dx'dy'} \ldots (3),$$

where x', y', z', have been written for x, y, z.

5.   Let us now refer the solid to the rectangular axes Ox, Oy, Oz, instead of Ox', Oy', Oz'. Let l, m, n, be the cosines of the angles $x'Ox$, $x'Oy$, $x'Oz$; let l', m', n', be the same for y', and l'', m'', n'', the same for z'. Then

$$\frac{d}{dx'} = l \frac{d}{dx} + m \frac{d}{dy} + n \frac{d}{dz}.$$

But we have also

$$x' = lx + my + nz,$$

and similar formulæ hold good with respect to y' and z'. Since symbols of differentiation combine with one another according to the same laws as factors, it follows that the right-hand member of equation (3) will be transformed exactly as if the symbols of differentiation were replaced by the corresponding coordinates. Hence there exists a system of rectangular axes, namely, the principal axes of the surface,

$$A'x'^2 + B'y'^2 + C'z'^2 + 2D'y'z' + 2E'z'x' + 2F'x'y' = 1 \ldots (4),$$

for which the equation of motion of heat takes the form

$$c\rho \frac{du}{dt} = A \frac{d^2u}{dx^2} + B \frac{d^2u}{dy^2} + C \frac{d^2u}{dz^2} \quad\dots\dots\dots\dots (5).$$

The system of axes of which the existence has just been established may conveniently be called the *thermic axes* of the crystal. Since the left-hand member of equation (4) is identical with $Ax^2 + By^2 + Cz^2$, it follows that not only do the principal axes of the surface (4) determine the directions of the thermic axes, but the constants A, B, C, are the squared reciprocals of the principal semi-axes of that surface.

6. Let us now take the thermic axes for axes of coordinates, and investigate the general expression for the flux of heat. The general expressions for f_x, f_y, f_z, being linear functions of the three differential coefficients of u with respect to x, y, z, will contain altogether nine arbitrary constants. Substituting the general expressions in (2), and comparing with (5), we find three relations between the constants, depending upon the choice of coordinate axes. These relations being introduced, the expressions may be put under the following form:

$$\left.\begin{aligned}
-f_x &= A \frac{du}{dx} - F_1 \frac{du}{dy} + E_1 \frac{du}{dz} \\
-f_y &= B \frac{du}{dy} - D_1 \frac{du}{dz} + F_1 \frac{du}{dx} \\
-f_z &= C \frac{du}{dz} - E_1 \frac{du}{dx} + D_1 \frac{du}{dy}
\end{aligned}\right\} \quad\dots\dots\dots (6).$$

I shall defer till towards the end of the paper a consideration of the reasons which make it probable that D_1, E_1, F_1 are necessarily equal to zero. For the present it may be observed that if the medium be symmetrical with respect to two rectangular planes, these constants must vanish. For the planes of symmetry must evidently contain the thermic axes; and on account of the symmetry supposed, if the planes of symmetry be taken for those of xz and yz, f_x must change sign with x, while f_y and f_z remain unchanged; and similarly when the sign of y is changed, f_y must change sign, while f_x and f_z remain unchanged. Referring to (6), we see that this requires the constants D_1, E_1, F_1 to vanish, so that

$$f_x = -A \frac{du}{dx}, \quad f_y = -B \frac{du}{dy}, \quad f_z = -C \frac{du}{dz} \quad\dots\dots\dots (7),$$

from whence the flux in any direction may be obtained by means of the formula (1). The formulæ (7) contain the expressions for the flux which result from the theory of M. Duhamel. The constants A, B, C, denote what may be called the *principal conductivities* of the crystal. The reader may suppose for the present that the following investigations are restricted to media which are symmetrical with respect to two rectangular planes.

7. It may be worth while to return to the coordinates x', y', z', which have a general direction, and examine the general expressions for the flux which correspond to the formulæ (7). Putting $f_{x'}$, $f_{y'}$, $f_{z'}$, for the fluxes across planes perpendicular to the axes of x', y', z', we get from (1) and (7)

$$\left. \begin{aligned} -f_{x'} &= A' \frac{du}{dx'} + F' \frac{du}{dy'} + E' \frac{du}{dz'} \\ -f_{y'} &= B' \frac{du}{dy'} + D' \frac{du}{dz'} + F' \frac{du}{dx'} \\ -f_{z'} &= C' \frac{du}{dz'} + E' \frac{du}{dx'} + D' \frac{du}{dy'} \end{aligned} \right\} \dots\dots\dots (8),$$

where
$$\left. \begin{aligned} A' &= l^2 A + m^2 B + n^2 C \\ D' &= l'l''A + m'm''B + n'n''C \end{aligned} \right\} \dots\dots\dots (9);$$

from whence the expressions for B', E', and C', F', may be written down by symmetry.

8. So long as we are only concerned with the succession of temperatures in an infinite solid, we have no occasion to consider the flux of heat, and the general equation (5) will enable us to perform all the requisite calculations. In this equation the *indestructibility** of heat is recognized, but not its *identity*. If we discard the latter idea, it is nonsense to talk of the heat gained, we will suppose, by a given element of the solid, as having come from this quarter rather than from that. If we denote by Δf_x, Δf_y, Δf_z the quantities by which the values of f_x, f_y, f_z given by (6) exceed those given by (7), we have

$$\frac{d\Delta f_x}{dx} + \frac{d\Delta f_y}{dy} + \frac{d\Delta f_z}{dz} = 0,$$

---

* According to the very important researches of Mr Joule, work is convertible into heat, from which there can be little doubt that conversely heat is convertible into work. As regards the present investigation, however, it is perfectly immaterial whether heat be indestructible, or only not destroyed, or rather whether it be not convertible into anything else, or only not converted.

which is analogous to the equation of continuity of an incompressible fluid.

If we suppose all possible systems of values assigned in succession to the constants D_1, E_1, F_1, the formulæ (6) will express all possible modes of transfer, consistent with our original assumption respecting the forms of f_x, f_y, f_z, by which the state of temperature of the solid at the end of the time t can pass into its state at the end of the time $t + dt$. Of course, if we suppose heat to be material, we cannot help attaching to it the idea of *individuality*. But if we suppose heat to consist in motion of some sort, which for my own part I regard as by far the more probable hypothesis, we require a definition of *sameness* of heat, supposing we find it convenient to treat the subject in this way. I am not now going to follow any further the subject which has just been broached; but I thought it might be worth while to point out in what manner the additional arbitrary constants found in the general expressions for the flux beyond what appeared in the equation of motion, or more properly *the equation of successive distribution*, corresponded to an attribute of heat which is necessarily involved in the idea of a flux, but which is not necessarily involved in the idea of the successive distributions of a given quantity of heat.

9. Besides the general equation (5), it is requisite to form the equation of condition which has to be satisfied at the surface when the solid radiates into a space at a different temperature. Let P' be a point in the surface, dS an element of the surface surrounding P', $P'N$ a normal drawn outwards at P', P a point in NP' produced, situated at the distance δ from P'. Consider the element of the solid bounded by dS, by a plane through P parallel to the tangent plane at P', and by a cylindrical surface circumscribing dS, and having its generating lines parallel to $P'N$; and suppose this element to be indefinitely diminished in such a manner that δ vanishes compared with the linear dimensions of dS, which is allowable if the curvature at P' be finite (that is, not infinitely great), as it must necessarily be in general. The quantity of heat which enters the element across the plane through P, as well as the quantity which escapes across dS, varies ultimately as dS. The area of the cylindrical surface varies as δ multiplied by the perimeter of dS, and there-

fore will vanish compared with dS, since δ vanishes compared with the linear dimensions of dS. Hence, even if the quantity of heat which entered by the cylindrical surface varied as the surface, it would vanish in the limit compared with the quantity which escapes by dS. In fact, however, even if we suppose δ comparable with the linear dimensions of dS, it may be shewn that the total quantity of heat which enters by the cylindrical surface is of the order δdS, because ultimately the quantity of heat which enters across that portion of the cylindrical surface for which the flux is positive is equal to the quantity which escapes across the remainder. Lastly, the gain or loss of heat by the element during a given time varies ultimately as the volume of the element. Hence, ultimately, the quantity of heat which enters the element across the plane through P, during the time dt, is equal to the quantity which escapes across dS. The former will be ultimately equal to $dSdt$ multiplied by the value of the flux obtained from the general formulæ (1) and (7) by taking x, y, z to denote the coordinates of P', and l, m, n, the direction-cosines of $P'N$. We may assume the latter to be proportional to the difference $u - v$ between the temperature u of the solid at the point P' and the temperature v of the surrounding space, and may accordingly express it by $h(u - v)dSdt$. Hence we have for the required condition,

$$lA\frac{du}{dx} + mB\frac{du}{dy} + nC\frac{du}{dz} + h(u-v) = 0\ldots\ldots(10).$$

The quantity h denotes the *exterior conductivity* of the solid. It is a certain function of l, m, n, the form of which M. Duhamel did not attempt to investigate, nor am I going to attempt the investigation myself. If however the crystal be covered with a thin coating of some other substance, sufficient to stop all direct radiation from the crystal into the surrounding space, h will depend upon the nature of the coating. In either case h will be constant throughout any plane face by which the crystal may be bounded.

10. Let us return to the consideration of the propagation of heat in the interior of the mass. Imagine the coordinates x, y, z, of any point altered in the ratios of $\sqrt{A}$ to $\sqrt{K}$, $\sqrt{B}$ to $\sqrt{K}$,

$\sqrt{C}$ to $\sqrt{K}$, where K is constant, and let ξ, η, ζ, be the results. The equation (5) becomes

$$c\rho \frac{du}{dt} = K \left(\frac{d^2u}{d\xi^2} + \frac{d^2u}{d\eta^2} + \frac{d^2u}{d\zeta^2} \right) \dots\dots\dots(11).$$

This will be true whatever be the value of K, but it will be convenient to suppose that

$$K^3 = ABC \dots\dots\dots\dots (12).$$

Now imagine a second solid formed from the first by altering all lines parallel to x in the ratio of $\sqrt{A}$ to $\sqrt{K}$, all lines parallel to y in the ratio of $\sqrt{B}$ to $\sqrt{K}$, and all lines parallel to z in the ratio of $\sqrt{C}$ to $\sqrt{K}$, nothing being as yet specified regarding the nature of the second solid, except that it is homogeneous. Imagine any number of points, lines, surfaces, or spaces, conceived as belonging to the first solid, and let the points, lines, &c. deduced from them by altering the coordinates in the ratios abovementioned, and conceived as belonging to the second solid, be said to *correspond* to the others. On account of the particular magnitude of K chosen, it is evident that the volumes of corresponding spaces will be equal. Let the second solid be called the *auxiliary solid*, and the operation of deducing either solid from the other, *derivation;* and suppose the temperatures equal at corresponding points of the two solids.

The equation (11) shews that the successive distributions of temperature in the interior of the auxiliary solid will take place as if this solid were an ordinary medium in which the interior conductivity bears to K the same ratio that the product of the specific heat and density bears to $c\rho$.

The first of equations (7) gives

$$f_x = -\sqrt{(AK)} \frac{du}{d\xi}.$$

If now we refer f_x, not to a unit of surface, but to that area of a plane perpendicular to the axis of x which is changed by derivation into a unit of surface, we must multiply the above expression by $\sqrt{(BC)}$ and divide it by K. Denoting the result by f_ξ, using f_η, f_ζ to denote for y, z, what f_ξ denotes for x, and taking account of (12), we get

$$f_\xi = -K \frac{du}{d\xi}, \ f_\eta = -K \frac{du}{d\eta}, \ f_\zeta = -K \frac{du}{d\zeta} \dots\dots (13).$$

It follows from these equations, that if we suppose not only the temperatures at corresponding points of the two solids to be always the same, but also equal quantities of heat to flow in equal times and in corresponding directions across corresponding surfaces, the flow of heat in the auxiliary solid is what would naturally belong to an ordinary medium having an interior conductivity K.

The density of the auxiliary solid being disposable, we may take it to be the same as that of the given solid, in which case corresponding spaces will contain equal quantities of matter. It is only necessary further to suppose the auxiliary solid to be an ordinary medium having a specific heat c, and an interior conductivity K, in order that the motion of heat in the interior of the two solids should precisely correspond.

11. It remains only to investigate the condition which must be satisfied relatively to the surface of the auxiliary solid, in order that the two solids should perfectly correspond in every respect. Retaining the notation of Art. 9, let $d\sigma$ be the element of surface which corresponds to dS; λ, μ, ν, the direction-cosines corresponding to l, m, n. The quantity of heat which escapes across dS during the time dt is ultimately equal to $h(u-v)\,dS\,dt$, and this must be equal to the quantity which escapes across $d\sigma$. Hence it is sufficient to attribute to the auxiliary solid an exterior conductivity k, such that

$$k = \frac{dS}{d\sigma}\,h \dots\dots\dots\dots\dots\dots(14).$$

But we have

$$\frac{\lambda}{\sqrt{A}\,.\,l} = \frac{\mu}{\sqrt{B}\,.\,m} = \frac{\nu}{\sqrt{C}\,.\,n} = \frac{1}{\sqrt{(Al^2 + Bm^2 + Cn^2)}}$$
$$= \sqrt{(A^{-1}\lambda^2 + B^{-1}\mu^2 + C^{-1}\nu^2)} \dots (15).$$

Also ldS, $\lambda d\sigma$, are the projections of dS, $d\sigma$, on the plane of yz, and these projections are proportional to $\sqrt{(BC)}$, K, or to $\sqrt{K}$, $\sqrt{A}$, whence we get from (15)

$$\frac{dS}{d\sigma} = \frac{\sqrt{K}}{\sqrt{(Al^2 + Bm^2 + Cn^2)}} = \sqrt{\{K(A^{-1}\lambda^2 + B^{-1}\mu^2 + C^{-1}\nu^2)\}} \dots(16).$$

The first or second of these expressions will be employed according as we suppose l, m, n, or λ, μ, ν, given.

If the crystal be covered by a thin coating of a given substance, h will be constant, and k will be a function of l, m, n, or of λ, μ, ν, which is determined by (14) and (16). These formulæ shew that in the case supposed k will have the same value for the opposite faces of a plate bounded by parallel surfaces.

By means of the auxiliary solid, we may reduce problems relating to the conduction of heat in crystals to corresponding problems relating to ordinary media; or, conversely, from a set of self-evident or known results relating to ordinary media, we may deduce a set of corresponding results relating to crystals.

12.   Let us first regard the crystal as infinite, in which case the auxiliary solid will be infinite likewise.

In an ordinary medium, if heat be introduced at one point according to any law, the isothermal surfaces will be a system of spheres, having the source of heat for their common centre, and the flow of heat at any point will take place in the direction of the radius vector drawn from the source. If the temperature be permanent, and the temperature at an infinite distance vanish, the temperature at any point will vary inversely as the distance from the source.

Hence, in a crystal, if heat be introduced at one point according to any law, the isothermal surfaces will be a system of similar and concentric ellipsoids, having their principal axes in the direction of the thermic axes drawn through the source, and proportional to the square roots of the principal conductivities. The flow of heat at any point will take place in the direction of the radius vector drawn from the source. If the temperature be permanent, and vanish at an infinite distance, the temperature along a given radius vector will vary inversely as the distance from the source.

It will frequently be convenient to refer to an ellipsoid constructed with its principal axes in the direction of the thermic axes, and equal to $2\sqrt{A}$, $2\sqrt{B}$, $2\sqrt{C}$, respectively. I shall call this ellipsoid the *thermic ellipsoid.*

13.   In an ordinary medium, whether finite or infinite, in which the temperature varies from point to point, and may be either constant or variable as regards the time, the flow of heat at any point takes place in the direction of the normal to the

isothermal surface passing through that point, that is, in a direction parallel to the radius vector drawn from the centre of the *thermic sphere* to the point of contact of a tangent plane drawn parallel to the isothermal surface at the point considered.

Now the tangency of two surfaces is evidently unchanged by derivation. Hence, in a crystal, if we have given the direction of the isothermal surface at any point, we may find that of the flow of heat by the following construction. In a direction parallel to the isothermal surface at the given point draw a tangent plane to the thermic ellipsoid, and join the centre with the point of contact: the flow of heat will take place in a direction parallel to this joining line. In other words, the flow of heat will take place in a direction parallel to the diameter which is conjugate to a plane parallel to the isothermal surface at the given point.

14. Conceive a plate bounded by parallel surfaces to be cut from a crystal, and heat to be applied towards its centre; and suppose the lateral boundaries sufficiently distant to produce no sensible influence on the result, so that we may regard the plate as infinite. In this case the auxiliary solid will likewise be an infinite plate bounded by parallel surfaces. Now if heat be supplied according to any law at one point of such a plate, or at any number of points situated in the same normal, the isothermal surfaces will be surfaces of revolution, having the normal drawn through the source or sources of heat for their axis, and the isothermal curves in which the parallel faces are cut by the isothermal surfaces will be circles, having their centres in the points in which the faces are cut by the normal abovementioned.

Hence, in a crystalline plate, if heat be supplied according to any law at one point, or at any number of points situated in a line parallel to the diameter of the thermic ellipsoid which is conjugate to the planes of the faces, (a line which for brevity I will call *the line of sources,*) any particular isothermal surface will be a surface generated by an ellipse which has its plane parallel to the faces, its centre in the line of sources, and its principal axes parallel and proportional to those of the ellipse in which the thermic ellipsoid is cut by a plane parallel to the faces. In particular, the isothermal curves on the two faces are ellipses of the kind just mentioned*. Hence

(1) If the plate be cut in a direction perpendicular to one of the thermic axes, the line joining the centres of a pair of ellipses which correspond on the two faces to a given temperature (such as that of melting wax) will be normal to the plate. The principal axes of the ellipses will be in the direction of the two remaining thermic axes, and will be proportional to the square roots of the corresponding conductivities.

(2) If the normal to the plate be not a thermic axis, the line joining the centres of the ellipses will be inclined to the normal, its direction being determined as above explained.

(3) If the plate be cut in a direction parallel to either of the circular sections of the thermic ellipsoid (the three principal conductivities being supposed unequal,) the isothermal curves on both faces will be circles, but the line joining the centres of the two systems of circles will be inclined to the normal.

If the heat be communicated uniformly along the line of sources, or if there be only a single source situated midway between the faces, or more generally if the sources be alike two

---

* The problem solved in this article forms a good example of the advantage of considering the auxiliary solid. In M. Duhamel's memoir the plate is regarded as extremely thin, so that the variation of temperature in passing from one point to another of the same normal may be considered insensible—and it is remarked that the second case (in which a normal to the plate is not a thermic axis) is much more difficult than the first; whereas here the plate is not necessarily thin, and both cases follow immediately from what with regard to an uncrystallized body is self-evident. M. Duhamel has shewn that the isothermal curves on the two faces are ellipses, having their principal axes parallel and proportional to those of the ellipses in which the thermic ellipsoid is cut by planes parallel to the faces of the plate: but his demonstration that the line joining the centres of the two systems of ellipses has the direction assigned in the text does not seem altogether satisfactory, because the analysis only applies to the case in which the thickness of the plate is regarded as indefinitely small; whereas the space by which the ellipse corresponding on one face to a given temperature overlaps the ellipse corresponding on the other face to the same temperature is a small quantity of the order of the thickness of the plate, and ought for consistency's sake to be neglected.

The results contained in the remaining part of this paper are not found in the memoirs of M. Duhamel. It may here be remarked, that the results arrived at by the consideration of the auxiliary solid, such for example as that of Art. 17, might have been obtained by referring the crystal to oblique axes parallel to a system of conjugate diameters of the thermic ellipsoid.

and two, those which belong to the same pair being situated at equal distances from the two faces respectively, the isothermal curves belonging to the same temperature in the two systems will be of equal magnitude, provided that the exterior conductivity h have the same value for the two faces. The last condition is satisfied, according to what has been already remarked, when the two faces are covered by a thin coating of the same substance, which regulates the exterior conductivity; but it is probable that it may be satisfied generally even if the faces be left bare, provided that they have the same degree of polish*.

The experiments of M. de Senarmont bear directly on the first two cases mentioned above. In the case of crystals which exhibited three different conductivities, it was found that when three plates were cut in the directions of the three principal planes, the ratio of the principal axes of the ellipses formed on one plate, as determined by observation, agreed very closely with the result calculated from the ratios which had previously been determined by observation from the other two plates. An interesting experiment bearing on the second case is described by M. de Senarmont in his second paper (p. 187). A rather thick plate of quartz, inclined to the axis at an angle of 45°, was drilled in a direction perpendicular to its plane, and heated by means of a wire inserted into the hole, after its two faces had been covered with wax. The curves marked out on the two faces approximated to the two bases of an elliptic cylinder, symmetrical with respect to the principal plane, and having its axis inclined towards the axis of the crystal, (which in quartz is the direction of greatest conductivity,) so as to cross the wire, which was perpendicular to the plate. The curves however were not elliptical but egg-shaped, having their axes of symmetry situated in the principal plane, the end at which the curvature was least being that which was nearest to the wire, so that the blunt ends on the two faces were turned in opposite directions, the curves being in other respects alike. It will be seen at once that the symmetry of the curves with respect to the principal plane, the obliquity of the line joining their centres,

---

* This result follows readily from the theory of molecular radiation, according to the suppositions usually made.

and their equality combined with *dissymmetry*, follow immediately from theory. We learn too from theory, that in order to procure ellipses it would be necessary to drill the hole in the direction of that diameter of the thermic spheroid which is conjugate to the plane of the plate.

15. Conceive a bar having a section with an arbitrary contour to be formed from an uncrystallized substance; let heat be applied in any manner at one or more places, and suppose the heat to escape again from the surface by radiation. Consider only those portions of the bar which are situated at a sufficient distance from the source or sources of heat to render insensible any irregularity arising from the mode in which the heat is communicated. If the bar be sufficiently slender, we may regard the temperature throughout a section drawn in a direction perpendicular to its length as approximately constant, without assuming thereby that the isothermal surfaces are perpendicular to the length. Let x be measured in the direction of the length, and consider the slice of the bar contained between the planes whose abscissæ are x and $x + dx$. Let u be the temperature of the bar at the distance of the first plane, p the perimeter, and Q the area of the section, h the exterior conductivity, or rather the mean of the exterior conductivities in case they should vary from one generating line to another; let c, ρ, K, be the same as before, and put $Q = ap$, so that $4a$ is the side of a square whose area divided by its perimeter is equal to Qp^{-1}.

The excess of the quantity of heat which enters during the time dt by the first of the plane ends of the slice over that which escapes by the second, is ultimately equal to $KQ\, d^2u/dx^2\, .\, dx\, dt$. The quantity which the slice loses by radiation is ultimately equal to $h\, p\, u\, dx\, dt$, if we take the temperature of the surrounding space, which is supposed to be constant, for the origin of temperatures. But the gain of heat by the slice is also equal to $c\rho Q\, du/dt\, .\, dx\, dt$. Hence we have

$$c\rho \frac{du}{dt} = K \frac{d^2u}{dx^2} - \frac{h}{a} u \quad \ldots\ldots\ldots\ldots\ldots (17).$$

If we suppose the heat to be continually supplied, and the temperature to have become stationary, we get from this equation

$$u = M e^{\sqrt{\frac{h}{Ka}}\,x} + N e^{-\sqrt{\frac{h}{Ka}}\,x} \quad \ldots\ldots\ldots\ldots (18),$$

where M and N are arbitrary constants. If now a be small, it follows from (18) that the lateral flux at the surface, which is equal to hu, as compared with the longitudinal flux, which is equal to $-K\,du/dx$, is a small quantity of the order $\sqrt{(ha/K)}$.

We may deduce the same consequence for the case of variable temperatures from the equation (17), without troubling ourselves with its solution. Conceive any number of bars of different sizes to be heated in a similar manner, and for greater generality, suppose the values of c, ρ, K, and h, as well as a, to be different for the different bars. Let x, x', x''... be corresponding lengths, and t, t', t'' corresponding times, relating to the several bars. The equation (17) shews that the temperatures at corresponding points and at the end of corresponding times may be the same in all the bars, provided

$$x \propto \sqrt{\frac{Ka}{h}}, \ t \propto \frac{c\rho a}{h} \ \dots\dots\dots\dots\dots \ (19).$$

These variations contain the definition of corresponding points and corresponding times. In order that the temperatures in the different bars should actually be the same at corresponding points and at the end of corresponding times, it is sufficient that the initial circumstances, or more generally the mode of communicating the heat, should be such as to give equal temperatures at the points and times defined by the variations (19), which in this point of view may be regarded as containing the definition of *similarity of heating*. Now, in comparing the longitudinal flux at corresponding points, if we take du the same, dx must vary as determined by (19), and therefore the flux will vary as $K\sqrt{(K^{-1}a^{-1}h)}$, or $\sqrt{(Ka^{-1}h)}$, and the ratio of the lateral flux at the surface to the longitudinal flux will vary as $\sqrt{(haK^{-1})}$; so that if we suppose a to decrease indefinitely, h and K being given, the ratio in question will be a small quantity of the order $\sqrt{(ha/K)}$ as before, and will ultimately vanish.

The second of the variations (19) shews that if we suppose the heat to be supplied to one bar in an irregular manner as regards the time, the fluctuations in the mode of communicating the heat must become more and more rapid as a decreases, in order that the similarity of temperatures may be kept up. If the fluctuations retain their original period, the motion of heat will tend indefinitely to become what we may regard at any instant as steady, and thus we fall back on the case first considered. We may conclude therefore generally, that if the bar be sufficiently slender, the direction of the maximum flux, even close to the surface, will sensibly coincide with that of the length of the bar; so that the isothermal surfaces, which are necessarily perpendicular to the direction of the flow of heat, will be planes perpendicular to the axis of the bar.

By supposing the bar to be the auxiliary solid belonging to a crystalline bar, we arrive at the following theorem. If a slender crystalline bar be heated at one end, and if we confine our attention to points of the bar situated at a sufficient distance from the source of heat to render insensible any irregularities attending the mode of communicating the heat, the isothermal surfaces will be sensibly planes parallel to the diametral plane of the thermic ellipsoid which is conjugate to the system of chords drawn parallel to the length of the bar. These planes will necessarily have an oblique position unless the direction of the length of the bar be a thermic axis of the crystal.

The same result might have been obtained without employing the auxiliary solid, by first shewing that when the bar is sufficiently slender the direction of the flow of heat sensibly coincides with that of the length of the bar. We should thus be led to a problem exactly the converse of that treated in Art. 13, namely, Given the direction of the flow of heat, to find that of the isothermal surface.

16. It may be shewn in a similar way, that if a thin plate be formed of an uncrystallized substance, and be heated at one or more places, or over a finite portion, if we consider only those parts of the plate which are situated at a sufficient distance from the sources of heat to render insensible any irregularities attending the mode in which the heat is communicated, the flow of heat will take place in a direction sensibly parallel to the plate, and therefore the isothermal surfaces will be cylindrical surfaces whose generating lines are perpendicular to the plate. It is here supposed that the lateral boundaries of the plate are situated at a sufficient distance to render their effect insensible.

Hence, in a thin crystalline plate heated in a similar manner, the isothermal surfaces, under similar restrictions, will be cylindrical surfaces whose generating lines are parallel to the diameter of the thermic ellipsoid which is conjugate to the plane of the plate.

17. The state of temperature, under given circumstances, of a rectangular parallelepiped formed of an uncrystallized substance, may be determined by certain known formulæ which it is not necessary here to describe.

Hence, the state of temperature of a parallelepiped cut from a crystal in such a manner that its edges are parallel to a system of conjugate diameters of the thermic ellipsoid may be determined by the same formulæ. This parallelepiped will of course be oblique-angled, except in the particular case in which its edges are parallel to the thermic axes. It may be remarked that a parallelepiped for which the state of temperature shall be determinable by the formulæ in question may be cut from a crystal in a manner quite as general as from an uncrystallized substance. In both cases the direction of the first edge is arbitrary, and, when it is fixed on, the plane of the other two edges is determined in direction. The direction of the second edge having been chosen arbitrarily in the plane above mentioned, that of the third edge is determined.

It does not seem worth while to notice the crystalline figures derived from spheres, &c., on account of the mechanical difficulty attending their execution. Besides, the derivation presents no theoretical difficulty.

Further consideration of the general expressions for the flux.

18. It has been already remarked, that if the crystal possess two planes of symmetry, the nine arbitrary constants which appear in the expressions for the flux in three rectangular directions, from which the flux in any other direction may be derived, reduce themselves to six, and the expressions for the flux take the form (8). I proceed now to consider what grounds we have for believing that these expressions, with only six arbitrary constants, are the most general possible.

In the first place, it may be observed that this result follows

readily from the theory of molecular radiation. In this theory
the extent of molecular radiation is supposed to be very great
compared with the mean interval between the molecules of a
body, so that the body may be treated as continuous. If E, E'
be any two elements of the body, situated sufficiently near one
another to render their mutual influence sensible, it is supposed
that, during the time dt, a quantity of heat proportional to Edt
radiates in all directions from E, whereof E' absorbs a portion
proportional to E'. On the other hand, E' emits and E absorbs
a quantity also proportional, so far as regards only the magnitudes of E, E', and dt, to $EE'dt$. The exchange of heat between
E and E' may therefore be expressed by $qEE'dt$. The quantity
q is supposed to be proportional, so far as regards its dependence
on the temperatures, to the small difference between the temperatures of E and E'. It will also depend upon the nature of
the body, upon the distance EE', and in the case of a crystalline body upon the direction of the line EE'; but we need not
now consider its dependence upon these quantities. If the length
$EE' = s$, and if we suppose the extent of internal radiation to be
very small, we may express the difference between the temperatures of E and E' by $du/ds \cdot s$. It follows then from the theory
we are considering, that the total flux of heat arises from the
exchange of heat between all possible pairs of elements, such as
E, E'; the exchange between any pair E, E' being proportional
to the rate of variation of temperature in the direction EE', and
accordingly independent of the variation of temperature in other
directions.

Now suppose the body referred to rectangular axes, and let P
be the mathematical point whose coordinates are x, y, z. Conceive the body divided into an infinite number of infinitely small
equal elements. Let E be the element which contains P, E' any
element in the neighbourhood of P, and consider the partial flux
in the neighbourhood of P which arises from the exchange of
heat between all pairs of elements which have the same relative
position as E and E'. Through P draw an elementary plane S,
which it will be convenient to consider as infinitely large compared with the dimensions of the elements such as E, and conceive S to assume in succession all possible directions by turning
round P. The partial flux across S will vary as the number of

points in which the lines, all equal and parallel to EE', which connect the pairs of elements, cut the plane S, or as the cosine of the angle between the normal to S and the direction EE'. Let $EE' = s$; let l', m', n', be the cosines of the angles which this line makes with the axes of x, y, z, and suppose S to be perpendicular to each of these axes in succession. We shall thus have for the partial fluxes f_x, f_y, f_z, quantities proportional to $l'\,du/ds$, $m'\,du/ds$, $n'\,du/ds$, or to

$$l'^2\frac{du}{dx} + l'm'\frac{du}{dy} + l'n'\frac{du}{dz}, \quad l'm'\frac{du}{dx} + m'^2\frac{du}{dy} + m'n'\frac{du}{dz},$$

$$l'n'\frac{du}{dx} + m'n'\frac{du}{dy} + n'^2\frac{du}{dz}.$$

Hence, the coefficient of du/dy in the expression for the partial flux f_x is equal to the coefficient of du/dx in the expression for the partial flux f_y, and the same applies to y, z, and to z, x. This being true for each partial flux, will be true likewise for the total flux, and therefore the general expressions for the flux in three rectangular directions, with nine arbitrary constants, will be reduced to the form (8), or the general expressions (6), referred to the thermic axes, to the form (7).

19. Let us further examine some of the consequences which would follow from the supposition that the expressions for the flux referred to the thermic axes have the general form (6). Conceive a crystalline mass, regarded as infinite, to be heated at one point according to any law, and let the source of heat be taken for origin. We have seen already that the succession of temperatures takes place in an infinite solid in exactly the same manner whether the expressions for the flux have the general form (6), or the more restricted form (7), and consequently, in the case supposed above, the temperature at a given time is some function of

$$A^{-1}x^2 + B^{-1}y^2 + C^{-1}z^2.$$

If x, y, z, be the coordinates of any point in a *line of motion*, or line traced at a given instant from point to point in the direction of the flow of heat, dx, dy, dz, will be proportional to f_x, f_y, f_z, which are given by (6), and in the present case du/dx, du/dy, du/dz, are proportional to $A^{-1}x$, $B^{-1}y$, $C^{-1}z$. Hence the differential equations of a line of motion are

$$\frac{dx}{x - F_1 B^{-1} y + E_1 C^{-1} z} = \frac{dy}{y - D_1 C^{-1} z + F_1 A^{-1} x}$$
$$= \frac{dz}{z - E_1 A^{-1} x + D_1 B^{-1} y} \dots (20).$$

Taking ξ, η, ζ, to denote the same quantities as in Art. 10, and putting for shortness

$$D_1 (BC)^{-\frac{1}{2}} = \omega', \quad E_1 (CA)^{-\frac{1}{2}} = \omega'', \quad F_1 (AB)^{-\frac{1}{2}} = \omega''' \dots (21),$$

we get

$$\frac{d\xi}{\xi - \omega''' \eta + \omega'' \zeta} = \frac{d\eta}{\eta - \omega' \zeta + \omega''' \xi} = \frac{d\zeta}{\zeta - \omega'' \xi + \omega' \eta} \dots (22).$$

Conceive an elastic solid to be fixed at the origin, and to expand alike in all directions and at all points with a velocity of expansion unity, so that a particle which at the end of the time t is situated at a distance r from the origin, at the end of the time $t + dt$ is situated at a distance $r(1 + dt)$. Conceive this solid at the same time to turn, with an angular velocity ω equal to $\sqrt{(\omega'^2 + \omega''^2 + \omega'''^2)}$, about an axis whose direction-cosines are $\omega' \omega^{-1}$, $\omega'' \omega^{-1}$, $\omega''' \omega^{-1}$. The direction of motion of any particle will represent the direction of the flow of heat in what we may still call the *auxiliary solid*, from whence the direction of the flow of heat in the given solid will be obtained by merely conceiving the whole figure differently magnified or diminished in three rectangular directions.

This *rotatory* sort of motion of heat, produced by the mere diffusion from the source outwards, certainly seems very strange, and leads us to think, independently of the theory of molecular radiation, that the expressions for the flux with six arbitrary constants only, namely the expressions (8), or the equivalent expressions (7), are the most general possible.

20. Let the auxiliary solid be referred to the rectangular axes of ξ', η', ζ', of which the last coincides with the axis to which ω refers. It may be seen immediately, without analytical transformation, that the differential equations to the lines of motion will be

$$\frac{d\xi'}{\xi' - \omega \eta'} = \frac{d\eta'}{\eta' + \omega \xi'} = \frac{d\zeta'}{\zeta'} \dots \dots \dots \dots (23).$$

Taking polar coordinates r, θ in the plane of ξ', η', we have

$$(\sin \theta + \omega \cos \theta)(\cos \theta \, dr - \sin \theta \, rd\theta)$$
$$= (\cos \theta - \omega \sin \theta)(\sin \theta \, dr + \cos \theta \, rd\theta);$$

whence $\qquad\qquad\qquad \omega dr = rd\theta$ .... (24),

the differential equation of a system of equiangular spirals in which the angle between the tangent and radius vector is equal to $\tan^{-1}\omega$. We have also from (23)

$$\frac{d\zeta'}{\zeta'} = \frac{(\xi' - \omega\eta')\,d\xi' + (\eta' + \omega\xi')\,d\eta'}{(\xi' - \omega\eta')^2 + (\eta' + \omega\xi')^2}$$

$$= (1 + \omega^2)^{-1} \left\{ \frac{\xi'd\xi' + \eta'd\eta'}{\xi'^2 + \eta'^2} + \omega \frac{\xi'd\eta' - \eta'd\xi'}{\xi'^2 + \eta'^2} \right\};$$

whence $\qquad (1 + \omega^2) \log \zeta' + \text{const} = \log r + \omega\theta$

$= \log r + \omega^2 \log r$ from (24). We have therefore

$$\zeta' = mr \qquad\qquad\qquad\qquad (25),$$

where m is an arbitrary constant.

Hence, in the plane of ξ', η', conceive an equiangular spiral described about the origin as pole, such that the angle between the tangent and the radius vector is equal to $\tan^{-1}\omega$. Let it assume all possible positions by turning once round the pole, and in each position let it be made the base of a cylinder whose generating lines are parallel to the axis of ζ'. Conceive also an infinite number of cones of revolution described with the origin for vertex and the axis of ζ' for axis. The system of curves of double curvature formed by the intersection of the cones with the cylinders will be the lines of motion in the auxiliary solid, on the supposition that the constant ω does not vanish. To obtain the lines of motion in the original solid, it will be sufficient to conceive the whole figure differently magnified or diminished in three rectangular directions, and we shall thus obtain a clearer idea of the form of the curves, which is the whole object of the investigation, than would have been derived from the rather complicated equations got from the integration of equations (20) in their original shape.

21. It may be observed, in conclusion, that even if there were reason to suppose that the constants D_1, E_1, F_1, were not necessarily equal to zero, it is only among crystals which possess

a peculiar sort of asymmetry, that we should expect to find traces of their existence. We have seen already that if the crystal possess two planes of symmetry, these constants can have no existence. But the crystalline form (taken as an index of the degree of symmetry of the internal structure) may indicate the non-existence of these constants even in cases in which the crystal does not possess a single plane of symmetry. Take for example quartz, which was one of the crystals employed by M. de Senarmont in his experiments. In this crystal, not only is there no plane of symmetry, but a peculiar kind of asymmetry is indicated by the occurrence of hemihedral faces, as well as by the optical properties of the crystal. Let three adjacent edges of the primitive rhombohedron meeting in one of the solid angles which is formed by three equal plane angles be denoted by G, H, I, and let the opposite edges be denoted by G', H', I'. If we suppose the interior structure to correspond to the crystalline form, whatever we can say of the structure with reference to the edges G, H, I, we can say with reference to the edges H, I, G, or I, G, H. This shews that the thermic ellipsoid must be an ellipsoid of revolution about the axis of the crystal, and that the line to which ω refers must coincide with the axis. But furthermore, whatever we can say of the structure with reference to G, H, I, we can say with reference to G', I', H', or I', H', G', or H', G', I'. This requires that $\omega = -\omega$, and therefore $\omega = 0$, and therefore $D_1 = 0$, $E_1 = 0$, $F_1 = 0$.

[From the *Transactions of the Royal Society of Edinburgh*, Vol. xx. p. 317.]

ON THE TOTAL INTENSITY OF INTERFERING LIGHT.

(Extracted from a letter addressed to Professor Kelland.)

[Communicated *January* 5, 1852. (*Proceedings R. S. E.*,
Vol. III. p. 98.)]

PEMBROKE COLLEGE, CAMBRIDGE.

MY DEAR SIR,

*   *   *   *

In reading your paper in the *Transactions of the Royal
Society of Edinburgh*, vol. xv., p. 315, some years ago, it occurred
to me to try whether it would not be possible to give a general
demonstration of the theorem, applying to apertures of all forms.
I arrived at a proof, which I wrote out, but have never published.
As I think it will interest you I will communicate it. You may
make any use you please of it.

CASE I. Aperture in front of a lens; light thrown on a screen
at the focus, or received through an eye-piece, through which the
luminous point is seen in focus.

The expression for the intensity is given in Airy's Tract,
Prop. 20. If the intensity of the incident light at the distance of
the aperture be taken for unity, and D be the quantity by which
any element of the area of the aperture must be divided in
forming the expression for the vibration, that expression becomes

$$\frac{1}{D} \iint \sin \frac{2\pi}{\lambda} \left(vt - B + \frac{px + qy}{b} \right) dx\, dy,$$

the integration being extended over the whole aperture. If it should be necessary to suppose a change of phase to take place in the act of diffraction, such change may be included in the constant B. If, then, I be the intensity,

$$D^2 I = \left(\iint \sin \frac{2\pi}{\lambda} \frac{px + qy}{b} \, dx \, dy \right)^2 + \left(\iint \cos \frac{2\pi}{\lambda} \frac{px + qy}{b} \, dx \, dy \right)^2 ;$$

and if $\bar{I}$ be the total illumination,

$$\bar{I} = \int_{-\infty}^{\infty} \int_{-\infty}^{\infty} I \, dp \, dq.$$

Now,

$$\left\{ \iint f(x, y) \, dx \, dy \right\}^2 = \iiiint f(x, y) f(x', y') \, dx \, dy \, dx' \, dy',$$

the limits of x', y' being the same as those of x, y. Hence,

$$D^2 I = \iiiint \cos \frac{2\pi}{b\lambda} (p \overline{x' - x} + q \overline{y' - y}) \, dx \, dy \, dx' \, dy'.$$

In the present shape of the integral, we must reserve the integration with respect to p and q till the end; but if we introduce the factor $\epsilon^{\mp \alpha p \mp \beta q}$, where the sign $-$ or $+$ is supposed to be taken according as p or q is positive or negative, we shall evidently arrive at the same result as before, provided we suppose in the end α and β to vanish. When this factor is introduced, we may, if we please, integrate with respect to p and q first. We thus get

$D^2 \bar{I} = $ limit of

$$\iiiiint \epsilon^{\mp \alpha p \mp \beta q} \cos \frac{2\pi}{b\lambda} (p \overline{x' - x} + q y' - y) \, dx \, dy \, dx' \, dy' \, dp \, dq.$$

Now,

$$\int_{-\infty}^{\infty} \epsilon^{\mp \alpha p} \cos (kp - Q) \, dp = \cos Q \int_{-\infty}^{\infty} \epsilon^{\mp \alpha p} \cos kp \, dp$$

$$+ \sin Q \int_{-\infty}^{\infty} \epsilon^{\mp \alpha p} \sin kp \, dp$$

$$= 2 \cos Q \int_{0}^{\infty} \epsilon^{-\alpha p} \cos kp \, dp = \frac{2\alpha \cos Q}{\alpha^2 + k^2}.$$

A similar formula holds good for q, whence

$$D^2 \bar{I} = \text{limit of} \iiiint \frac{4\alpha\beta}{\left\{\alpha^2 + \left(\frac{2\pi (x' - x)}{b\lambda}\right)^2\right\} \left\{\beta^2 + \left(\frac{2\pi (y' - y)}{b\lambda}\right)^2\right\}} dx\, dy\, dx'\, dy'.$$

Let now

$$\frac{2\pi (x' - x)}{b\lambda} = \alpha u,$$

whence

$$dx' = \frac{b\lambda\alpha}{2\pi} du,$$

and the limits of u are ultimately $-\infty$ and $+\infty$, since α ultimately vanishes. Hence

$$\text{limit of} \int \frac{2\alpha dx'}{\alpha^2 + \left(\frac{2\pi (x' - x)}{b\lambda}\right)^2} = \frac{b\lambda}{\pi} \int_{-\infty}^{\infty} \frac{du}{1 + u^2} = b\lambda.$$

A similar formula holds good for y', and we have, therefore,

$$D^2 \bar{I} = b^2\lambda^2 \iint dx\, dy = b^2\lambda^2 A,$$

if A be the whole area of the aperture or apertures.

Now $\bar{I}$ ought to be equal to A, and, therefore,

$$D = b\lambda.$$

CASE II.   Aperture in front of a screen.

The formula for the illumination is given in Airy's Tract, Art. 73. We have as before,

$$D^2 \bar{I} = \text{limit of} \iiiint\!\!\iint \epsilon^{\mp \alpha p \mp \beta q} \cos \frac{\pi (a + b)}{\lambda ab} \left\{ \left(x' - \frac{ap}{a + b}\right)^2\right.$$

$$\left. - \left(x - \frac{ap}{a + b}\right)^2 + \left(y' - \frac{aq}{a + b}\right)^2 - \left(y - \frac{aq}{a + b}\right)^2 \right\} dx\, dy\, dx'\, dy'\, dp\, dq$$

$$= \text{limit of} \iiiint\!\!\iint \epsilon^{\mp \alpha p \mp \beta q} \cos \left\{ \frac{\pi (a + b)}{\lambda ab} [x'^2 - x^2 + y'^2 - y^2] \right.$$

$$\left. - \frac{2\pi p}{\lambda b} (x' - x) - \frac{2\pi q}{\lambda b} (y' - y) \right\} dx\, dy\, dx'\, dy'\, dp\, dq$$

$$= \text{limit of} \iiiint \frac{2\alpha}{\alpha^2 + \left(\frac{2\pi(x'-x)}{\lambda b}\right)^2} \cdot \frac{2\beta}{\beta^2 + \left(\frac{2\pi(y'-y)}{\lambda b}\right)^2} \cdot$$

$$\cos \frac{\pi(a+b)}{\lambda ab}(x'^2 - x^2 + y'^2 - y^2)\, dx\, dy\, dx'\, dy'.$$

Now, when α vanishes, the whole of the integral

$$\int_{-\infty}^{\infty} \frac{2\alpha\, dx'}{\alpha^2 + \left(\frac{2\pi(x'-x)}{\lambda b}\right)^2}$$

is ultimately comprised between limits for which x' is infinitely close to x, and similarly with respect to y'; so that ultimately

$$\cos \frac{\pi(a+b)}{\lambda ab}(x'^2 - x^2 + y'^2 - y^2) = 1$$

within the limits for which the quantity under the integral sign does not vanish. Hence, passing to the limit, we get

$$D^2 \bar{I} = \lambda^2 b^2 \iint dx\, dy = \lambda^2 b^2 A,$$

as before.

CASE III. Everything the same as in Case II., except that the phase of vibration is retarded by ρ, where ρ is some function of x and y.

This case is very general. It includes, as particular cases, those numbered I. and II. The experiment with Fresnel's mirrors or a flat prism is also included as a particular case*.

From what precedes, it is plain that we should have in this case

$$D^2 \bar{I} = \text{limit of} \iiiint \frac{2\alpha}{\alpha^2 + \left(\frac{2\pi(x'-x)}{\lambda b}\right)^2} \cdot \frac{2\beta}{\beta^2 + \left(\frac{2\pi(y'-y)}{\lambda b}\right)^2} \cdot$$

$$\cos \left\{ \frac{\pi(a+b)}{\lambda ab}[x'^2 - x^2 + y'^2 - y^2] - \rho' + \rho \right\} dx\, dy\, dx'\, dy',$$

* Thus, in the case of the flat prism, if P, Q be the virtual images corresponding to the halves AB, BC, if we produce AB to D, we may suppose the light which falls on BC, instead of coming from Q, to come from P, and to have been accelerated by the passage through the wedge DBC of air instead of the same wedge of glass.

where ρ' is the same function of x' and y' that ρ is of x and y. The same reasoning as before leads to the same result.

I do not regard the preceding demonstration of a result which you were the first to announce, as of any physical interest after what you have yourself done. Still it may not seem wholly uninteresting, in an analytical point of view, to demonstrate the proposition for any form of aperture.

Of course, by comparing the result $\lambda^2 b^2 A$ with that obtained, in particular cases, by integrating in the straightforward way, we may arrive at the values of various definite integrals.

I am, dear Sir,

Yours very truly,

G. G. STOKES.

[From the *Transactions of the Cambridge Philosophical Society*, Vol. IX., p. 399 (read *Feb.* 16 and *March* 15, 1852).]

ON THE COMPOSITION AND RESOLUTION OF STREAMS OF POLARIZED LIGHT FROM DIFFERENT SOURCES.

WHEN a stream of polarized light is decomposed into two streams which, after having been modified in a slightly different manner, are reunited, the mixture is found to have acquired properties which are quite distinct from those of the original stream, and give rise to a number of curious and apparently complicated phenomena. These phenomena have now, however, through the labours of Young and Fresnel, been completely reduced to law, and embraced in a theory, the wonderful simplicity of which is such as to bear with it the stamp of truth. But when two polarized streams from different sources mix together, the mixture possesses properties intermediate between those of the original streams, and none of the curious phenomena depending upon the interference of polarized light are manifested. The properties of such mixtures form but an uninviting subject of investigation; and accordingly, though to a certain extent they are obvious, and must have forced themselves upon the attention of all who have paid any special attention to the physical theory of light, they do not seem hitherto to have been studied in detail.

Were the only object of such a study to enable us to calculate with greater facility the results obtained by means of certain complicated combinations, the subject might deservedly be deemed of small importance. For the object of the philosopher is not to complicate, but to simplify and analyze, so as to reduce phenomena to laws, which in their turn may be made the stepping-stones for ascending to a general theory which shall embrace them all; and when such a theory has been arrived at, and thoroughly verified, the task of deducing from it the results which ought to be observed under a combination of circumstances which has nothing to recommend it for consideration but its complexity, may well be abandoned for new and more fertile fields of research. But in the present case certain difficulties seem to have arisen respecting the connexion between common and elliptically polarized light which it needed only a more detailed study of the laws of combination of polarized light to overcome; and accordingly the subject may be deemed not wholly devoid of importance.

The early part of the following paper is devoted to a demonstration of various properties of elliptically polarized light, and of oppositely polarized streams. When two streams of light are called oppositely polarized, it is meant that, so far as relates to its state of polarization, one stream is what the other becomes when it is turned in azimuth through 90°, and has its nature reversed as regards right-handed and left-handed. Most, if not all, of these properties have doubtless already occurred to persons studying the subject, but I am not aware of any formal demonstrations of them which have been published; and indeed some artifices were required in order to avoid being encumbered in the demonstrations with long analytical expressions. The combination of several independent polarized streams is next considered, and with respect to this subject a proposition is proved which may be regarded as the capital theorem of the paper. It is as follows.

When any number of independent polarized streams, of given refrangibility, are mixed together, the nature of the mixture is completely determined by the values of four constants, which are certain functions of the intensities of the streams, and of the azimuths and eccentricities of the ellipses by which they are respectively characterized; so that any two groups of polarized streams which furnish the same values for each of these four constants are optically equivalent.

It is a simple consequence of this theorem, that any group of polarized streams is equivalent to a stream of common light combined with a stream of elliptically polarized light from a different source. The intensities of these two streams, as well as the azimuth and eccentricity of the ellipse which characterizes the latter, are determined by certain formulæ, which will be found in their place.

The general principles established in this paper bear on two questions of physical interest. Strong reasons are adduced in favour of the universality of the law, that the two polarized pencils which a doubly refracting medium of any nature is capable of propagating independently in a given direction are polarized oppositely. In strictness, we ought to speak of two series of waves rather than two pencils; for it is the fronts of the waves, not the rays, which are supposed to have a common direction. The other point alluded to relates to the distinction between common, and elliptically polarized light. It is shewn that the changes which are continually taking place in the mode of vibration may be of any nature, and that there is no occasion, in the case of common light, to suppose the transition from a series of vibrations of one kind to a series of another kind to be abrupt.

At the end of the paper the general formulæ are applied to the case of some actual experiments, but these applications are not of sufficient importance to deserve separate mention.

1. Consider a stream of light polarized in the most general way, that is, elliptically polarized, and propagated through the free ether. Let the medium be referred to the rectangular axes of x, y, z, the axis of z being measured in the direction of propagation. Let α and $\alpha + 90°$ be the azimuths of the principal planes, that is, the planes of maximum and minimum polarization, azimuths being measured about the axis of z from x towards y. Let the rectangular components of the displacements of the ether be represented by lines drawn in the planes of polarization of the plane-polarized streams which these components, taken separately, would constitute. I make this assumption to avoid entering into the question whether the vibrations of plane-polarized light are parallel or perpendicular to the plane of polarization. If we adopt the former theory, the actual lines in the figures which we are to suppose drawn will represent in magnitude and direction the ethereal displacements; if we adopt the latter, the same will still be the case if we first suppose all our figures turned round the axis of z, in a given direction, through 90°.

Let the co-ordinates x', y' be measured in the principal planes whose azimuths are α, $\alpha + 90°$; let β be the angle whose tangent is equal to the ratio of the axes of the ellipse described, the numerical value of β being supposed not to lie beyond the limits 0 and 90°; let v be the velocity of propagation, t the time, λ the length of a wave, and put for shortness,

$$\frac{2\pi}{\lambda}(v t - z) = \phi \dots\dots\dots\dots\dots (1).$$

Then about the time t, and at no very great distance from a given point, suppose the origin, we may represent the displacements belonging to the given stream of elliptically polarized light by

$$x' = c \cos \beta \sin (\phi + \epsilon), \quad y' = c \sin \beta \cos (\phi + \epsilon)\dots\dots(2).$$

If the light be convergent or divergent, c will depend upon ϕ, but for our present object any variation of c arising from this cause will not enter into account. The value of α, which determines the direction in which x' is measured, as well as that of β, is given by the nature of the polarization. The polarization is right-handed or left-handed according to the sign of β. As to c and ϵ, the phenomena of optics oblige us to suppose that they are constant, or sensibly constant, for a great number of consecutive undulations, but that they change in an irregular manner a great number of times in the course of one second. The known rapidity of the luminous vibrations allows abundant scope for such a supposition, since c and ϵ may be constant for millions of consecutive undulations, and yet change millions of times in a second. This series of changes, rapid with respect to the duration of impressions on the retina, but slow compared with the periodic changes in the motion of the ethereal particles, is exactly what we might have expected beforehand from a consideration of the circumstances under which light is produced, so far at least as its sources are accessible to us; and thus in this point, as in so many others, the theory of undulations commends itself for its simplicity.

If c were constant c^2 would be a measure of the intensity, so long as we were only comparing different streams having the same refrangibility. But since c is liable to the changes just mentioned, if we wish to express ourselves exactly, avoiding conventional abbreviations, we must say that the intensity is measured, not by c^2, but by the mean value of c^2, which may conveniently be represented by $\mathfrak{m}(c^2)$.

2. Let us examine now whether it be always possible to resolve the given disturbance into two which, taken separately, would correspond to two elliptically polarized streams of given nature. For the sake of clear ideas, it may be supposed that the azimuths and eccentricities of the ellipses belonging to these two streams are given and invariable, while the azimuth and eccentricity of the ellipse belonging to the first stream are given for that stream, but vary from one to another of a set of streams which we wish to consider in succession.

Let x_1, c_1, &c. be for the first, and x_2, c_2, &c. be for the second stream of the pair, what x', c, &c. were for the original stream; and resolve all the displacements along the principal axes of the latter stream. Then, in order that the original disturbance may be equivalent to the pair, we must have, independently of ϕ,

$$\left.\begin{aligned}
x_1 \cos(\alpha_1 - \alpha) - y_1 \sin(\alpha_1 - \alpha) + x_2 \cos(\alpha_2 - \alpha) \\
- y_2 \sin(\alpha_2 - \alpha) = x'; \\
x_1 \sin(\alpha_1 - \alpha) + y_1 \cos(\alpha_1 - \alpha) + x_2 \sin(\alpha_2 - \alpha) \\
+ y_2 \cos(\alpha_2 - \alpha) = y'.
\end{aligned}\right\} \quad \dots\dots (3).$$

Conceive x_1, y_1, x_2, y_2, x', and y' expressed in terms of ϕ by the formulæ (2) and the similar formulæ whereby x_1, y_1, &c. are expressed, and then let the sines and cosines of $\phi + \epsilon$, $\phi + \epsilon_1$, and $\phi + \epsilon_2$ be developed. In order that equations (3) may be satisfied independently of ϕ, the coefficients of $\sin \phi$ and $\cos \phi$ must separately be equal to zero, so that each of these equations will split into two. We shall thus have four equations to determine the four unknown quantities c_1, c_2, ϵ_1, and ϵ_2. For the sake of shortness, let

$$c \cos \epsilon = g, \qquad\qquad c \sin \epsilon = h,$$

whether the letters be or be not affected with suffixes; and further put

$$\alpha_1 - \alpha = \gamma_1, \qquad\qquad \alpha_2 - \alpha = \gamma_2;$$

then our four equations become

$$
\begin{aligned}
\cos\beta_1\cos\gamma_1 . g_1 + \sin\beta_1\sin\gamma_1 . h_1 + \cos\beta_2\cos\gamma_2 . g_2 \\
+ \sin\beta_2\sin\gamma_2 . h_2 = \cos\beta . g, \\
\cos\beta_1\cos\gamma_1 . h_1 - \sin\beta_1\sin\gamma_1 . g_1 + \cos\beta_2\cos\gamma_2 . h_2 \\
- \sin\beta_2\sin\gamma_2 . g_2 = \cos\beta . h, \\
\cos\beta_1\sin\gamma_1 . g_1 - \sin\beta_1\cos\gamma_1 . h_1 + \cos\beta_2\sin\gamma_2 . g_2 \\
- \sin\beta_2\cos\gamma_2 . h_2 = -\sin\beta . h, \\
\cos\beta_1\sin\gamma_1 . h_1 + \sin\beta_1\cos\gamma_1 . g_1 + \cos\beta_2\sin\gamma_2 . h_2 \\
+ \sin\beta_2\cos\gamma_2 . g_2 = \sin\beta . g.
\end{aligned}
\Bigg\} \dots (4).
$$

Multiplying the first and second of these equations by 1, $\sqrt{(-1)}$, and adding, then multiplying the third and fourth by $-\sqrt{(-1)}$, 1, and adding, and putting generally

$$ g + \sqrt{-1}\, h = G \dots\dots\dots\dots\dots (5), $$

we have

$$
\begin{aligned}
(\cos\beta_1\cos\gamma_1 - \sqrt{-1}\sin\beta_1\sin\gamma_1)\, G_1 + (\cos\beta_2\cos\gamma_2 \\
- \sqrt{-1}\sin\beta_2\sin\gamma_2)\, G_2 = \cos\beta . G, \\
(\sin\beta_1\cos\gamma_1 - \sqrt{-1}\cos\beta_1\sin\gamma_1)\, G_1 + (\sin\beta_2\cos\gamma_2 \\
- \sqrt{-1}\cos\beta_2\sin\gamma_2)\, G_2 = \sin\beta . G;
\end{aligned}
\Bigg\} \dots (6),
$$

which two equations are equivalent to the four (4).

Putting for shortness p_1, p_2, q_1, q_2 for the coefficients in the left-hand members of equations (6), we have

$$ \frac{G_1}{q_2\cos\beta - p_2\sin\beta} = \frac{G_2}{p_1\sin\beta - q_1\cos\beta} = \frac{G}{p_1 q_2 - p_2 q_1} \dots (7). $$

On substituting for p_1, q_2, &c. their values, we find

$$ p_1 q_2 - p_2 q_1 = \cos(\gamma_1 - \gamma_2)\sin(\beta_2 - \beta_1) $$
$$ + \sqrt{-1}\sin(\gamma_1 - \gamma_2)\cos(\beta_1 + \beta_2). $$

Now the equations (6) cannot be incompatible or identical unless the above quantity vanish. But this can only take place when

$$ \sin(\beta_2 - \beta_1) = 0 \ \text{ and } \ \sin(\gamma_1 - \gamma_2) = 0, $$

or else

$$ \cos(\beta_1 + \beta_2) = 0 \ \text{ and } \ \cos(\gamma_1 - \gamma_2) = 0, $$

or lastly

$$ \sin(\beta_2 - \beta_1) = 0 \ \text{ and } \ \cos(\beta_1 + \beta_2) = 0. $$

The first case gives $\beta_2 = \beta_1$, $\gamma_1 - \gamma_2 = \alpha_1 - \alpha_2 = 0$ or $\pm 180°$, so that the two streams into which it was proposed to resolve the first are of the same nature. The second case gives $\beta_2 = 90° - \beta_1$, $\gamma_1 - \gamma_2 = \alpha_1 - \alpha_2 = \pm 90°$, which shews that the two streams are identical in their nature, only the first and second principal planes of the first of these streams are accounted respectively the second and first of the second stream. The third case gives $\beta_2 = \beta_1 = \pm 45°$, so that the two streams are circularly polarized and of the same kind, which is a particular instance of the first case.

Hence, universally, a stream of elliptically polarized light may be resolved into two streams of elliptically polarized light in which the polarizations are of any kind that we please, but different from one another.

Substituting for q_2, &c. their values in (7), and replacing γ_1, γ_2 by $\alpha_1 - \alpha$, $\alpha_2 - \alpha$, for which they had been temporarily written, we find

$$\left.\begin{array}{l}\{\sin(\beta_2 - \beta)\cos(\alpha_2 - \alpha) \\ \qquad - \sqrt{-1}\cos(\beta_2 + \beta)\sin(\alpha_2 - \alpha)\}^{-1}\, G_1 \\ = \{\sin(\beta - \beta_1)\cos(\alpha - \alpha_1) \\ \qquad - \sqrt{-1}\cos(\beta + \beta_1)\sin(\alpha - \alpha_1)\}^{-1}\, G_2 \\ = \{\sin(\beta_2 - \beta_1)\cos(\alpha_2 - \alpha_1) \\ \qquad - \sqrt{-1}\cos(\beta_2 + \beta_1)\sin(\alpha_2 - \alpha_1)\}^{-1}\, G \end{array}\right\} \dots (8).$$

3. Among these various modes of resolution there is one which possesses several peculiar properties, any one of which might serve to define it. Let us in the first place examine under what circumstances the intensity of the stream made up of the two components is independent of any retardation which the phase of vibration of one component may have undergone relatively to the phase of vibration of the other previously to the recomposition.

For this purpose there is evidently no occasion to consider the manner in which a_1, b_1, ϵ_1, a_2, b_2, ϵ_2 are made up of a, b, ϵ, but we may start with the components. Let ρ_1, ρ_2 be the retardations of phase which take place before recomposition, and resolve the disturbances along the axes of x, y. We shall have for the resolved parts

$$x = S\{a_1 \cos \alpha_1 \sin (\phi + \epsilon_1 - \rho_1) - b_1 \sin \alpha_1 \cos (\phi + \epsilon_1 - \rho_1)\}$$
$$y = S\{a_1 \sin \alpha_1 \sin (\phi + \epsilon_1 - \rho_1) + b_1 \cos \alpha_1 \cos (\phi + \epsilon_1 - \rho_1)\},$$

where S denotes the sum of the expression written down and that formed from it by replacing the suffix 1 by 2. To form the expression for the intensity, or rather what would be the intensity if the quantities c and ϵ were absolutely constant, not merely constant for a great number of successive undulations, we must develope the expressions for x and y so as to contain the sine and cosine of $\phi + \kappa$, and take the sum of the squares of the coefficients. κ is here a constant quantity which may be chosen at pleasure, and which it will be convenient to take equal to $\epsilon_1 - \rho_1$. If I be the intensity, in the sense above explained, or as it may be called the *temporary intensity*, we find, putting δ for $\epsilon_2 - \rho_2 - \epsilon_1 + \rho_1$,

$$I = \{a_1 \cos \alpha_1 + a_2 \cos \alpha_2 \cos \delta + b_2 \sin \alpha_2 \sin \delta\}^2$$
$$+ \{- b_1 \sin \alpha_1 + a_2 \cos \alpha_2 \sin \delta - b_2 \sin \alpha_2 \cos \delta\}^2$$
$$+ \{a_1 \sin \alpha_1 + a_2 \sin \alpha_2 \cos \delta - b_2 \cos \alpha_2 \sin \delta\}^2$$
$$+ \{b_1 \cos \alpha_1 + a_2 \sin \alpha_2 \sin \delta + b_2 \cos \alpha_2 \cos \delta\}^2$$
$$= a_1^2 + b_1^2 + a_2^2 + b_2^2 + 2(a_1 a_2 + b_1 b_2) \cos (\alpha_2 - \alpha_1) \cos \delta$$
$$+ 2(a_1 b_2 + a_2 b_1) \sin (\alpha_2 - \alpha_1) \sin \delta.$$

On putting for a, b their values $c \cos \beta$, $c \sin \beta$, this expression becomes

$$I = c_1^2 + c_2^2 + 2c_1 c_2 \{\cos (\alpha_2 - \alpha_1) \cos (\beta_2 - \beta_1) \cos \delta$$
$$+ \sin (\alpha_2 - \alpha_1) \sin (\beta_2 + \beta_1) \sin \delta\} \ldots (9).$$

In order that I may be independent of the difference of phase $\rho_2 - \rho_1$, and therefore of δ, we must have either

$$\cos (\alpha_2 - \alpha_1) = 0, \qquad \sin (\beta_2 + \beta_1) = 0, \ldots\ldots(10),$$
$$\text{or } \sin (\alpha_2 - \alpha_1) = 0, \qquad \cos (\beta_2 - \beta_1) = 0, \ldots\ldots(11).$$

The equations (10) give $\alpha_2 - \alpha_1 = \pm 90°$, $\beta_2 = -\beta_1$, so that the ellipses described in the case of the two streams are similar, their major axes perpendicular to each other, and the direction of revolution in the one stream contrary to that in the other. It will be easily seen that the equations (11) differ from (10) only in this, that what are regarded as the first and second principal planes of the second stream when equations (10) are satisfied, are accounted respectively the second and first when (11) are satisfied.

Two streams thus related may be said to be *oppositely polarized*. Two streams of plane-polarized light in which the planes of polarization are at right angles to each other, and two streams of circularly polarized light, one right-handed and the other left-handed, are particular cases of streams oppositely polarized.

In the reasoning of this article, nothing depends upon the precise relation between the two polarized streams and the original stream. All that it is necessary to suppose is, that the two polarized streams came originally from the same polarized source, so that the changes in epoch and intensity, that is, the changes in the quantities ϵ, c, are the same for the two streams. Nothing depends upon the precise nature of these changes, which may be either abrupt or continuous, but must be sufficiently infrequent if abrupt, or sufficiently gradual if continuous, to allow of our regarding c and ϵ as constant for a great number of successive undulations. Our results will apply just as well to the disturbance produced by the union of two neighbouring streams coming originally from the same polarized source, but having had their polarizations modified, as to that produced by the union, after recomposition, of the components of a single polarized stream. Since the resulting intensity is independent of δ, it follows that two oppositely polarized streams coming originally from the same polarized source are incapable of interfering, but two streams polarized otherwise than oppositely necessarily interfere, to a greater or less degree, when the difference in their retardation of phase is sufficiently small. Of course the interference here spoken of means only that which is exhibited without analyzation.

4. Two interfering streams may be said to interfere perfectly when the fluctuations of intensity are the greatest that the difference in the intensities of the interfering streams admits of, so that in case of equality the minima are absolutely equal to zero. Referring to (9), we see that in order that this may be the case the maximum value of the coefficient of $2c_1c_2$ must be equal to 1. Now the maximum value of $A \cos \delta + \beta \sin \delta$ is $\sqrt{(A^2 + B^2)}$, and therefore we must have

$$\cos^2(\alpha_2 - \alpha_1) \cos^2(\beta_2 - \beta_1) + \sin^2(\alpha_2 - \alpha_1) \sin^2(\beta_2 + \beta_1) = 1$$
$$= \cos^2(\alpha_2 - \alpha_1) + \sin^2(\alpha_2 - \alpha_1),$$

whence,

$$\cos^2(\alpha_2 - \alpha_1)\sin^2(\beta_2 - \beta_1) + \sin^2(\alpha_2 - \alpha_1)\cos^2(\beta_2 + \beta_1) = 0,$$

which leads to the very same conditions that have been already discussed in Art. 2. Hence two polarized streams coming from the same polarized source are capable of interfering perfectly if the polarizations are the same, not at all if the polarizations are opposite, and in intermediate cases of course in intermediate degrees.

5. When a stream of polarized light is resolved into two oppositely polarized streams, which are again compounded after their phases have been differently altered, we have from (9), taking account of (10) or (11),

$$I = c_1^2 + c_2^2 \quad \dots\dots\dots\dots\dots\dots (12),$$

so that the intensity of the resultant is equal to the sum of the intensities of the components, and is therefore constant, that is, independent of $\rho_1 - \rho_2$, and is accordingly equal to what it was at first, when ρ_1 and ρ_2 were each equal to zero, that is, equal to the intensity of the original stream.

It may be readily proved from the formulæ (8) that it is only in the case in which the polarizations of the two components of the original polarized stream are opposite that the intensity of the original stream, whatever be the nature of its polarization, is equal to the sum of the intensities of the component streams. For, changing the sign of $\sqrt{(-1)}$ in these formulæ, multiplying the resulting equations, member for member, by the equations (8), and observing that if G' be what G becomes, $GG' = g^2 + h^2 = c^2$, we find

$$\begin{aligned} &\{\sin^2(\beta_2 - \beta)\cos^2(\alpha_2 - \alpha) + \cos^2(\beta_2 + \beta)\sin^2(\alpha_2 - \alpha)\}^{-1}c_1^2 \\ &= \{\sin^2(\beta - \beta_1)\cos^2(\alpha_1 - \alpha) + \cos^2(\beta + \beta_1)\sin^2(\alpha_1 - \alpha)\}^{-1}c_2^2 \\ &= \{\sin^2(\beta_2 - \beta_1)\cos^2(\alpha_2 - \alpha_1) + \cos^2(\beta_2 + \beta_1)\sin^2(\alpha_2 - \alpha_1)\}^{-1}c^2 \end{aligned} \quad (13).$$

In order that $\mathfrak{m}(c^2)$ may be equal to $\mathfrak{m}(c_1^2) + \mathfrak{m}(c_2^2)$, it is necessary that c^2 be equal to $c_1^2 + c_2^2$, because, whatever fluctuations c_1 and c_2 may undergo in a moderate time, such as the tenth part of a second, c_1 and c_2 are always proportional to c. Hence the sum of the quantities whose reciprocals are the coefficients of c_1^2 and c_2^2 must be equal to that whose reciprocal is the coefficient of c^2. Since this has to be true independently of

β, let the quantities $\sin^2(\beta_2 - \beta)$, &c. be replaced by sines and cosines of multiple arcs, and let our equation be put under the form

$$A + B \cos 2\beta + C \sin 2\beta = 0.$$

Then A, B, C must be separately equal to zero, or

$$\left.\begin{array}{c} \sin^2(\beta_2 - \beta_1)\cos^2(\alpha_2 - \alpha_1) + \cos^2(\beta_2 + \beta_1)\sin^2(\alpha_2 - \alpha_1) = 1 ; \\ \cos 2\beta_2 \cos 2(\alpha_2 - \alpha) + \cos 2\beta_1 \cos 2(\alpha_1 - \alpha) = 0 ; \\ \sin 2\beta_2 + \sin 2\beta_1 = 0. \end{array}\right\} \quad (14).$$

Replacing unity in the right-hand member of the first of these equations by

$$\cos^2(\alpha_2 - \alpha_1) + \sin^2(\alpha_2 - \alpha_1),$$

we find

$$\cos^2(\beta_2 - \beta_1)\cos^2(\alpha_2 - \alpha_1) + \sin^2(\beta_2 + \beta_1)\sin^2(\alpha_2 - \alpha_1) = 0 ;$$

whence $\beta_2 = -\beta_1$, $\alpha_2 = \alpha_1 + 90°$, or else β_2 and β_1 differ by $90°$, and $\alpha_2 = \alpha_1$, except in the particular case in which $\beta_1 = \pm 45°$, when $\beta_2 = \mp 45°$ satisfies the equation independently of α_2. Hence the streams must be polarized oppositely, a condition which may always be expressed by

$$\beta_2 = -\beta_1, \quad \alpha_2 = \alpha_1 + 90°,$$

which equations satisfy the second and third of equations (14) independently of α, as it might have been foreseen that they would, since it has been already shewn that the condition (12) is satisfied in the case of oppositely polarized streams. It now appears that it is only in the case of such streams that this is satisfied.

6. The properties of oppositely polarized pencils which have been proved, render it in a high degree probable that it is a general law that in a doubly refracting medium the two polarized pencils transmitted in a given direction are oppositely polarized. Were this not the case, the two pencils, polarized otherwise than oppositely, into which a polarized pencil is resolved on entering into the medium, would at emergence compound a pencil of which the intensity would depend upon the retardations of phase of one pencil relatively to the other, so that such a medium, when examined with polarized light, ought to exhibit rings or colours without the employment of an analyzer. It is here supposed

that the light enters the medium at the first surface, and leaves it at the second, in a direction normal to the surface, or very nearly so, and likewise that the refracting power of the medium for the two pencils is not very different, so that the effect of re-flexion at the two surfaces may be disregarded, the only sensible effect being to diminish the intensity of each of the two pencils in the same proportion, without affecting its state of polariza-tion. These views would lead us to scrutinize very carefully any experimental evidence brought forward which would lead to the conclusion, that the two polarized pencils which a doubly re-fracting medium was capable of propagating in a given direction were polarized otherwise than oppositely.

7. In ordinary doubly refracting crystals, whether uniaxal or biaxal, and in doubly refracting liquids, such as syrop of sugar, it is generally admitted that the two pencils transmitted in a given direction are oppositely polarized. In the former case the two pencils are polarized in rectangular planes, in the latter they are circularly polarized, one being right-handed and the other left-handed. The same is the case with quartz for pencils transmitted in the direction of the axis; but in following out the researches in which he so successfully connected the pheno-mena, identical with those of a doubly refracting liquid, which quartz exhibits in the direction of the axis, with the phenomena which it exhibits as a uniaxal crystal, Mr Airy met with an ex-perimental result which seemed to shew that while the ellipses which characterize the two streams transmitted in a given direc-tion, oblique to the axis, have their major axes situated, one in a principal plane, and the other perpendicular to the principal plane, and the directions of revolution are opposite, the eccen-tricities of the ellipses, though nearly, are not quite equal*. This conclusion depended upon the result of certain experiments made by means of a Fresnel's rhomb. The nature of the ex-periments seemed to eliminate the effect of an error in the rhomb, that is, a deviation from 90° in the retardation of phase which it produced in a pencil polarized in the plane of reflexion relatively to a pencil polarized in a plane perpendicular to the former. The effect of a possible index error in the plane of reflexion seemed also to be eliminated; and the quantities on

* *Cambridge Philosophical Transactions*, Vol. IV. p. 204.

which the results depended, though small, seemed to be beyond mere errors of pointing. Being impressed however with a strong conviction that the result depended in some way on the mode of observation, I was led to scrutinize the different steps of the process, and it occurred to me that the apparent inequality of eccentricities was probably due to defects of annealing in the rhomb. It is next to impossible to procure a piece of glass of such a size free from defects of that nature, for which reason I believe that even a good Fresnel's rhomb is not to be trusted for minute quantities, except in the case of merely differential observations.

The same views lead to the conclusion that the two pencils transmitted through magnetized glass in a direction oblique to the lines of magnetic force are oppositely polarized. Some theoretical investigations in which I was engaged some time ago led me to the result that these polarized streams are circularly polarized, as well as those transmitted along the line of magnetic force, and that the difference between the wave-velocities varies as the cosine of the inclination of the wave-normal to the line of magnetic force. I hope at some future time to bring these researches before the notice of this Society.

8. After these preliminary investigations respecting the nature of opposite polarization, which indeed contain, it is probable, but little that has not already occurred to persons who have studied the subject, it is time to come to the more immediate object of this paper, which relates to the combination of independent streams. But first it will be convenient to state explicitly a principle which is generally recognized.

When any number of polarized streams from different sources mix together, after having been variously modified by reflexion, refraction, transmission through doubly refracting media, tourmalines, &c., the intensity of the mixture is equal to the sum of the intensities due to the separate streams.

The reason of this law may be easily seen. The components whereby the disturbance due to any one stream is originally expressed have to be resolved, their components resolved again, and so on; and of these partial disturbances the phases of vibration have to be altered by quantities independent of the time, and the coefficients in some cases diminished in given ratios, and in some cases suppressed altogether. Each stream has to be treated in a similar way. The final disturbance being resolved in any two rectangular directions, each component must be put under the form $U \cos \phi + V \sin \phi$, and the sum of the squares of U and V must be taken to form the expression for the temporary intensity. All the quantities such as U and V will evidently be linear functions of $c \cos \epsilon$, $c \sin \epsilon$, $c' \cos \epsilon'$, $c' \sin \epsilon'$, &c., where c, c' ... and ϵ, ϵ' ... refer to the different streams, so that U for instance will be of the form

$$A c \cos \epsilon + B c \sin \epsilon + A'c' \cos \epsilon' + B'c' \sin \epsilon' + \ldots$$

where A, B, A', B' ... are independent of the time. The temporary intensity will involve U^2, but the actual intensity will involve $\mathfrak{m}(U^2)$, or

$$\mathfrak{m}\Sigma (A c \cos \epsilon + B c \sin \epsilon)^2 + 2\mathfrak{m}\Sigma \{(A c \cos \epsilon + B c \sin \epsilon)$$
$$(A'c' \cos \epsilon' + B'c' \sin \epsilon')\}.$$

Now the products such as $\cos \epsilon \cos \epsilon'$, $\cos \epsilon \sin \epsilon'$, &c. will have a mean value zero, since the changes in ϵ and those in ϵ' have no relation to each other, and therefore the expression for $\mathfrak{m}(U^2)$ becomes

$$\mathfrak{m}\Sigma (A c \cos \epsilon + B c \sin \epsilon)^2, \text{ or } \Sigma\mathfrak{m} (A c \cos \epsilon + B c \sin \epsilon)^2,$$

that is, the sum of the quantities by which it would be expressed were the different streams taken separately.

Two streams which come from different sources, or which, though in strictness they come from the same source, are such that the changes of epoch and intensity in the one have no relation to the changes of epoch and intensity in the other, may be called *independent*.

9. Suppose that there are any number of independent polarized streams mixing together; let the mixture be resolved in any manner into two oppositely polarized streams, and let us examine the intensity of each.

Let us take one stream first. The intensities of its components are given by the formulæ (13), which become somewhat simpler in the case of opposite polarizations, since $\beta_2 = -\beta_1$, and $\alpha_2 = 90° + \alpha_1$. Hence

$$c_1^2 = \{\sin^2 (\beta_1 + \beta) \sin^2 (\alpha_1 - \alpha) + \cos^2 (\beta_1 - \beta) \cos^2 (\alpha_1 - \alpha)\} c^2;$$

whence we find

$$\mathfrak{m}\,(c_1^{\,2}) = \tfrac{1}{2}\,\{1 + \sin 2\beta_1 \sin 2\beta + \cos 2\alpha_1 \cos 2\alpha \cos 2\beta_1 \cos 2\beta$$
$$+ \sin 2\alpha_1 \sin 2\alpha \cos 2\beta_1 \cos 2\beta\}\, \mathfrak{m}\,(c^2)\ \ldots(15).$$

There will be no occasion to write down the value of $\mathfrak{m}\,(c_2^{\,2})$, since c_1 may be taken to refer to either component.

Let us pass now to the consideration of a group consisting of any number of independent polarized streams. Let

$$\Sigma \mathfrak{m}\,(c^2) = A, \quad \Sigma \sin 2\beta \mathfrak{m}\,(c^2) = B, \quad \Sigma \cos 2\alpha \cos 2\beta \mathfrak{m}\,(c^2) = C, \atop \Sigma \sin 2\alpha \cos 2\beta \mathfrak{m}\,(c^2) = D, \Bigg\}\,(16),$$

and let c_1 now refer to one of the components of the whole group; then

$$2\mathfrak{m}\,(c_1^{\,2}) = A + B \sin 2\beta_1 + C \cos 2\alpha_1 \cos 2\beta_1 + D \sin 2\alpha_1 \cos 2\beta_1 \ \ldots(17).$$

It follows that if there are two groups of independent polarized streams which are such as to give the same values to each of the four quantities A, B, C, D defined by (16), if the groups be resolved in any manner whatsoever, which is the same for both, into two oppositely polarized streams, the intensities of the components of the one group will be respectively equal to the intensities of the components of the other group. Conversely, if two groups of oppositely polarized streams are such that when they are resolved in any manner, the same for both, into two oppositely polarized streams, the intensities of the components of the one group are respectively equal to the intensities of the components of the other group, the quantities A, B, C, D must be the same for the two groups. For, if we take accented letters to refer to the second group, the second member of equation (17), and the expression thence derived by accenting A, B, C, D, must be equal independently of α_1 and β_1, which requires that A', B', C', D' be respectively equal to A, B, C, D.

DEFINITION. Two such groups will be said to be *equivalent*.

10. The theoretical definition of equivalence which has just been given agrees completely with the experimental tests of equivalence. One of the most ready as well as delicate modes of detecting minute traces of polarization, and at the same time determining qualitatively the nature of the polarization, consists in viewing the light to be examined through a plate of calcareous

spar or other crystal, cut for shewing rings, followed by a Nicol's prism. The plane-polarized pencils respectively stopped by and transmitted through the Nicol's prism consisted, on entering the crystal, of pencils elliptically polarized in opposite ways; and the nature of this elliptic polarization changes in every possible manner from one point to another of the field of view. If these two streams of light be equivalent according to the definition given in the preceding article, they will present exactly the same appearance on being viewed through a crystal followed by a Nicol's prism or other analyzer.

11. THEOREM. Let a polarized stream be resolved into two oppositely polarized streams; let the phase of vibration of one of the streams be altered by a given quantity relatively to that of the other, and let the streams be then compounded. If the polarization of the original stream be now changed to its opposite, the polarization of the final stream will also be changed to its opposite.

The straightforward mode of demonstrating this theorem, by making use of the general expressions, would lead to laborious analytical processes, which are wholly unnecessary. For the formulæ which determine the components of a given stream are expressed by simple equations, so that the results are unique, and accordingly whenever we can foresee what the result will be, it is sufficient to shew that the formulæ themselves, or the geometrical conditions of which the formulæ are merely the expressions, are satisfied.

For shortness' sake call the original stream X, and its components O, E. Let ρ be the given quantity, positive or negative, by which the phase of vibration of O is retarded relatively to that of E. Let o, e denote the streams O, E after the changes of phase, and Y the stream resulting from their reunion. Conceive now all the vibrations with which we are concerned to be turned in azimuth through $90°$. This will not affect the geometrical relations connecting components and resultants. Let X', O', E', o', e', Y' be the streams which X, O, E, o, e, Y thus become. The streams O', E' are evidently polarized in the same manner respectively as E, O, except that *right-handed* is changed into *left-handed*, and *vice versa;* and in passing from O' to o' the phase is retarded by ρ. Now conceive the direction of motion of a given particle reversed, the motions of all other particles being derived from that of the first according to the general law of wave propagation. The relations between components and resultants will evidently not be violated; and if X'', O'', E'', o'', e'', Y'' denote the streams into which X', O', E', o', e', Y' are thus changed, it is evident that the polarizations of X'', O'', E'', o'', e'', Y'' are respectively opposite to those of X, O, E, o, e, Y. But on account of the reversion in direction of motion it is plain that there is reversion as regards acceleration and retardation of phase, so that in passing from the pair O'', E'' to the pair o'', e'' the phase of O'' is *accelerated* by ρ relatively to the phase of E''.

Hence the stream X'', polarized oppositely to X, is resolved into two E'', O'', polarized in the same manner respectively as O, E, which are recompounded after the phase of the one polarized in the same manner as O has been retarded by ρ relatively to the phase of the other, and the result is Y'', a stream polarized in a manner opposite to Y, which proves the theorem.

12. This theorem may be applied to the case of light transmitted through a slice of a doubly refracting crystal, and shews that two streams going in oppositely polarized come out oppositely polarized. Also, since nothing in the demonstration depends upon the order in which the decompositions and recompositions take place, it is immaterial whether X, X'' denote a pair of oppositely polarized incident streams, which give rise to the emergent streams Y, Y'', or X, X'' denote a pair of oppositely polarized emergent streams, which came from the incident streams Y, Y''. A particular case of this theorem was assumed in the preceding article, when it was stated that the pencils, polarized in perpendicular planes, which on coming out of a crystal of calcareous spar are respectively stopped by and transmitted through a Nicol's prism, went into the crystal oppositely polarized.

The theorem will evidently be true of a train consisting of any number of crystalline plates, each possessing the property of resolving the incident light into two oppositely polarized streams, which are propagated within the medium with different velocities. For two oppositely polarized streams incident on the first plate give rise to two emergent streams which fall oppositely polarized on the second plate, and so on. Since the number

of plates may be supposed to increase and their thickness to decrease indefinitely, while at the same time the steps by which
the doubly refracting nature of the plates alters from one to
another become separately insensible, the theorem will be true
if the whole train, or part of it, consist of a substance of which
the doubly refracting nature alters continuously, as for example
a piece of strained or unannealed glass. If, however, the train
contain a member which performs a partial analysis of the light,
as for example a plate of smoky quartz, or a plate of glass inclined to the incident light at a considerable angle, it will no
longer be true that two pencils going in oppositely polarized will
come out oppositely polarized.

13. THEOREM. If two equivalent groups of polarized streams
be resolved in any manner, which is the same for both, into two
oppositely polarized groups, and these be recombined after the
phase of one of the components has been retarded by a given
quantity relatively to that of the other, the two groups of resultant
streams will be equivalent.

Let the groups of resultants be each resolved in any manner
into two oppositely polarized streams, and call these O, E. By
Art. 11, if O', E' be the streams which furnish O, E respectively,
O', E' are oppositely polarized. Now by Art. 3, the intensities
of O, E are the same respectively as those of O', E'; but these
are the same for the one group as for the other, by the definition of equivalence. Therefore the intensities of O, E are the
same for the one group as for the other; but O, E are any two
oppositely polarized components of the resultant groups; therefore
these groups are equivalent.

Hence, if two equivalent groups be transmitted through a
crystalline plate, the emergent groups will be equivalent; and
by the same reasoning as in Art. 12 the theorem may be extended to an optical train consisting of any number of crystalline
plates, pieces of unannealed glass, &c.

14. THEOREM. If two equivalent groups be resolved in any
manner, the same for both, into two polarized streams, the intensities of the components of the one group will be respectively
equal to the intensities of the components of the other group.

The proof of this theorem is very easy. It is sufficient to
treat the general expressions (13) exactly as in Art. 9, only that

no relations are to be introduced between α_2, β_2, and α_1, β_1. Since the coefficient of c^2 in (13) is constant, if we consider as variable such quantities only as may change in passing from the one group to the other, it will be easily seen that the intensity of either component depends on the same four constants A, B, C, D as before; and since these have the same values for equivalent groups, it follows that in the most general mode of resolution the components of any two such groups have the same intensities respectively.

15. THEOREM. If two equivalent groups be each resolved in the same manner into two oppositely polarized streams, and these be recombined after their vibrations have been diminished in two different given ratios, the resultant groups will be equivalent.

Let each of the resultant groups be resolved into two oppositely polarized streams O, E, and let O', E' denote the streams which furnished O, E respectively. It is easily seen that the streams O', E' are both polarized, though not in general oppositely. Were there any occasion to determine the nature of the polarizations, it might easily be done by following a process the reverse of that by which the modified are deduced from the original groups. Thus, if it were required to determine the polarization of O', we should resolve O into streams oppositely polarized in the manner originally given, augment the vibrations in ratios the inverse of those in which they are actually diminished, and recombine the streams so obtained. For our present purpose, however, it is sufficient to observe that the polarizations of O', E' depend only on the mode of resolution, and not on the nature of the original group, and that therefore they are the same for the one group as for the other. The intensity of O' is not the same as that of O, as in the case considered in Art. 13, but bears to it a ratio depending only on the mode of resolution, and therefore the same for each of the two original groups. The same is true of the intensity of E' compared with that of E. Now by Art. 14 the intensities of O', E' are the same for the two original groups, and the intensities of O, E bear to those of O', E' respectively, ratios the same for the two groups: therefore the intensities of O, E are the same for the two final groups; but O, E are any two oppositely polarized components of these groups; therefore these groups are equivalent.

16. It follows from this theorem that no partial analysis of light, such, for example, as would be produced by reflexion from the surface of glass or metal, or by transmission through a doubly absorbing medium, can from equivalent groups produce groups which are not equivalent to each other; and we have seen already that this cannot be done by means of the alteration of phase accompanying double refraction. It follows, therefore, that equivalent groups are optically undistinguishable.

In proving this property of equivalent groups, it has been supposed that the polarizations of the two streams into which any group was resolved were opposite, such being the case in nature. But did a medium exist such that the two streams of light which it transmitted independently were polarized otherwise than oppositely, it would still not enable us to distinguish between equivalent groups.

17. The experimental definition of common light is, light which is incapable of exhibiting rings of any kind when examined by a crystal of Iceland spar and an analyzer, or by some equivalent combination. Consequently, a group of independent polarized streams will together be equivalent to common light when, on being resolved in any manner into two oppositely polarized pencils, the intensities of the two are the same, and of course equal to half that of the original group. Accordingly, in order that the group should be equivalent to common light, it is necessary and sufficient that the constants B, C, D should vanish.

18. Let us now see under what circumstances two independent streams of polarized light can together be equivalent to common light.

Let α, β refer to the first, and α', β' to the second stream, and let the intensities of the two streams be as 1 to n; then we get from the formulæ (16)

$$\sin 2\beta + n \sin 2\beta' = 0\;;$$
$$\cos 2\alpha \cos 2\beta + n \cos 2\alpha' \cos 2\beta' = 0\;;$$
$$\sin 2\alpha \cos 2\beta + n \sin 2\alpha' \cos 2\beta' = 0.$$

Transposing, squaring, and adding, we find $n^2 = 1$, and therefore $n = 1$, since n is essentially positive. Since β and β' are supposed not to lie beyond the limits $-90°$ and $+90°$, we get from the first equation $\beta' = -\beta$, or $\beta' = \pm 90° - \beta$, $+$ or $-$ according as β is positive or negative. Now it is plain that any one solution must be expressed analytically in two ways, in which the values of β' are complementary, and the values of α' differ by $90°$, since either principal axis of the ellipse belonging to the second stream may be that whose azimuth is α'. Accordingly, we may reject the second solution as being nothing more than the first expressed in a different way, and may therefore suppose $\beta' = -\beta$. The second and third equations then give $\cos 2\alpha' = -\cos 2\alpha$, $\sin 2\alpha' = -\sin 2\alpha$, and therefore α and α' differ by $90°$. The equations are indeed satisfied by $\beta = -\beta' = \pm 45°$, but this solution is only a particular case of the former.

It follows therefore that common light is equivalent to any two independent oppositely polarized streams of half the intensity; and no two independent polarized streams can together be equivalent to common light, unless they be polarized oppositely, and have their intensities equal.

19. We have seen that the nature of the mixture of a given group of independent polarized streams is determined by the values of the four constants A, B, C, D. Consider now the mixture of a stream of common light having an intensity J, and a stream, independent of the former, consisting of elliptically polarized light having an intensity J', and having α' for the azimuth of its plane of maximum polarization, and $\tan \beta'$ for the ratio of the axes of the ellipse which characterizes it.

By the preceding article, the stream of common light is equivalent to two independent streams, plane-polarized in azimuths $0°$ and $90°$, having each an intensity equal to $\frac{1}{2}J$. Hence, applying the formulæ (16) and (17) to the mixture, we have

$$2\mathfrak{m}\,(c_1)^2 = J + J' + J' \sin 2\beta' \sin 2\beta_1 + J' \cos 2\alpha' \cos 2\beta' \cos 2\alpha_1 \cos 2\beta_1$$
$$+ J' \sin 2\alpha' \cos 2\beta' \sin 2\alpha_1 \cos 2\beta_1 ;$$

and this mixture will be equivalent to the original group of polarized streams, provided

$$\left. \begin{array}{ll} J + J' = A ; & J' \sin 2\beta' = B ; \\ J' \cos 2\alpha' \cos 2\beta' = C ; & J' \sin 2\alpha' \cos 2\beta' = D. \end{array} \right\} \dots (18).$$

These equations give

$$J' = \sqrt{(B^2 + C^2 + D^2)}; \qquad J = A - \sqrt{(B^2 + C^2 + D^2)}; \\ \left.\sin 2\beta' = \frac{B}{\sqrt{(B^2 + C^2 + D^2)}}; \qquad \tan 2\alpha' = \frac{D}{C}.\right\} \quad \ldots(19).$$

These formulæ can always be satisfied, and therefore it is always possible to represent the given group by a stream of common light combined with a stream of elliptically polarized light independent of the former. Moreover, there is only one way in which the group can be so represented. For, though the third of equations (19) gives two values for β' complementary to each other, these values, as before explained, lead only to two ways of expressing the same result. If we choose that value of β which is numerically the smaller, then among the different values of α', differing by 90°, which satisfy the fourth equation, we must choose one which gives to $\cos 2\alpha'$ the same sign as C.

20. Let us now apply the principles and formulæ which have just been established to a few examples. And first let us take one of the fundamental experiments by which MM. Arago and Fresnel established the laws of interference of polarized light, or rather an analogous experiment mentioned by Sir John Herschel. The experiment selected is the following.

Two neighbouring pencils of common light from the same source are made to form fringes of interference. A tourmaline, carefully worked to a uniform thickness, is cut in two, and its halves interposed in the way of the two streams respectively. It is found that when the planes of polarization of the two tourmalines are parallel the fringes are formed perfectly; but as one of the tourmalines is turned round in azimuth the fringes become fainter, and at last, when the planes of polarization become perpendicular to each other, the fringes disappear.

Let the planes of polarization of the tourmalines be inclined at an angle α, and let it be required to investigate an expression for the intensity of the fringes. Since common light is equivalent to two independent streams, of equal intensity, polarized in opposite ways, let the original light be represented by two independent streams, having each an intensity equal to unity, polarized in planes respectively parallel and perpendicular to the plane of polarization of the first tourmaline. If c, c' be the coefficients of vibration in the two streams respectively, $c \cos \alpha$, $c' \sin \alpha$ will be the coefficients of the resolved parts in the direction of the vibrations transmitted by the second tourmaline. Hence we shall have, mixing together, two independent streams, in one of which the temporary intensity fluctuates between the limits $(c \pm c \cos^2 \alpha)^2 + (c \cos \alpha \sin \alpha)^2$ as the interval of retardation changes in passing from one point to another in the field of view. The temporary intensity of the other stream being $(c' \sin \alpha)^2$, the intensity at different points will fluctuate between the limits

$$\mathfrak{m} \, (c^2 \pm 2c^2 \cos^2 \alpha + c^2 \cos^2 \alpha) + \mathfrak{m} \, (c' \sin \alpha)^2.$$

It is needless to take account of the absorption which takes place even on the pencils which the tourmalines do transmit, because it affects both pencils equally. Since $\mathfrak{m} \, (c^2) = \mathfrak{m} \, (c'^2) = 1$, we have for the limits of fluctuation of the intensity $2 \pm 2 \cos^2 \alpha$. When $\alpha = 0$ these limits become 4 and 0, and the interference is perfect. When $\alpha = 90°$ the limits coalesce, becoming each equal to 2, and there are no fringes. As α increases from 0 to 90°, the superior limit continually decreases, and the inferior increases, and consequently the fringes become fainter and fainter.

21. It is a well-known law of interference that if two rays of common light from the same source be polarized in rectangular planes, and afterwards be brought to the same plane of polarization, or in other words, analyzed so as to retain only light polarized in a particular plane, they will not interfere, but if the light be primitively polarized in one plane they will interfere. This law seems to have presented a difficulty to some, because, it would be argued, the most general kind of vibrations are elliptic, so that we must suppose the vibrations of the ether in the case of common light to be of this kind; and yet the phenomena of interference are exhibited perfectly well if the light be at first elliptically polarized instead of plane-polarized. For my own part, I never could see the difficulty, but on the other hand it seems to me that it would be an immense difficulty were the law anything else than what it is. For, if we consider the rectangular components of the vibrations which make up common light, these components being measured along any two rectangular axes perpendicular to the ray, we must suppose them to be independent of each other, or at least to

have no fixed relation to each other so far as regards the changes in the mode of vibration, which we must suppose to be taking place continually, though slowly, it may be, in comparison with the time of a luminous vibration. To suppose otherwise would be contrary to the idea of common light, in which it is implied that on the average whatever we can say of one plane passing through the ray, we can say of another: whatever we can say of the direction one way round we can say of the other way round.

At the end of his excellent Tract on the Undulatory Theory, Mr Airy has shewn how the simple supposition of the existence, in common light, of successive series of undulations, in which the vibrations of one series have no relation to those of another, would account at the same time for the interference of common light and the non-interference of the pencils, polarized in rectangular planes, into which common light may be conceived to be decomposed. But he has, I think, introduced a gratuitous difficulty into the subject, by asserting that it is necessary to suppose the transition from one series into another to be abrupt, and that a gradual change in the nature of the vibrations is inadmissible. This assertion, which seems to have led others to conceive that there was here a difficulty with which the undulatory theory had to contend, seems to have resulted from an investigation from which it appeared that common light could not be represented by an indefinite series of elliptic vibrations, in which the major axis of the ellipse was supposed to revolve uniformly, rapidly, with regard to the duration of impressions on the retina, though slowly with regard to the time of a luminous vibration. I have elsewhere pointed out on what grounds I conceive that the instance of the revolving ellipse is not a case in point, namely, that it is not a fair representation of common light, because it gives a preponderance on the average to one direction of revolution over the opposite, which is contrary to the idea of common light*. Let us now apply the general formulæ (16) and (17) to this case.

Let $c \cos \beta$, $c \sin \beta$ be the semi-axes of the ellipse, α the azimuth of the first axis at a given time. So far as relates to

---

* Report of the meeting of the British Association at Swansea in 1848. Transactions of the Sections, p. 5.

the stream for which the azimuth of the first axis lies between α and $\alpha + d\alpha$, we have $\mathfrak{m}\,(c^2) = (2\pi)^{-1}\,c^2 d\alpha$; and in the application of the formulæ (16) the summation, to which Σ refers, will of course pass into an integration. We have therefore

$$A = \frac{c^2}{2\pi} \int_0^{2\pi} d\alpha = c^2 \,;$$

$$B = \frac{c^2}{2\pi} \sin 2\beta \int_0^{2\pi} d\alpha = c^2 \sin 2\beta \,;$$

$$C = 0 \,; \qquad\qquad D = 0 \,;$$

whence we get from the formulæ (19), supposing β positive,

$$J' = c^2 \sin 2\beta \,; \qquad J = c^2 (1 - \sin 2\beta) \,; \qquad \beta' = 45^\circ \,;$$

while α' remains indeterminate. Hence the mixture is equivalent, not to common light alone, but to a stream of common light having an intensity equal to $c^2 (1 - \sin 2\beta)$, combined with a stream of circularly polarized light, independent of the former, having an intensity equal to $c^2 \sin 2\beta$, and being of the same character as regards right-handed or left-handed as the original stream would be were the ellipse stationary. The result of supposing β negative is here assumed as obvious.

22. Suppose that a polarizing prism and a mica plate, which produce elliptic polarization, are made to revolve together with great rapidity. The stream of light thus produced will be equivalent to the former. The only difference is that in the former case c was supposed constant, whereas in the case of actual experiment it will be subject to the fluctuations mentioned at the beginning of this paper; but the mean values represented by $\mathfrak{m}$ will not be affected when these fluctuations are taken into account, and therefore the same formulæ will continue to apply. Hence, if the polarization be circular the rotation will make no difference; if it be plane, the light will appear completely depolarized; in intermediate cases the result will be intermediate, and the light will be equivalent to a mixture of common light and circularly polarized light. The reader may compare these conclusions of theory with some experiments by Professor Dove*.

23. As a last example, let light polarized by transmission through a Nicol's prism be transmitted through a second Nicol's

* See *Philosophical Magazine*, Vol. xxx. (1847) p. 465.

prism, which is made to revolve uniformly and rapidly while the first remains fixed.

Let α be the azimuth of the plane of polarization of the second Nicol's prism, measured from that of the first, c the coefficient of vibration in the stream transmitted through the first prism. The stream passing through the second prism is made up of an infinite number of independent streams such as that whose intensity is $(2\pi)^{-1}\cos^2\alpha\, d\alpha$ multiplied by the mean value of c^2. Hence we have from the formulæ (16)

$$A = \tfrac{1}{2}\mathfrak{m}\,(c^2); \qquad B = 0; \qquad C = \tfrac{1}{4}\mathfrak{m}\,(c^2); \qquad D = 0;$$

whence, taking the intensity of the original stream as unity, we have

$$J' = \tfrac{1}{4}; \qquad J = \tfrac{1}{4}; \qquad \beta' = 0; \qquad \alpha' = 0;$$

or the light is equivalent to a mixture of common light having an intensity $\tfrac{1}{4}$, and light independent of the former, of the same intensity, polarized at an azimuth zero. This result may be compared with one of Professor Dove's experiments.

If a rotating crystalline plate, cut from a doubly refracting crystal in a direction not nearly coinciding with the optic axis or one of the optic axes, and not sufficiently thin to exhibit colours in polarized light, be substituted for the rotating Nicol's prism, since the plate is too thick to allow of the exhibition of any phenomena depending on the interference of the oppositely polarized pencils, the effect will be just the same as in the case of the Nicol's prism, only the intensity of each stream will be doubled.

In applying the formulæ of this paper to experiments in which one part of an optical train is made to revolve rapidly, it must be understood that the other parts of the train are at rest, or at least do not revolve with a velocity nearly equal to the former. Otherwise, particular phenomena will be exhibited depending on the simultaneous movements of two or more parts of the train, as appeared in Professor Dove's experiments; and in the calculation of these phenomena it will not be allowable to substitute for the stream of light emerging from the polarizer, or first revolving piece whatever it be, the streams of common and elliptically polarized light to which, for general purposes, it is equivalent.

[From the *Proceedings of the Royal Society*, Vol. vi., p. 195.]

ABSTRACT OF A PAPER "ON THE CHANGE OF REFRANGIBILITY
OF LIGHT*."

[Read *May* 27, 1852.]

THE author was led into the researches detailed in this paper
by considering a very singular phenomenon which Sir John
Herschel had discovered in the case of a weak solution of sulphate
of quinine, and various other salts of the same alkaloid. This
fluid appears colourless and transparent, like water, when viewed
by transmitted light, but exhibits in certain aspects a peculiar
blue colour. Sir John Herschel found that when the fluid was
illuminated by a beam of ordinary daylight, the blue light was
produced only throughout a very thin stratum of fluid adjacent to
the surface by which the light entered. It was unpolarized. It
passed freely through many inches of the fluid. The incident
beam, after having passed through the stratum from which the
blue light came, was not sensibly enfeebled nor coloured, but yet
it had lost the power of producing the usual blue colour when
admitted into a solution of sulphate of quinine. A beam of light
modified in this mysterious manner was called by Sir John Herschel
epipolized.

Several years before Sir David Brewster had discovered in the
case of an alcoholic solution of the green colouring matter of
leaves a very remarkable phenomenon, which he has designated

* [As the paper in question is of considerable length, and contains a good deal
of detail, it seemed desirable in this case to reprint the abstract as well as the
paper itself. In the abstract the reader will find in short compass the more
important of the results contained in the paper.]

as *internal dispersion*. On admitting into this fluid a beam of sunlight condensed by a lens, he was surprised by finding the path of the rays within the fluid marked by a bright light of a blood-red colour, strangely contrasting with the beautiful green of the fluid itself when seen in moderate thickness. Sir David afterwards observed the same phenomenon in various vegetable solutions and essential oils, and in some solids. He conceived it to be due to coloured particles held in suspension. But there was one circumstance attending the phenomenon which seemed very difficult of explanation on such a supposition, namely, that the whole or a great part of the dispersed beam was unpolarized, whereas a beam reflected from suspended particles might be expected to be polarized by reflexion. And such was, in fact, the case with those beams which were plainly due to nothing but particles held in suspension. From the general identity of the circumstances attending the two phenomena, Sir David Brewster was led to conclude that epipolic was merely a particular case of internal dispersion, peculiar only in this respect, that the rays capable of dispersion were dispersed with unusual rapidity. But what rays they were which were capable of affecting a solution of sulphate of quinine, why the active rays were so quickly used up, while the dispersed rays which they produced passed freely through the fluid, why the transmitted light when subjected to prismatic analysis showed no deficiencies in those regions to which, with respect to refrangibility, the dispersed rays chiefly belonged, were questions to which the answers appeared to be involved in as much mystery as ever.

After having repeated some of the experiments of Sir David Brewster and Sir John Herschel, the author could not fail to take a most lively interest in the phenomenon. The firm conviction which he felt that two portions of light were not distinguishable as to their nature otherwise than by refrangibility and state of polarization, left him but few hypotheses to choose between, respecting the explanation of the phenomenon. In fact, having regarded it at first as an axiom that dispersed light of any particular refrangibility could only have arisen from light of the same refrangibility contained in the incident beam, he was led by necessity to adopt hypotheses of so artificial a character as to render them wholly improbable. He was thus compelled to adopt the other alternative, namely, to suppose that in the process of internal dispersion the refrangibility of light had been changed. Startling as such a supposition might appear at first sight, the ease with which it accounted for the whole phenomenon was such as already to produce a strong probability of its truth. Accordingly the author determined to put this hypothesis to the test of experiment.

The experiments soon placed the fact of a change of refrangibility beyond all doubt. It would exceed the limits of an abstract like the present to describe the various experiments. It will be sufficient to mention some of the more remarkable results.

A pure spectrum from sunlight having been formed in air in the usual manner, a glass vessel containing a weak solution of sulphate of quinine was placed in it. The rays belonging to the greater part of the visible spectrum passed freely through the fluid, just as if it had been water, being merely reflected here and there from motes. But from a point about half-way between the fixed lines G and H to far beyond the extreme violet the incident rays gave rise to light of a sky-blue colour, which emanated in all directions from the portion of the fluid which was under the influence of the incident rays. The anterior surface of the blue space coincided of course with the inner surface of the vessel in which the fluid was contained. The posterior surface marked the distance to which the incident rays were able to penetrate before they were absorbed. This distance was at first considerable, greater than the diameter of the vessel, but it decreased with great rapidity as the refrangibility of the incident rays increased, so that from a little beyond the extreme violet to the end the blue space was reduced to an excessively thin stratum adjacent to the surface by which the incident rays entered. It appears therefore that this fluid, which is so transparent with respect to nearly the whole of the visible rays, is of an inky blackness with respect to the invisible rays more refrangible than the extreme violet. The fixed lines belonging to the violet and the invisible region beyond were beautifully represented by dark planes interrupting the blue space. When the eye was properly placed, these planes were of course projected into lines. The author has made a sketch of these fixed lines, which accompanies the paper. They may be readily identified with the fixed lines represented in M. Becquerel's map of the fixed lines of the chemical spectrum. The last line seen in a solution of sulphate of quinine appears to be the line next beyond the last represented in M. Becquerel's map. Under very favourable circumstances two dusky bands were seen still further on. Several circumstances led the author to conclude that in all probability fixed lines might be readily seen corresponding to still more refrangible rays, were it not for the opacity of glass with respect to those rays of very high refrangibility.

It is very easy to prove experimentally that the blue dispersed light corresponding to any particular part of the incident spectrum is not homogeneous light, having a refrangibility equal to that of the incident rays, and rendered visible in consequence of its complete isolation; but that it is in fact heterogeneous light, consisting of rays extending over a wide range of refrangibility, and not passing beyond the limits of refrangibility of the spectrum visible under ordinary circumstances. To show this it is sufficient to isolate a part of the incident spectrum, and view the narrow beam of dispersed light which it produces through a prism held to the eye.

In Sir David Brewster's mode of observation, the beam of light which was of the same nature as the blue light exhibited by a solution of sulphate of quinine was necessarily mixed with the beam due merely to reflexion from suspended particles; and in the case of vegetable solutions, a beam of the latter kind almost always exists, to a greater or less degree. But in the method of observation employed by the author, to which he was led by the discovery of the change of refrangibility, the two beams are exhibited quite distinct from one another. The author proposes to call the two kinds of internal dispersion just mentioned *true internal dispersion* and *false internal dispersion*, the latter being nothing more than the scattering of light which is produced by suspended particles, and having, as is now perfectly plain, nothing to do with the remarkable phenomenon of true internal dispersion.

Now that the nature of the latter phenomenon is better known, it is of course possible to employ methods of observation by which it may be detected even when only feebly exhibited. It proves to be almost universal in vegetable solutions, that is, in solutions made directly from various parts of vegetables. When vegetable products are obtained in a state of isolation, their solutions sometimes exhibit the phenomenon and sometimes do not, or at least exhibit it so feebly that it is impossible to say whether what they do show may not be due to some impurity. Among fluids which exhibit the phenomenon in a high degree, or according to the author's expression are highly *sensitive*, may be mentioned a weak decoction of the bark of the horse-chestnut, an alcoholic extract from the seeds of the *Datura stramonium*, weak tincture of turmeric, and a decoction of madder in a solution of alum. In these cases the general character of the dispersion resembles that exhibited by a solution of sulphate of quinine, but the tint of the dispersed light, and the part of the spectrum at which the dispersion begins, are different in different cases. In the last fluid, for example, the dispersion commences somewhere about the fixed line *D*, and continues from thence onwards far beyond the extreme violet. The dispersed light is yellow, or yellowish orange.

In the case of other fluids, however, some of them sensitive in a very high degree, the mode in which light is dispersed internally presents some very remarkable peculiarities. One of the most singular examples occurs in the case of an alcoholic solution of the green colouring matter of leaves. This fluid disperses a rich red light. The dispersion commences abruptly about the fixed line *B*, and continues from thence onwards throughout the visible spectrum and a little beyond. The dispersion is subject to fluctuations intimately connected with the singular absorption bands exhibited by this medium.

In order that a medium should be capable of changing the refrangibility of light incident upon it, it is not necessary that the medium should be a fluid, or a clear solid. Washed papers and other opaque substances produce the same effect, but of course the mode of observation must be changed. The author has observed the change of refrangibility in various ways. It will be sufficient to mention here that which was found most generally useful, which he calls the method of observing by a *linear spectrum*. The method is as follows.

A series of prisms and a lens are arranged in the usual manner for forming a pure spectrum, but the slit by which the light enters, instead of being parallel, is placed in a direction perpendicular to the edges of the prisms. A linear spectrum is thus formed at the focus of the lens, consisting of an infinite succession of images of the slit arranged one after the other in the order of refrangibility, and of course overlapping each other to a certain extent. The substance to be examined is placed in the linear spectrum, and the line of light seen upon it is viewed through a prism held to the eye. In this way it is found that almost all common organic substances, such as wood, cork, paper, calico, bone, ivory, horn, wool, quills, feathers, leather, the skin of the hand, the nails, are sensitive in a greater or less degree. Organic substances which are dark-coloured are frequently found to be insensible, but, on the other hand, scarlet cloth and various other dyed articles are highly sensitive. By means of a linear spectrum the peculiar dispersion of a red light produced by chlorophyll, or some of its modifications, may be observed not only in a solution, but in a green leaf, or on a washed paper, or in a sea-weed.

The highly sensitive papers obtained by washing paper with tincture of turmeric, or a solution of sulphate of quinine, or some other highly sensitive medium, display their sensibility in a remarkable manner when they are examined in a linear spectrum. In these cases, however, the paper produces a very striking effect when merely held so as to receive a pure spectrum formed in the usual manner, that is, with a slit parallel to the edges of the prisms. Such a paper may be used as a screen for showing the fixed lines belonging to the invisible rays, though they are not thus shown quite so well as by using a solution. The extraordinary prolongation of the spectrum seen when it is received on turmeric paper has been already observed by Sir John Herschel, by whom it was attributed to a peculiarity in the reflecting power of that substance. Of course it now appears that the true explanation is very different.

A high degree of sensibility appears to be rather rare among inorganic compounds. Certain specimens of fluor spar, as is already known, give a copious internal dispersion of a deep blue light; but this is plainly due to some foreign ingredient, the nature of which is at present unknown. But there is one class of inorganic compounds which are very remarkable for their sensibility, namely, certain compounds of peroxide of uranium, including the ornamental glass called canary glass, and the natural mineral yellow uranite. In these compounds the dispersed light is found on analysis to consist of bright bands arranged at regular intervals. A very remarkable system of absorption bands is also found among these compounds, which is plainly connected with the system of bright bands seen in the spectrum of the dispersed light. The connection between the absorption and internal dispersion exhibited by these compounds is very singular, and is of a totally different nature from the connection which has been already mentioned as occurring in solutions of the green colouring matter of leaves.

There is one law relating to the change of refrangibility which appears to be quite universal, namely, that the refrangibility of light is *always lowered* by internal dispersion.   The incident rays being homogeneous, the dispersed light is found to be more or less composite. Its colour depends simply on its refrangibility, having no relation to the colour of the incident light, or to the circumstance that the incident rays were visible or invisible.   The dispersed light appears to emanate in all directions, as if the solid or fluid were self-luminous while under the influence of the incident rays.

The phenomenon of the change of refrangibility of light admits of several important applications.   In the first place it enables us to determine instantaneously the transparency or opacity of a solid or fluid with respect to the invisible rays more refrangible than the violet, and that, not only for these rays as a whole, but for the rays of each refrangibility in particular.   For this purpose it is sufficient to form a pure spectrum with sun-light as usual, employing instead of a screen a vessel containing a decoction of the bark of the horse-chestnut, or a slab of canary glass, or some other highly sensitive medium, and then to interpose the medium to be examined, which, if fluid, would have to be contained in a vessel with parallel sides of glass.   Glass itself ceases to be transparent about the region corresponding to the end of the author's map, and to carry on these experiments with respect to invisible rays of still higher refrangibility would require the substitution of quartz for glass.   The reflecting power of a surface with respect to the invisible rays may be examined in a similar manner.

The effect produced on sensitive media leads to interesting information respecting the nature of various flames.   Thus, for example, it appears that the feeble flame of alcohol is extremely brilliant with regard to invisible rays of very high refrangibility. The flame of hydrogen appears to abound in invisible rays of still higher refrangibility.

By means of the phenomena relating to the change of refrangibility, the independent existence of one or more sensitive substances may frequently be observed in a mixture of various compounds. In this way the phenomenon seems likely to prove of value in the separation of organic compounds. The phenomena sometimes also afford curious evidence of chemical combinations; but this subject cannot here be further dwelt upon.

The appearance which the rays from an electric spark produce in a solution of sulphate of quinine, shows that the spark is very rich in invisible rays of excessively high refrangibility, such as would plainly put them far beyond the limits of the maps which have hitherto been made of the fixed lines in the chemical part of the solar spectrum. These rays are stopped by glass, but transmitted through quartz. These circumstances render it probable that the phosphorogenic rays of an electric spark are nothing more than rays of the same nature as those of light, but which are invisible, and not only so, but of excessively high refrangibility. If so, they ought to be stopped by a very small quantity of a substance known to absorb those rays with great energy. Accordingly the author found that while the rays from an electric spark which excite the phosphorescence of Canton's phosphorus pass freely through water and quartz, they are stopped on adding to the water an excessively small quantity of sulphate of quinine.

At the end of the paper the author explains what he conceives to be the cause of the change of refrangibility, and enters into some speculations to account for the law according to which the refrangibility of light is always lowered in the process of internal dispersion.

[From the *Philosophical Transactions* for 1852, p. 463.]

ON THE CHANGE OF REFRANGIBILITY OF LIGHT*.

[Read *May* 27, 1852.]

1. THE following researches originated in a consideration of the very remarkable phenomenon discovered by Sir John Herschel in a solution of sulphate of quinine, and described by him in two papers printed in the *Philosophical Transactions* for 1845, entitled "On a Case of Superficial Colour presented by a Homogeneous Liquid internally colourless," and "On the Epipolic Dispersion of Light." The solution of quinine, though it appears to be perfectly transparent and colourless, like water, when viewed by transmitted light, exhibits nevertheless in certain aspects, and under certain incidences of the light, a beautiful celestial blue colour. It appears from the experiments of Sir John Herschel that the blue colour comes only from a stratum of fluid of small but finite thickness adjacent to the surface by

---

* [In choosing the title of this paper, all that was intended was, to express the previously unsuspected origin of peculiar coloured lights which were exhibited by certain bodies, and which were in some cases matters of common observation, and had been subjected to examination by physicists. The capital fact now brought to light was that the colours in question were due to rays incident upon the body which were of a different refrangibility from those which constituted the colour observed. The title was meant to express a fact of observation, not any theory. It must have been from misapprehension on this point that a view as to the nature of the phenomenon has been attributed to me by more than one writer, a view which I have even been said to persist in, according to which the change of refrangibility takes place in the act of reflexion from the molecules of the body. Such a view is not mine, and never was, and is directly opposed to the views stated in the paper. The abstract, which was published some months before the paper, contains nothing about theoretical views, but for these refers to the paper. I can only conjecture that the view erroneously, though without hesitation, attributed to me was assumed to be mine in consequence of the title of the paper.]

which the light enters. After passing through this stratum, the incident light, though not sensibly enfeebled nor coloured, has lost the power of producing the same effect, and therefore may be considered as in some way or other qualitatively different from the original light. The dispersion which takes place near the surface of this liquid is called by Sir John Herschel *epipolic*, and he applies the term *epipolized* to a beam of light which, having been transmitted through a quiniferous solution, has been thereby rendered incapable of further undergoing epipolic dispersion. In one experiment, in which sun-light was used, a feeble blue gleam was observed to extend to nearly half an inch from the surface. As regards the dispersed light itself, when analysed by a prism it was found to consist of rays extending over a great range of refrangibility : the less refrangible extremity of the spectrum was however wanting. On being analysed by a tourmaline, it showed no signs of polarization. A special experiment showed that the dispersed light was perhaps incapable, at any rate not peculiarly susceptible, of being again dispersed.

2. In a paper " On the Decomposition and Dispersion of Light within Solid and Fluid Bodies," read before the Royal Society of Edinburgh in 1846, and printed in the 16th volume of their *Transactions,* as well as in the *Philosophical Magazine* for June 1848, Sir David Brewster notices these results of Sir John Herschel's, and states the conclusions, in some respects different, at which he had arrived by operating in a different way. The phenomenon of internal dispersion had been discovered by him some years before, and is briefly noticed in a paper read before the Royal Society of Edinburgh in 1833[*]. It is described at length, as exhibited in the particular case of fluor-spar, in a paper communicated to the British Association at Newcastle in 1838[†]. In Sir David Brewster's experiments the sun's light was condensed by a lens, and so admitted into the solid or fluid to be examined; which afforded peculiar facilities for the study of the phenomena. On examining in this way a solution of sulphate of quinine, it was found that light was dispersed, not merely close to the surface, but at a long distance within the fluid : and Sir David Brewster was led to conclude that the dispersion produced

---

[*] *Edinburgh Transactions*, Vol. XII. p. 542.

[†] Eighth Report.—Transactions of the Sections, p. 10.

by sulphate of quinine was only a particular case of the general phenomenon of internal dispersion. On analysing the blue beam by a rhomb of calcareous spar, it was found that a considerable portion of it, consisting chiefly of the less refrangible rays, was polarized in the plane of reflexion, while the more refrangible of its rays, constituting an intensely blue beam, had a different polarization.

3. On repeating some of Sir John Herschel's experiments, I was immediately satisfied of the reality of the phenomenon, notwithstanding its mysterious nature, that is to say, that an epipolized beam of light is in some way or other qualitatively different from the light originally incident on the fluid. On making the observation in the manner of Sir David Brewster, it seemed no less evident that the phenomenon belonged to the class of internal dispersion*. Nevertheless, the singular phenomenon discovered by Sir John Herschel manifested itself even in this mode of observation. If indeed the vessel containing the solution were so placed that the image of the sun in the focus of the lens lay a little way inside the fluid, the phenomenon was masked, because the increase of intensity due to an increase of concentration in approaching the focus made up for the decrease of intensity due to passing out of the blue band. But when the vessel was moved so that the focus of the lens fell either further inside the fluid or else outside the vessel, the narrow blue band adjacent to the surface was seen *as well as* the blue beam which shot far into the fluid. Light which has been "epipolized" by transmission through a moderate thickness of the solution is indeed capable of undergoing further dispersion, but not *epipolic* dispersion, *if that term be restricted* to the dispersion by which the narrow blue band is produced. It was no doubt of great importance to assign to the phenomenon its true place as a member of the class of phenomena of internal dispersion. Never-

---

* By this, I merely mean that, to take a particular example, the exhibition of a blue light by a solution of sulphate of quinine appeared to be a phenomenon of the same nature as the exhibition of a red light by a solution of the green colouring matter of leaves, although the latter does not manifest the same singular concentration as the former in the neighbourhood of the surface by which the light enters; and the latter had already been observed by Sir David Brewster, and the phenomenon designated as *internal dispersion*. I make this remark because Sir David Brewster has applied this same term to another class of phenomena which are totally different.

theless the mystery was by no means cleared up; rather, we were prepared to expect something of the same sort in other instances of internal dispersion. In fact, the mystery consisted, not in the narrowness of the stratum from which most of the blue light came, but in the circumstance that it was possible for light, by passing across such a stratum, to be deprived of the power of producing the same effect again, without, apparently, being altered in any other respect.

4. To one who regards light as a subtle and mysterious agent, of which the laws indeed are in a good measure known to us, but respecting the nature of which we are utterly ignorant, the phenomenon might seem merely to make another striking addition to the modes of decomposition with which we were already acquainted. But in the mind of one who regards the theory of undulations as being for light what the theory of universal gravitation is for the motions of the heavenly bodies, it was calculated to excite a much more lively interest. Whatever difficulty there might be in explaining how the effect was produced, we ought at least to be able to say what the effect was that had been produced; wherein, for example, epipolized light differed from light which had not undergone that modification.

In speculating on the nature of the phenomenon, there is one point which deserves especial attention. Although the passage through a thickness of fluid amounting to a small fraction of an inch is sufficient to purge the incident light from those rays which are capable of producing epipolic dispersion, the dispersed rays themselves traverse many inches of the fluid with perfect freedom. It appears therefore that the rays producing dispersion are in some way or other of a different nature from the dispersed rays produced. Now, according to the undulatory theory, the nature of light is defined by two things, its period of vibration, and its state of polarization. To the former corresponds its refrangibility, and, so far as the eye is a judge of colour, its colour*. To a change,

---

* It has been maintained by some philosophers of the first eminence that light of definite refrangibility may still be compound, and though no longer decomposable by prismatic refraction might still be so by other means. I am not now speaking of compositions and resolutions depending upon polarization. It has even been suggested by the advocates of the undulatory theory, that possibly a difference of properties in lights of the same refrangibility might correspond to a difference in the law of vibration, and that lights of given refrangibility may differ

then, either in the refrangibility or in the state of polarization we are to look for an explanation of the phenomenon.

5.   Regarding it at first as an axiom that the dispersed light of any given refrangibility could only have arisen from light of the same refrangibility contained in the incident beam, I was led to look in the direction of polarization for the required change in the nature of the light.   Since a fluid has no axes, circular polarization is the only kind which can here come into play.   As some fluids are doubly refracting, transmitting right-handed and left-handed circularly polarized light with different velocities, so, it might be, this fluid was doubly absorbing, absorbing say right-handed circularly polarized light of certain refrangibilities with great energy, and freely transmitting left-handed.   The right-handed light, absorbed, in the sense of withdrawn from the incident beam, might have been more strictly speaking scattered, and thereby depolarized.   The common light so produced would be equivalent to two streams, of equal intensity, one of right-handed, and the other of left-handed circularly polarized light. Of these the latter would be freely transmitted, while the former would be scattered anew, and so on.   Yet this hypothesis, sufficiently improbable already, was not enough.   New suppositions were still required, to account for the circumstance that an epipolized beam, when subjected to prismatic analysis with a low magnifying power, exhibited no bands of absorption in the region to which, as regards their refrangibility, the dispersed rays principally belong ; so that altogether this theory bore not the slightest semblance of truth.

6.   I found myself thus fairly driven to suppose that the change of nature consisted in a change of refrangibility.   From the time of Newton it had been believed that light retains its refrangibility through all the modifications which it may undergo.

in tint, just as musical notes of given pitch differ in quality.   Were it not for the strong conviction I felt that light of definite refrangibility is in the strict sense of the word homogeneous, I should probably have been led to look in this direction for an explanation of the remarkable phenomena presented by a solution of sulphate of quinine.   It would lead me too far from the subject of the present paper to explain the grounds of this conviction.   I will only observe that I have not overlooked the remarkable effect of absorbing media in causing apparent changes of colour in a pure spectrum ; but this I believe to be a subjective phenomenon, depending upon contrast,

Nevertheless it seemed to me less improbable that the refrangibility should have changed, than that the undulatory theory should have been found at fault. And when I reflected on the extreme simplicity of the whole explanation if only this one supposition be admitted, I could not help feeling a strong expectation that it would turn out to be true. In fact, we have only to suppose that the invisible rays beyond the extreme violet give rise by internal dispersion to others which fall within the limits of refrangibility between which the retina of the human eye is affected, and the explanation is obvious. The narrowness of the blue band observed by Sir John Herschel would merely indicate that the fluid, though highly transparent with regard to the visible rays, was nearly opaque with regard to the invisible. According to the law of continuity, the passage from almost perfect transparency to a high degree of opacity would not take place abruptly; and thus rays of intermediate refrangibilities might produce the blue gleam noticed by Sir John Herschel, or the blue cylinder, or rather cone, observed by Sir David Brewster. We should thus, too, have an immediate explanation of a remarkable circumstance connected with the blue band, namely that it can hardly be seen by strong candle-light, though readily seen by even weak daylight. For candle-light, as is well known, is deficient in the chemical rays situated beyond the extreme violet.

7.   My first experiments were made with coloured glasses. A test tube was about half filled with a solution consisting of disulphate of quinine dissolved in 200 times its weight of water acidulated with sulphuric acid. The tube, having been first covered with black paper, with the exception of a hole by which the light might enter, was placed in a vertical position in front of a window, the hole being turned towards the light. On looking down from above, in a direction nearly parallel to the surface of the glass, a blue arc was well seen, extending only a very short distance into the fluid, and situated immediately behind the hole. As this arc, though extremely distinct, was not of course what could be called brilliant, I did not at first venture, for the experiment I had in view, to use any but pale glasses. Having no direct means of determining which were opaque with regard to the invisible rays situated beyond the extreme violet, I sought among a collection of orange, yellow, and brown glasses, which, from transmitting mainly the less refrangible rays, seemed

the most likely to absorb the chemical rays. I presently found a pale smoke-coloured glass, which, when placed immediately in front of the hole, prevented the formation of the blue arc, although when placed immediately in front of the eye it transmitted a large proportion of the light of which the arc consisted. The colour of the arc was of course modified, and rendered more nearly white.

On trying other pale glasses, I found one of a puce colour, which, when placed in front of the hole, allowed the arc to be formed, though it absorbed it when placed in front of the eye. A yellow, and likewise a yellowish green glass allowed the arc to be seen in both positions; but its colour was decidedly different according as the glass was placed in front of the hole or in front of the eye. The breadth, too, of the arc was differently affected by different coloured glasses placed in front of the hole, some causing the light to be more, and others less concentrated towards the surface of the test tube than when the incident light was unimpeded.

8. The sun's light was next reflected horizontally into a darkened room, and allowed to pass through a hole in a vertical board which was placed in the window. The hole contained a lens of rather short focus. On placing a test tube containing the solution, in a vertical position, in front of the lens, at such a distance that the focus lay some way inside the fluid, the narrow blue band described by Sir John Herschel and the blue beam mentioned by Sir David Brewster were seen independently of each other. On trying different coloured glasses, which were placed, first in front of the fluid, and then in front of the eye, it was found that the blue beam, as had previously proved to be the case with the narrow band, was for the most part differently affected according as the glass was placed so as to intercept the incident or the dispersed light. Moreover, the long blue beam and the narrow band did not behave in the same manner under the action of the same coloured glass.

9. To my own mind these experiments were conclusive as to the fact of a change of refrangibility. Admitting that the effect of a coloured glass is simply to stop a certain fraction of the incident light, that fraction being a function of the refrangibility, it is plain that the results can be explained in no other way. It

must be confessed however that these results are merely an extension of that which precisely constitutes the peculiarity of the phenomenon. For, take the case of the narrow blue band formed by ordinary daylight. Imagine a glass vessel with parallel sides to be filled with a portion of the solution, and placed so as to intercept, first the incident, and then the dispersed light. In the first position the light incident on the fluid under examination would be "epipolized" by transmission through the fluid contained in the vessel, and therefore the blue band would be cut off, whereas when the vessel was held in front of the eye the blue band would be freely transmitted. Hence the effects of the coloured glasses are analogous to, but less striking than, the effect of a stratum of the solution of sulphate of quinine in the imaginary experiment above described. There is to be sure one important difference in the two cases, namely, that in the case of the stratum of fluid the epipolic dispersion which is prevented in the fluid under examination is produced near the first surface of the stratum, whereas no such dispersion is produced, or at any rate necessarily produced, in the coloured glasses. Whatever the reader may think of the results obtained with coloured glasses, the next experiment it is presumed will be deemed conclusive.

10.   The board in the window containing the lens having been replaced by a pair of boards adapted to form a vertical slit, the sun's light was reflected horizontally through the slit, and transmitted through three Munich prisms placed one after the other close to it. A tolerably pure spectrum was thus formed at the distance of some feet from the slit. A test tube containing the solution was then placed vertically a little beyond the extreme red of the spectrum, and afterwards gradually moved horizontally through the colours. Throughout nearly the whole of the visible spectrum the light passed through the fluid as it would have done through so much water; but on arriving nearly at the violet extremity a ghost-like gleam of pale blue light shot right across the tube. On continuing to move the tube, the blue light at first increased in intensity and afterwards gradually died away. It did not however cease to appear until the tube had been moved far beyond the violet extremity of the spectrum visible on a screen. Before disappearing, the blue light was observed to be confined to an excessively thin stratum of fluid adjacent to the surface by which the light entered, whereas when it first appeared, namely

when the tube was placed a little short of the extreme violet, the blue light had extended completely across it. It was certainly a curious sight to see the tube instantaneously lighted up when plunged into the invisible rays: it was literally *darkness visible*. Altogether the phenomenon had something of an unearthly appearance.

11. Since the fluid is so intensely opaque with regard to rays of extreme refrangibility, it might be expected, that, though it appears transparent and colourless when examined merely by viewing a white object through it, it would yet exhibit a very sensible absorbing action with regard to the extreme violet rays when subjected to prismatic analysis. To try whether such were really the case, I reflected the sun's light horizontally through a slit, at which was placed a test tube filled with the liquid, and analysed the line of light by a prism, the eye being defended by a deep blue glass. I was barely able to make out the fixed line H, that is, the less refrangible band of the pair, although in similar circumstances I can generally see about as far beyond the more refrangible band as it is beyond H. However, to make the result more decisive by using a greater thickness, as well as to render the observation strictly differential, I placed a tumbler filled with water behind the slit, the blue glass before it, and then viewed the slit through the prism. I saw as far as usual into the violet. The water was then poured out and replaced by the solution of sulphate of quinine, which, when viewed by transmitted light, appeared as transparent as the water which it had replaced. When the tumbler was now placed behind the slit, the blue beam of dispersed light was observed to extend quite across it, a distance of about three inches, and would evidently have gone much further. On viewing the slit through the prism, the spectrum was found to be cut off about half-way between the fixed lines G and H. The termination was pretty definite, which indicates that, at least for that part of the spectrum, the absorbing energy of the fluid rapidly increased with the refrangibility of the light; there, however, an evident diminution of intensity produced by the fluid, extending from the termination of the spectrum to near G.

12. There could no longer be any doubt, either as to the fact of a change of refrangibility, or as to the explanation thereby of the remarkable phenomenon exhibited by sulphate of quinine.

18—2

Epipolized light is merely light which has been purged of the invisible, or at most feebly illuminating rays more refrangible than the violet; and the term itself, which in fact was only adopted provisionally by Sir John Herschel, and which has now served its purpose, may henceforth be discarded, especially as it is calculated to convey a false impression respecting the cause of the phenomenon. It remained to examine other instances of internal dispersion, of which, according to Sir David Brewster's observations, the dispersion produced by sulphate of quinine is only a particular case; to endeavour to make out the laws according to which a change of refrangibility takes place; and, if possible, to account for these laws on mechanical principles.

13. In giving an account of my further experiments, I think it best to describe in detail the phenomena observed in some of the more remarkable instances of internal dispersion before attempting to draw any general conclusions. It will save repetition to explain in the first instance the methods of observation employed, which on the whole may very fairly be divided into four, though occasionally it was convenient to employ intermediate methods, or a combination of two of them. Of course I frequently availed myself of Sir David Brewster's method of observation, in which the effect of the incident light is studied as a whole; but the methods here referred to relate to an investigation of the separate offices of the portions of light of different degrees of refrangibility which are found in the incident beam. As my researches proceeded, new methods of observation suggested themselves, but these will be described in their place.

Methods of Observation employed.

FIRST METHOD.—The sun's light was reflected horizontally through a small lens, which was fixed in a hole in a vertical board. The cone of emergent rays was allowed to enter the solid or fluid examined. A coloured glass or other absorbing medium was then placed, first so as to intercept the incident rays, and then between the substance examined and the eye. For shortness' sake these positions will be designated as *the first* and *the second*. Sometimes a coloured glass was allowed to remain in front of the hole, and a second glass was added, first in front of the hole and then in front of the eye.

SECOND METHOD.—The sun's light, reflected as before, was transmitted through a series of three or four Munich prisms placed one immediately after the other, and each nearly in the position of minimum deviation. It was then transmitted through a small lens in a board close to the last prism, and so allowed to enter the body to be examined, which was generally placed so that the first surface coincided, or nearly so, with the focus of the lens. The diameter of the lens was much smaller than the breadth or height of the prisms, so that the lens was completely filled with white light, the component parts of which however entered in different directions. Regarding the image of the sun in the focus of the small lens as a point, we may conceive the light incident on the body under examination as consisting of a series of cones, corresponding to different refrangibilities, the axes of which lay in a horizontal plane and intersected in the centre of the lens, the vertices being arranged in a horizontal line near the surface of the body examined.

THIRD METHOD.—The sun's light was reflected horizontally through a vertical slit, and received on the prisms, which were arranged as before, but placed at the distance of several feet from the slit. A large lens of rather long focus was placed immediately after the last prism, with its plane perpendicular, or nearly so, to the beam of light which had passed through the prisms, and with its centre about the middle of this beam. The body examined was placed at the distance of the image of the slit, or nearly so.

FOURTH METHOD.—Everything being arranged as in the third method, a board with a small lens of short focus was placed at the distance of the image of the slit, or between that and the image of the sun, which was a little nearer to the prisms, inasmuch as the focal length of the large lens commonly employed, though much smaller, was not incomparably smaller than the distance of the lens from the slit. A second slit was generally added immediately in front of the small lens. The body examined was placed at the focus of the small lens. The dispersed light was viewed from above, and analysed by a prism, being refracted sideways.

The object of these several arrangements will appear in the course of the paper. The prisms employed consisted, three of

them of flint glass and one of crown. The refracting angles of
the former were about 43°, 33°, and 24°, and that of the latter
about 45°. The refracting faces of the smallest of the prisms
(the flint of 43°) were 1·35 inch high and 1·60 long. The small
lens used was one or other of a pair of which the apertures were
0·34 inch and 0·22 inch, and the focal lengths 0·75 inch and
0·50 inch. The focal length of the large lens generally used was
about twelve inches. Once or twice a lens was tried which had
a focal length about three times as great, but the light proved
too faint for most purposes. In the third method it was sometimes convenient to employ a lens of only 6½ inches focal length,
but the 12-inch lens was employed in the fourth method, except
on a few occasions, when the lens of 36 inches focal length was
used. With the 12-inch lens the length of the spectrum from the
fixed line B to H was usually about an inch and a quarter.

It will be convenient for the purposes of this paper to employ
certain terms in a particular sense, but as some of these terms
relate to phenomena which have not yet been described, it will
be well previously to relate in detail what was observed in one
remarkable instance of internal dispersion.

Solution of Sulphate of Quinine.

14. The effects of some pale coloured glasses in the case
of this fluid have already been mentioned. But there is one glass
of which the effect is still more striking. It is well known that
a deep cobalt blue glass is highly transparent with regard to the
chemical rays. Accordingly I found that a blue glass, so deep
that only the brighter objects in a room could be seen through
it, produced but very little effect when placed so as to intercept
the light incident on the fluid. When placed immediately in
front of the eye, at first everything disappeared except the light
reflected from the convexities of the glass tube; but when the
eye became a little accustomed to the darkness it was possible
to make out the existence of the band. The contrast between
the effects of this glass and of the pale brown glass already
mentioned was most striking.

15. When the fluid was examined by the second method,
the dispersed light was found to consist of two beams, separated

from each other at their entrance into the fluid, that is, at the vertical surface of separation of the fluid and the containing vessel, and afterwards still further separated by divergence. Of course each beam must have been made up of a series of cones having their axes diverging from the centre of the lens, and their vertices situated at its focus. The first beam, or that which was produced by light of less refrangibility, consisted of the brighter colours of the spectrum in their natural order. It had a discontinuous, sparkling appearance, and was plainly due merely to motes which were suspended in the fluid. On being viewed from above through a Nicol's prism, it was found to consist chiefly of light polarized in the plane of reflexion. Taken as a whole, it served as a fiducial line to which to refer the position of the second beam, and thereby judge of the refrangibility of the rays by which it was produced.

This second beam was a good deal the brighter of the two. Its colour was a beautiful sky-blue, which was nearly the same throughout, but just about its first border, that is, where it arose from the least refrangible of those rays which were capable of producing it, the colour was less pure. It had a perfectly continuous appearance. When viewed from above through a doubly refracting achromatic prism of quartz, which allowed a direct comparison of the two images, it offered no traces of polarization. It was produced by light polarized in a vertical or horizontal plane as well as by common light, and in that case, as well as in the former, manifested no traces of polarization *.

The short distance that the more refrangible rays were able to penetrate into the fluid might readily be perceived in this experiment, but the second method of observation was not adapted to bring out this part of the phenomenon.

16. On examining the fluid by the third method, the result was very striking, although of course only what might have been

---

* These two results, namely, that the blue beam which constitutes the greater part of the light dispersed by a solution of sulphate of quinine is unpolarized, or according to his expression possesses a *quaquaversus* polarization, and that that still remains the case when the incident light is polarized, have been already announced by Sir David Brewster, who appears to have been led to attend to the polarization of the light from Sir John Herschel's observation, that the blue light arising from epipolic dispersion in a solution of sulphate of quinine was unpolarized. It seemed important however to repeat the observation on the blue beam obtained in a state of isolation.

anticipated. The principal fixed lines of the violet, and of the chemical parts of the spectrum beyond, were seen with beautiful distinctness as dark planes interrupting an otherwise perfectly continuous mass of blue light. To see any particular fixed line with most distinctness, it was of course necessary to hold the eye in the corresponding plane, when the dark plane was foreshortened into a dark line. From the red end of the spectrum, as far as the line G, or thereabouts, the light passed freely through the fluid, or at least was only reflected here and there from motes held in mechanical suspension. About G the dispersion just commenced to be sensible, and there were traces of that line seen as a dark plane interrupting a mass of continuous but excessively faint light. For some distance further on the dispersed light remained so faint that it might have been passed over if not specially looked for. It was about half-way between G and H, or a little before, that it first became so strong as to arrest attention, and a little further on it became very conspicuous, the tint meanwhile changing to a pale sky-blue. The light was very copious about the two broad bands of the group H and for some distance from H towards G. Some of the fixed lines less refrangible than H were very plain, and beyond H a good number were visible, which will presently be further described. The whole system of fixed lines thus visible as interruptions in the dispersed light had a resolvable appearance; but with a very narrow slit and a lens of long focus at the prisms the light would have been too faint for convenient observation.

The dispersed light about G, and for some distance further on, was so very faint that I might have overlooked it had it not arrested my attention when observing by the fourth method; indeed, I have sometimes specially looked for it in the third arrangement without having been able to see it. Practically speaking, the dispersion might be said to commence about half-way between G and H.

17. On refracting the whole system sideways through a prism of moderate angle held in front of the eye, the fixed lines became confused, and the finer ones disappeared. The edges of the broad bands H were tinged with prismatic colours, like the edges of two slips of black velvet placed on a sheet of pale blue paper, and viewed through a prism. This experiment exhibits the

compound character of the dispersed light, notwithstanding the perfect homogeneity of the incident light.

18.   The third method of observation is well adapted to bring into view the variation in the absorbing energy of the medium corresponding to a variation in the refrangibility of the incident rays.   When the eye is placed vertically over the vessel containing the solution, so that the dark planes corresponding to the fixed lines of the spectrum are projected into dark lines, of which the length is not exaggerated by obliquity, the boundary of the dispersed light is projected into a curve, which serves to represent to the eye the relation between the absorbing power of the medium and the refrangibility of the incident light.   This curve is not exactly that which Sir John Herschel has treated of in the theory of absorption, and considered as the type of the absorbing medium to which it is applied, but nevertheless it serves much the same purpose.   It is true, that, independently of any change in the absorbing energy of the medium, an increasing faintness in the dispersed light would produce to a certain extent an approximation of the curve to its axis; but practically, in the case of sulphate of quinine, as well as in a great many others, the appearance is such as to leave no doubt as to the existence of a most intense absorbing energy on the part of the medium with respect to rays of very high refrangibilities*.

In the case of a solution of sulphate of quinine of the strength of one part of the disulphate to 200 parts of acidulated water, it has been already stated that a portion of the rays which are capable of producing dispersed light passed across a thickness of 3 inches.   On forming a pure spectrum, the fixed line H was traced about an inch into the fluid.   On passing from H towards G, the distance that the incident rays penetrated into the fluid increased with great rapidity, while on passing in the contrary direction it diminished no less rapidly, so that from a point situated at no great distance beyond H to where the light ceased, the dispersion was confined to the immediate

---

* I should here remark, that, after the researches described in this paper had far advanced, I met accidentally with a passage in the *Comptes Rendus*, tom. xvii. p. 883, in which M. Ed. Becquerel mentions a solution of acid sulphate of quinine as a medium eminently remarkable for its absorbing power with respect to the rays more refrangible than H.

neighbourhood of the surface. When the solution was diluted so as to be only one-tenth of the former strength, a conspicuous fixed line, or rather band of sensible breadth, situated in the first group of fixed lines beyond *H*, was observed to penetrate about an inch into the fluid. On passing onwards from the band above-mentioned in the direction of the more refrangible rays, the distance that the incident rays penetrated into the fluid rapidly decreased, and thus the rapid increase in the absorbing energy of the fluid was brought into view in a part of the spectrum in which, with the stronger solution, it could not be so conveniently made out, inasmuch as the posterior surface of the space from which the dispersed light came almost confounded itself with the anterior surface of the fluid.

The high degree of opacity with regard to rays of great refrangibility which the addition of so small a proportion of sulphate of quinine is sufficient to produce in water is certainly very remarkable; nevertheless it is only what I have constantly observed while following out these researches.

19. In observing by the fourth method, the part of the spectrum to which the incident light belonged was determined sometimes by the colour, sometimes by means of the fixed lines of the spectrum. It almost always happened that there were motes enough suspended in the fluid to cause a portion of the dispersed beam to consist merely of light which had undergone ordinary reflexion. When the whole dispersed beam was analysed by a prism, the beam which consisted of light reflected from motes was separated from the rest; it was in general easily recognised by its sparkling appearance, but at any rate was known by its consisting almost wholly of light polarized in the plane of incidence, whereas the truly dispersed light was un-polarized. It consisted of course of light of definite refrangibility, the same as that of the incident light, and thus served as a fiducial line to which to refer by estimation the refrangibilities of the component parts of the dispersed light. Of course this part of the observation was possible only when the incident rays belonged to the visible part of the spectrum.

On moving the lens horizontally through the colours of the spectrum, in a direction from the red to the violet, it was found that the dispersion was first perceptible in the blue. When the dispersed light was separated by a prism from the light reflected from motes, it was found to consist of an exceedingly small quantity of red; further on some yellow began to enter into its composition; further still, perhaps about the junction of the blue and indigo, the dispersed beam began to grow brighter, and was found on analysis to contain some green in addition to the former colours. In the indigo it got still brighter, and when viewed as a whole was somewhat greenish. Further still it became something of a pale slaty blue, and was found on analysis to contain some indigo, or at least highly refrangible blue. On proceeding further the dispersed light became first of a deeper blue and then, a little short of the fixed line *H*, whiter. At a considerable distance beyond *H* the dispersed light was if anything a shade more nearly white.

By this method of observation the dispersion can be detected earlier in the spectrum than by the third method, and moreover the change in the colour of the dispersed light is much more easily perceived; indeed the most striking part of this change takes place while the dispersed light is so very faint that it can hardly be seen in observing by the third method; moreover, even in the bright part of the dispersed beam, it is not at all easy by the latter method to make out the change of tint corresponding to a change in the refrangibility of the incident rays, because the tint changes so gradually and so slightly that the eye glides from one part of the dispersed beam to another without noticing any change.

20. It has been already mentioned that the blue beam of dispersed light seen in a solution of sulphate of quinine was produced whether the incident light was polarized in or perpendicularly to the plane of reflexion, or more properly plane of dispersion, that is, the plane containing the incident ray and that dispersed ray which enters the eye. A question naturally presents itself, whether the intensity of the dispersed light is strictly the same in the two cases. By combining a lens of rather short focus and a doubly refracting prism with the four prisms, I satisfied myself that the difference of intensity, if there were any, was not great, but the experiment presented some practical difficulties. However, the result of the following experiment appeared to be as decisive as a negative result could well be.

The arrangement being the same as in the third method, but the lens in front of the prisms having a focal length of only 6·5 inches, the incident light was polarized in a vertical plane previously to passing through the slit, by transmission through a pile of plates. The two beams of light were seen as usual in the fluid, namely, the blue beam due to internal dispersion, and the fainter coloured beam due to motes. The former of these, which was quite separate from the latter, exhibited the principal fixed lines belonging to the highly refrangible part of the spectrum. A plate of selenite was then interposed immediately in front of the vessel, so as to modify the polarization of the light entering the fluid. This plate was obtained by an irregular natural cleavage, and was cemented with Canada balsam between two discs of glass. When examined by polarized light it exhibited a succession of beautiful and varied tints, according to the various thicknesses of the different parts. Now when the plate was moved about in front of the vessel, without altering its perpendicularity to the incident light, different portions of the beam due to motes were observed to disappear and reappear, or at least to become faint and then bright again, so that a person ignorant of the cause, and not looking at the disc, might have supposed that the observer had been holding in front of the vessel a piece of dirty glass, having the dirt laid on in patches; but in whatever manner the disc was moved in its own plane without rotation, or turned round an axis perpendicular to its plane, not the slightest perceptible change was produced in any part of the blue beam.

Explanation of Terms.

21. In all the experiments described in this paper in which a spectrum was formed for the sake of examining the separate action of portions of light of different refrangibilities, the length of the spectrum was horizontal, so that the fixed lines were vertical. Nevertheless it will be convenient, for the sake of shortness, to use the prepositions *above* and *below* to signify respectively *on the more refrangible side of* and *on the less refrangible side of*.

The principal fixed lines of the visible spectrum will be denoted by letters in accordance with Fraunhofer's admirable map. These lines are now too well known to need description.

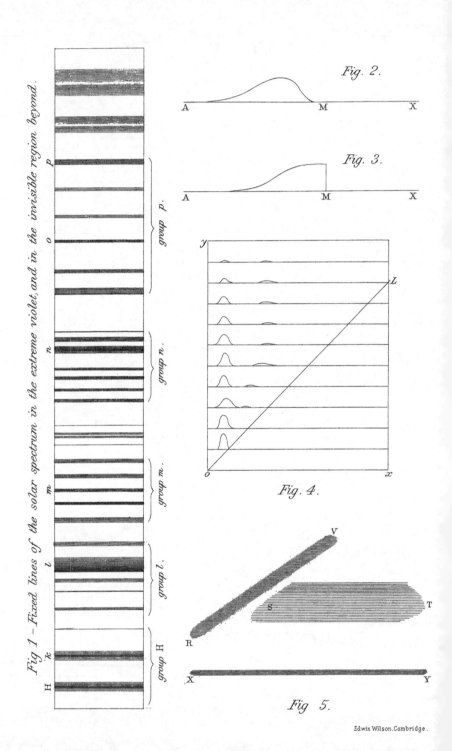

Fig 1 – Fixed lines of the solar spectrum in the extreme violet, and in the invisible region beyond.

Fig. 2.

Fig. 3.

Fig. 4.

Fig 5.

Edwin Wilson. Cambridge.

The only map of the fixed lines of the chemical spectrum which I had for a good while after these researches were commenced is Professor Draper's, which will be found in the twenty-second volume of the *Philosophical Magazine* (1843). Of course this map cannot be compared for accuracy of detail with Fraunhofer's map of the visible spectrum, nor does it profess to give more than some of the most conspicuous lines selected from among a great multitude. The suppression of so many lines, without any representation by shading of their general effect, renders it difficult to identify those which are laid down, at least if I may judge from my own observations; besides, Professor Draper's spectrum was so much purer than the one with which I found it most convenient to work, that the two are not comparable with each other.

22. I have made a sketch of the fixed lines from H to the end, which accompanies this paper (see Plate). The fixed lines of the visible spectrum are so well known that I thought it unnecessary to begin before H. A solution of sulphate of quinine is a very good medium for showing the lines, but a yellow glass, which will be mentioned presently, is quite as good, or rather better. The map represents the spectrum as seen with the lens of 12 inches focal length in front of the prisms. The breadth of the slit was not always quite the same: it may be estimated at about the $\frac{1}{20}$th of an inch. The map contains 32 fixed lines or bands more refrangible than H, which is the utmost that I have been able on different occasions to see with this lens, though with a lens of longer focus and a narrower slit the number of fixed lines which might be counted was, as might be expected, a good deal larger. As I have not yet identified these lines, except in certain cases, with those which had previously been represented by means of photographic impressions, I have thought it advisable not to attempt an identification, but to attach letters to the more conspicuous lines in my map without reference to former maps. As the capitals L, M, N, O, P have already been appropriated to designate certain fixed lines, I have made use of the small letters l, m, n, o, p, to prevent confusion.

In drawing the map, I have endeavoured to preserve the character of the lines with respect to blackness or faintness, sharpness or diffuseness. The distances were not laid down by measurement, except here and there, and they are not, I fear, quite so accurate as might be desired; still, I feel assured that no one viewing the actual object would feel any difficulty in identifying the lines with those in my map, provided the circumstances under which his spectrum was formed at all approached to those under which mine was seen when the arrangement as to focal length of the lens, &c. was that most convenient for general purposes.

The more conspicuous lines in the part of the spectrum represented in the map may conveniently be arranged in five groups, which I will call the groups H, l, m, n, p. The group H consists chiefly of the well-known pair of bands of which the first contains Fraunhofer's line H; the second band I have marked k, in accordance with Professor Draper's map. The most conspicuous object in the next group consists of a broad dark band, l. This band is between once and twice as broad as H, and is darker in the less refrangible half than in the other. With a lens of 3 feet focal length and a narrow slit it was resolved into lines, which is probably the reason why it is altogether omitted in Professor Draper's map, while the first three lines of the group (if I do not mistake as to the identification) are represented, forming his group L. Under the circumstances to which the accompanying map corresponds, the band l appears as a very striking object, perhaps, with the exception of the bands H, k, the most conspicuous in the whole spectrum. With a still lower power it appears as a very black and conspicuous line. A double line beyond l completes the group l, after which comes another remarkable group m, consisting of five lines or bands. Of these the first is rather shady, though sharply cut off on its more refrangible side, but the others, and especially I think the second and third, are particularly dark and well-defined. I have marked the middle line m, not because it is more conspicuous than its neighbours, but on account of its central situation. After a very faint group, consisting apparently of four lines, comes another very conspicuous group n, consisting of two pairs of dark bands followed by another pair of bands which are broad and very dark. The first of these is a good deal broader than the second, but is not so broad as the band H; the second is followed by a fine line. This is as far as it is easy to see; but when the sunshine is clear, and the arrangements are made with a little care, a group of six lines is seen much further on. Of these, the first two are only moderately dark, and the first is rather diffuse; they stand off a little from the others, and are a little closer together than the other four. Of the latter, the first, marked o, is very strong, considering the faintness of the light which it interrupts; the second and third are faint, and difficult to see; the fourth, marked p, is black like the first, and a good deal broader. The line p was situated, by measurement, as far beyond H as H beyond b. Once or twice in the height of summer, and under the most favourable circumstances, I have observed two broad dusky bands still further on. The first of these had the appearance of being resolvable into two. The excessively faint light seen beyond the second seemed to end rather abruptly at the distance represented by the border of the accompanying plate, as if there were there the edge of another dark band beyond which nothing could be seen. In order to see the dusky bands last mentioned, and even to see the group p to most advantage, it was necessary to allow the central part of the beam incident on the prisms to pass through them close to their edges, so that evidently a great deal of light was lost by passing by the prisms altogether. This circumstance, combined with others which I have observed, convinces me that the great obstacle to seeing the fixed lines in this part of the spectrum consists in the opacity of glass. Were glass as transparent with respect to the invisible rays of very high refrangibility as it is with respect to the rays belonging to the visible spectrum, I know not how much further I might have been able to see.

I have endeavoured to identify the fixed lines in my map with the fixed lines represented in M. Silbermann's map of the chemical spectrum, with a copy of which my friend Professor Thomson has kindly furnished me. I am still uncertain respecting the identification. M. Silbermann's map is so very much more detailed than my own, and must have been made with so much purer a spectrum, that the two systems of lines are not directly comparable.

23. From the difficulty of identification some persons might be disposed to imagine that the chemical rays, and those which produced the blue light in a solution of quinine, were of a different nature, and had each a system of fixed lines of its own. For my own part, I was too well acquainted with the Protean character of fixed lines to regard the difficulty of identification as any valid argument in support of such a view. And that this difficulty

arose from nothing more than the different degrees of purity of the spectra is now put past dispute, for my friend Mr Kingsley of Sidney Sussex College, to whom I recently showed some of the experiments mentioned in this paper, has kindly taken for me some photographs of spectra having nearly the same degree of extent and purity as those with which I worked, and these show the fixed lines just as they appeared in a solution of sulphate of quinine and in other media*.

24. The position of a point in the spectrum which does not coincide with one of the principal fixed lines, will be denoted by referring it to two of those lines, in a manner which will be most easily explained by an example. Thus $\frac{1}{2}GH$, $G\frac{1}{2}H$, $GH\frac{1}{2}$ will be used to denote respectively a point situated at a distance below G equal to half the interval from G to H, a point midway between G and H, and a point situated at the same distance above H. In using this notation, the letters denoting fixed lines will be written in the order of refrangibility, and the fraction expressing the part of the interval between these lines, which must be conceived to be measured off in order to reach the point whose position it is required to express, will be written before, between, or after the letters, according as the measurement is to be taken from the first line in the negative direction, from the first line in the positive direction, or from the second line in the positive direction, the positive direction being that of increasing refrangibility.

25. From the experiments already described, it appears that the beam of dispersed light which was observed in the experiments of Sir David Brewster consisted of two very distinct portions, one arising merely from light reflected from motes, and the other having a far more remarkable origin. It will be convenient to have names for these two kinds of dispersion, and I shall accordingly call them respectively *false internal dispersion* and *true internal dispersion*, or simply *false dispersion* and *true dispersion* when the context sufficiently shows that internal dispersion is spoken of. When dispersion is mentioned without qualification, it is to be understood of true dispersion. Now that it appears that the mere reflexion of light from solid particles held in mechanical suspension has nothing to do with that remarkable

kind of internal dispersion which is characterized by the "*quaqua-versus* polarization," the phenomenon of false dispersion ceases to be of much interest in an optical point of view; while on the other hand the phenomenon of true dispersion, which had always been very remarkable, is now calculated to excite a great additional interest. It will be convenient to mention here the principal characters by which true and false dispersion may be distinguished, although it will be anticipating in some measure the results of observations yet to be described.

26. In true dispersion the dispersed light has a perfectly continuous appearance. In false dispersion, on the other hand, it has generally more or less of a sparkling appearance, and on close inspection is either wholly resolved into bright specks, or so far resolved as to leave on the mind the impression that if the resolution be not complete it is only for want of a sufficient magnifying power.

In true dispersion the dispersed light is perfectly unpolarized. In false dispersion, on the contrary, at a proper inclination the light is almost perfectly polarized in the plane of reflexion.

In false dispersion, which is merely a phenomenon of reflexion, the dispersed light has of course the same refrangibility as the incident light. In true dispersion heterogeneous dispersed light arises from a homogeneous beam incident on the body by which the dispersion is produced.

27. In those bodies, whether solid or liquid, which possess in a high degree the power of internal dispersion, the colour thence arising may be seen by exposing the body to ordinary daylight, looking at it in such a direction that the regularly reflected light does not enter the eye, and excluding transmitted light by placing a piece of black cloth or velvet behind, or by some similar contrivance. It has been usual to speak of the colour so exhibited as displayed by reflexion. As however the cause now appears to be so very different from ordinary reflexion, it seems objectionable to continue to use that term without qualification, and I shall accordingly speak of the phenomenon as *dispersive reflexion**. Thus

---

* I confess I do not like this term. I am almost inclined to coin a word, and call the appearance *fluorescence*, from fluor-spar, as the analogous term *opalescence* is derived from the name of a mineral.

dispersive reflexion is nothing more than internal dispersion considered as viewed in a particular way.

28. The tint exhibited by dispersive reflexion is modified in a peculiar manner by the absorbing power of the medium. In the first place, the light which enters the eye in a given direction is made up of portions which have been dispersed by particles situated at different distances from the surface at which the light emerges. The word *particle* is here used as synonymous, not with *molecule*, but with *differential element*. If we consider any particular particle, the light which it sends into the eye has had to traverse the medium, first in reaching the particle, and then in proceeding towards the eye. On account of the change of refrangibility which takes place in dispersion, the effect of the absorption of the medium is different for the two portions of the whole path within the medium, so that this effect may be regarded as a function of two independent variables, namely, the lengths of the path before and after dispersion; whereas, had the light been merely reflected from coloured particles held in suspension, the effect of absorption would have been a function of only one independent variable, namely, the length of the entire path within the medium.

29. When false dispersion abounds in a fluid, it may be detected at once by the eye, without having recourse to any of the characters already mentioned whereby it may be distinguished from true dispersion. When a fluid is free from false dispersion it appears perfectly clear, when viewed by transmitted light, although it may be highly coloured, and may even possess to such an extent the property of exhibiting true internal dispersion as to display, when properly viewed, a copious dispersive reflexion. On the contrary, when false dispersion abounds, the fluid, if not plainly muddy, has at least a sort of opalescent appearance when viewed by transmitted light, which, after a little experience, the eye in most cases readily recognizes. In viewing the phenomenon of dispersive reflexion, as exhibited in a fluid, it might be supposed that the fluid was water, or else some clear though coloured liquid, holding in suspension a water colour in a state of extreme subdivision. But on holding the fluid before the eye, so as to view it by transmitted light, or rather view a bright well-defined object through it, the illusion is instantly dispelled. The reason of this

difference appears to admit of easy explanation, and will be noticed further on.

30. Light will be spoken of in this paper as *active* when it is considered in its capacity of producing other light by internal dispersion. A medium will be said to be *sensitive* when it is capable of exhibiting dispersed light under the influence of light (visible or invisible) incident upon it. In the contrary case it will be called *insensible*.

I shall now return to the description of the appearances exhibited by some of the media most remarkable for their sensibility.

Decoction of the Bark of the Horse-Chestnut (Æsculus hippocastanum).

31. In Sir John Herschel's second paper it is stated that esculine possesses in perfection the peculiar properties which had been found to belong to quinine. Having tried without success to procure the former alkaloid*, I was content to let this substance pass, till I found how admirably a mere decoction or infusion of the bark of the tree answered for all purposes of observation.

This medium is even more sensitive than a solution of sulphate of quinine, and disperses like it a blue light. The description of the mode of dispersion in the latter medium will apply in almost all points to the former : the principal difference consists in the circumstance that in the horse-chestnut solution the dispersion begins earlier in the spectrum than in the solution of quinine. In a solution of sulphate of quinine of convenient strength, we have seen that the dispersion came on at about $G\frac{1}{2}H$, the excessively faint dispersion which was exhibited earlier being left out of consideration, whereas in a decoction of the bark of the horse-chestnut, diluted so as to be of a convenient strength, it came on a little before G. This explains the reason of an observation of Sir David Brewster's, who has remarked that "a beam of light that has passed through the esculine solution disperses blue light, but not copiously, when transmitted through the quinine solution ; but the beam that has passed through quinine is copiously dispersed when transmitted through esculine†."

---

* [It is a glucoside.]

† *Philosophical Magazine*, Vol. xxxii. (June 1848), p. 406.

Green Fluor-Spar from Alston Moor.

32. It is well known that some specimens of fluor-spar exhibit a sort of double colour. In particular, a variety found at Alston Moor, which is green when seen by transmitted light, appears when viewed in a certain manner of a beautiful deep blue. This blue colour seems to have been considered by Sir John Herschel as merely superficial. It has been shown however by Sir David Brewster to arise from light dispersed in the interior of the crystal, and to have no particular relation to the surface.

The crystal with which the following observations were made was of a fine but not intense green when viewed by transmitted light. On viewing a pure spectrum through it, there was found to be a dark band of absorption in the red. This band was narrow, and by no means intense. The crystal exhibited a copious deep blue by dispersive reflexion.

33. On admitting into the crystal a cone of sunlight formed by a lens of short focus, and then analysing the dispersed beam, it was found to consist of a very little red followed by a dark interval, then green, faintly fringed below with less refrangible colours down perhaps to the orange, then blue, or bluish-green, followed by a great deal of indigo or violet. Independently of the gap in the red, the spectrum was not quite continuous, for a band of bluish-green, not very broad, was separated by dusky bands from the green below and the indigo above. The separate red band and the two dusky bands were all so faint as to be difficult to see.

The dispersed beam was readily proved to be truly dispersed, for it was unpolarized, and a pale brown glass cut it off when placed in the first position, although it transmitted it in a great measure when placed in the second.

34. When the crystal was examined by the third method, the general result closely resembled that produced by sulphate of quinine. The dispersion commenced about half-way between G and H, and continued from thence onwards far beyond H. It was strongest about H. The fixed lines were seen with beautiful distinctness as dark planes in the crystal. The groups H, l, m were quite evident, and n might be seen without difficulty. I have

even seen some of the fixed lines of the group p. The tint of the dispersed light appeared as nearly as possible uniform throughout. The distance to which this light could be traced from the surface did not at all diminish so rapidly in this crystal, with an increase in the refrangibility of the incident light, as it had done in the case of a solution of sulphate of quinine. Indeed, it was difficult to say how far the decrease in the depth to which the incident rays could be traced, by means of the dispersed light which they produced, was due merely to the increasing faintness of the light, and how far it indicated a real increase in the absorbing energy of the crystal; whereas in the case of sulphate of quinine the appearance presented unequivocally indicated a very rapid increase of absorbing power.

35. On examining the crystal by the second method, the general appearance was the same as in the case of sulphate of quinine, but the beam of falsely dispersed light was absent. In addition to the copious beam of deep blue light dispersed by the most refrangible rays, there was however a faint beam of red or reddish light dispersed by rays of low refrangibility. This beam was too faint to be seen by the third method of examination. It will be remembered that the prismatic analysis of the transmitted light gave a band of absorption in the red. Another crystal of a pale colour, which did not give a similar band of absorption in the red, exhibited nothing but the blue beam of dispersed light when examined by the second method.

36. On examining the crystal by the fourth method, the extreme red proved inactive. The activity commenced about the most refrangible limit of the red transmitted by a deep blue glass, when the dispersed light was red, but extremely faint. On moving the lens onwards through the spectrum, the dispersed light rapidly became brighter, and then died away. When at its brightest, although even then it was almost too faint for prismatic examination, it appeared to consist of not quite homogeneous light a little lower in refrangibility than the active light. For a considerable distance further on there was no sensible dispersion produced. The dispersed light became again perceptible when the active light belonged to the greenish yellow, or not till the blue, according to the intensity of the incident light. As the lens moved on

the dispersed light remained faint for a considerable time. It was first reddish and then brownish, with a refrangibility answering to its colour. When the active light was at $G\frac{1}{2}H$, or thereabouts, the dispersed light rapidly grew much brighter, and became of a fine blue. On analysis it was found to consist of rays the refrangibility of which ranged within wide limits. The red rays were, however, almost wholly wanting, while the rays belonging to the more refrangible part of the spectrum resulting from the analysis of the dispersed beam were particularly copious. The most refrangible limit of the dispersed light did not quite reach in refrangibility the active light. The dispersed light was most copious when the active light belonged to the neighbourhood of H. As the lens moved on the dispersed light grew less bright, and gradually died away.

Solution of Guaiacum in Alcohol.

37. This is one of the media mentioned by Sir David Brewster, who remarks that it "disperses, by the stratum chiefly near its surface, a beautiful violet light."

When this fluid is examined by the third or fourth method, it is found to exhibit a copious internal dispersion, which begins to be conspicuous much lower down in the spectrum than in the cases already described. In observing by the third method, the true dispersion appeared to commence about the end of the green, the dispersed light being reddish-brown. By the fourth method the dispersion could be traced as low down as $D\frac{1}{3}b$, the dispersed light being reddish. As the lens moved onwards, in a direction from the red to the violet, the more refrangible colours entered in succession into the dispersed beam, and it became successively brownish, yellowish, greenish, and bluish. In whatever part of the spectrum the lens might be, it was found that the most refrangible part of the dispersed beam was of lower refrangibility than the active light. This could be easily determined by means of the beam of falsely dispersed light, which was always visible so long as the active light belonged to the visible part of the spectrum.

38. With the third arrangement the fixed lines were seen as before by means of the dispersed light, but in this fluid they

could be seen much lower down in the spectrum than in the solution of sulphate of quinine. The group H was seen on a greenish ground. About the group l the ground was still greenish, but the dispersed light was not very copious. The beautiful violet light mentioned by Sir David Brewster is produced only by rays of extremely high refrangibility, and is found to extend from the beginning of the group m to the end of the group n, and even further. This part of the dispersion is best seen with a rather dilute solution.

39. In a solution of guaiacum, just as in the solution of sulphate of quinine, the absorbing power of the medium increases very rapidly with the refrangibility of the light. This is shown by the rapid decrease in the distance from the surface to which the dispersed light can be traced. The reason why the violet dispersed light is confined to a very thin stratum adjacent to the surface by which the light enters, is simply that the medium is so nearly opaque with regard to the invisible rays beyond the extreme violet that all such rays are absorbed by the time the light has passed through a very thin stratum of the fluid.

40. If the solution be strong the colour is of considerable depth. In all such cases it is necessary to take the precaution, mentioned by Sir David Brewster, of transmitting the incident beam as near as possible to the upper surface, so as just to graze it. The absorption of the medium would otherwise modify the tint of the dispersed beam.

41. The solutions of quinine and guaiacum present a striking contrast with respect to the change of tint of the dispersed beam. In the former solution the change is but slight, if we except that part of the dispersion which is very faint; whereas in the latter, the prismatic colour which makes the nearest match to the composite tint of the dispersed beam runs through nearly the entire spectrum, as the refrangibility of the active light changes from that of the green rays to that of invisible rays situated far beyond the extreme violet.

Tincture of Turmeric.

42. This fluid is very sensitive, and exhibits a pretty copious dispersive reflexion of a greenish light. In its mode of internal

dispersion it strongly resembles a solution of guaiacum, but the final tint of the dispersed light does not correspond to so high a mean refrangibility. When the fluid was examined by the third method, the true dispersion appeared to commence about b. The absorbing power was so great for the rays of high refrangibility, that from a little above F (in the case of tincture not diluted with alcohol) to the end the dispersed light seemed to be confined to the mere surface. By the fourth method the dispersion was as usual traced a little lower down in the spectrum. When the dispersed beam was first perceived it was nearly homogeneous, and its refrangibility was only a very little less than that of the active light. As the refrangibility of the active light increased, new colours, in the order of their refrangibility, entered into the dispersed beam, which became more and more composite, while at the same time its upper limit became distinctly separated from the beam of falsely dispersed light, which, when the whole dispersed beam was analysed by a prism, was always found in advance of the other. The tint of the dispersed beam passed from orange through yellow to yellowish green, which was its final tint. Tincture of turmeric is well adapted for exhibiting the fixed lines in the invisible part of the spectrum, though perhaps not quite so well as a solution of sulphate of quinine.

Alcoholic Extract from the Seeds of the Datura Stramonium.

43. This fluid, which I was led to try in consequence of Sir David Brewster's paper, proved to be remarkably sensitive, and exhibited a copious dispersive reflexion of a pale but lively green. The general phenomena are so nearly the same as in a solution of sulphate of quinine that there is no need of a separate description. The principal difference consists in the tint, which is green instead of blue. In the present case, however, the fluid, in addition to its dispersion of green, dispersed a red beam under the influence of certain red rays. As the lens employed in the fourth method of examination was moved from the extreme red onwards, the light was at first inactive, but when the lens reached a certain point of the spectrum, a red beam of truly dispersed light suddenly appeared, which disappeared with almost equal suddenness as the lens moved on. In this mode of observation the refrangibility of the dispersed could hardly be distinguished from that of

the active light; but on combining the first and third methods, by removing the lens, placing the vessel truly in focus, and holding a blue glass alternately in front of the vessel and in front of the eye, I satisfied myself that the truly dispersed beam, taken as a whole, was of lower refrangibility than the light by which it was produced. The utility of the blue glass depended upon the circumstance that the upper extremity of the extreme red which it transmitted nearly coincided with the point of the spectrum at which the red beam occurred. This red beam was doubtless due to the presence of a small quantity of chlorophyll, or one of its modifications. The light transmitted by the fluid exhibited on prismatic analysis the absorption band in the red which is so characteristic of that substance.

The colour of the solution was a pale brownish yellow; it would no doubt have been still paler, and perhaps nearly colourless, had the sensitive principle to which the green dispersion was due been present in equal quantity but in a state of purity. As it was, the fluid was pale enough to exhibit well, when poured into a test tube and held in front of a window, a narrow arc on the side of the incident light, like sulphate of quinine, only in this case the arc was green instead of blue.

Frequency of the occurrence of true internal dispersion having the same general character as that which takes place in the cases above described.

44. If we except the red dispersed beam produced by red rays in the crystal of fluor-spar and in the stramonium extract, a strong similarity may be observed in the mode of internal dispersion which takes place in the cases hitherto described. As the refrangibility of the incident light continually increases, the rays are at first inactive. At a certain point of the spectrum, varying according to circumstances, the true dispersion begins to be sensible, but is faint at first. After remaining faint for some distance it presently becomes more copious. It remains very conspicuous through the whole of the violet and beyond, and then gradually dies away. It consists at first of light of comparatively low refrangibility, and then new colours in the order of their refrangibility enter into it. Frequently the greater part of the

change of prismatic composition takes place while the dispersed light is very faint, so that practically speaking we may almost say that the tint is uniform. Sometimes, when the dispersion just commences, the dispersed light is nearly homogeneous, and has a refrangibility so nearly equal to that of the active light that the beams due to true and false dispersion can hardly be separated.

45. Now this, so far as I have observed, is much the commonest kind of true internal dispersion, although sometimes the phenomenon presents very striking singularities. In the paper in which Sir David Brewster first announced the discovery of internal dispersion, he remarks "that it is a phenomenon which occurs almost always in vegetable solutions, and almost never in chemical ones or in coloured glasses *." For my own part, I have rarely met with a vegetable solution which did not exhibit more or less the phenomenon of *true* internal dispersion. Its existence may in general be easily detected in the following manner. The sun's light being reflected horizontally through a lens, a deep blue glass is left in such a position as to intercept the light incident on the vessel containing the fluid, which is placed at the focus of the lens. A pale brown glass of the proper kind is then placed so as to intercept, first the incident, and then the dispersed light. A vessel with flat sides filled with a solution of sulphate of quinine would be better, and then the placing of the medium in the second position might be dispensed with, the medium being sensibly transparent. Sometimes it is useful to have recourse to analysis through a doubly refracting prism, or a rhomb of calcareous spar. In this way true internal dispersion may often be detected in a fluid which is actually muddy, in which case, were the effect of the incident light observed as a whole, the true would be masked by the enormous quantity of false dispersion which such a medium would offer.

46. The fluids obtained by treating the leaves and other parts of plants with alcohol or hot water are almost always sensitive, so far as I have observed. The solutions in water presently ferment, and are frequently highly sensitive in the early stages of fermentation; they are usually more or less sensitive in all stages.

* *Edinburgh Transactions*, Vol. xii. p. 542.

Different kinds of fungus furnish very sensitive solutions. When aqueous solutions become muddy by decomposition, other clear and often highly sensitive liquids may be obtained from them by various chemical processes. Port and sherry are decidedly sensitive. In such cases the fluid is a mixture of several substances, of which some may be sensitive and others insensible. When vegetable substances are isolated they are frequently insensible, or else so very slightly sensitive when examined under great concentration of the highly refrangible rays, that it is quite impossible to say whether the sensibility thus exhibited may not be due to some impurity: thus, several solutions containing sugar, salicine, morphine, or strychnine were found to be insensible. A solution of veratrine in alcohol proved to be sensitive in a pretty high degree, dispersing internally a bluish light. Sir David Brewster has remarked that a solution of sulphate of strychnine in alcohol dispersed light *after it had stood for some days*. This observation I have verified with reference to *true* dispersion, which the solution exhibits, though not very copiously, after it has been made some time. There can be little doubt that the sensitive principle in this case is not strychnine, but some product of its decomposition. I now come to some instances of internal dispersion which are far more striking.

Solution of Leaf-Green in Alcohol.

47. It was in this very remarkable fluid that the phenomenon of internal dispersion was first discovered by Sir David Brewster, while engaged in researches relating to absorption. The character of the internal dispersion of a solution of leaf-green is no less remarkable than the character of its absorption. On account of the close connexion which seems to exist between the two phenomena, it will be requisite first to say a few words about the latter.

When green leaves are treated with alcohol, a fluid is obtained which is of a beautiful emerald-green in moderate thicknesses, but red in great thicknesses, and which has a very remarkable effect on the spectrum. A good number of the following observations on the internal dispersion of leaf-green were made with a solution obtained from the leaves of the common nettle, by first boiling them in water and then treating them with cold alcohol, the leaves having previously been partially dried by pressing them between sheets of blotting paper. Nettle was chosen partly because it stands boiling without losing its green colour, and partly for other reasons. My object in boiling the leaves was to obtain the green colouring matter more nearly in a state of isolation, but it seems to have the additional advantage of giving a solution less liable to decomposition. Indeed, this fluid seemed disposed to remain permanently unchanged when kept in the dark; but a small portion of it which was exposed to strong light had its colour rapidly discharged.

48. When fresh leaves are left in contact with alcohol in the dark, or in only weak light, the colour of the fluid changes by degrees, and it seems to approximate (making allowance for impurities) to a type which is nearly represented by the fluid obtained in this manner from laurel leaves, or that obtained by treating with alcohol tea leaves from which a good deal of brown colouring matter has first been extracted by water. This type was rather ideal than actual, being derived from a comparison of different cases, until it seemed to be realized in the case of a fluid obtained by re-dissolving in alcohol a crust* which had formed itself at the bottom of a test tube containing leaf-green. The principle to which the peculiar absorption and internal dispersion of such a fluid seems due may be called modified leaf-green. The fluid itself is not green but olive-coloured, becoming red at great thicknesses.

49. When solutions of leaf-green, and of its various modifications, are examined in different thicknesses by the light of a candle, there are five bands of absorption which may be observed

[* From the spectrum of its solution, this crust must have been the product of decomposition by acids of the principal red-absorbing and red-fluorescing constituent of the chlorophyll mixture. This mixture consists (in land plants) mainly of two red-fluorescing and red-absorbing substances and a yellow non-fluorescent substance, all showing characteristic bands of absorption. They are all three, especially the first two, easily altered by acids. The "modified leaf-green" is the mixture with the first two substances decomposed as if by the minute quantity of acid obtained from the extract of the leaves. As the relative proportion of the three substances in the extract is liable to vary with the circumstances of the manipulation, and as the first two are very easily altered by the least trace of acid, there is some uncertainty in endeavouring (as in § 49) to identify the bands of absorption described by different observers who worked with the natural mixture.]

in the spectrum. These will be called, in the order of their refrangibility, Nos. 1, 2, 3, 4 and 5, the bright bands below the respective dark bands being also numbered in the same manner. Of the dark bands, Nos. 1, 2, 3 and 5, are the first four in Sir David Brewster's plate*. No. 4 is mentioned in the memoir, but not represented in the plate, which corresponds to a thickness not sufficient to bring out this band. The last band in the plate could not be seen without strong light. The dark bands Nos. 1 and 2 are situated in the red, No. 3 about the yellow or greenish yellow, No. 4 in the green, and No. 5 early in the blue. Of these, No. 1 is in small thicknesses by far the most intense, and it may be readily seen even in a very dilute solution; it might apparently be used as a chemical test of chlorophyll, or one of its modifications. The test would be of very easy application, since it would be sufficient to hold a test tube with the liquid at arm's length before a candle at a little distance, and view the linear image of the flame through a prism applied to the eye.

50.   Fresh and modified leaf-green differ much in the order in which the bright bands are absorbed, and in the degree to which the dark bands are developed before they cease to be visible by the absorption of the part of the spectrum in which they are situated. In the green fluid, the dark band No. 5 is not usually seen, because the spectrum is there cut off, unless a very small thickness be used. With a moderate thickness, Nos. 2 and 3, especially the former, are well seen, and No. 1 is very intense. As the absorption goes on, the bright bands Nos. 2 and 3 are absorbed, and there is left the red band No. 1, and a double green band, consisting of the bright bands Nos. 4 and 5, separated by the dark band No. 4, which by this time has come out. In modified leaf-green, the dark bands Nos. 4 and 5 are much more conspicuous than in the green fluid, but No. 3 is wanting, or all but wanting. With a thickness by which the absorption is well developed, the conspicuous bright bands are in this case Nos. 1 and 3, and next to them No. 2, whereas in the green fluid Nos. 2 and 3 were quickly absorbed, or at least the whole of No. 2, and the greater part of No. 3.

51.   It seems worthy of remark, that, especially in the case of the green fluid, the absorbing power alters with the refrangibility

of the light at a very different rate on the two sides of the intense dark band No. 1. This might be inferred from the order in which the bright bands disappear; but it was rendered visible to the eye by the following easy experiment. A narrow test tube was partly filled with a solution of leaf-green, and then a few drops of alcohol were added, which remained at the top, and there diluted the solution. The tube was then held before a candle, and the linear image of the flame was viewed through a prism. In the under part the dark band No. 1 was broad, the bright band No. 2 being narrow, and almost obliterated, but in the upper part the dark band No. 1 was very narrow. Now on tracing upwards the sides of this dark band, it was found that the less refrangible side was almost straight, and the diminution in the breadth of the band was produced by the encroachment of the bright band No. 2. Speaking approximately, we may say that in proceeding from the extreme red onwards, at a certain point of the spectrum the fluid passes abruptly from transparent to opaque, and then gradually becomes almost transparent again.

52. It may here be remarked, that although the absorption produced by leaf-green is best studied in a solution, its leading characters may be observed very well by merely placing a green leaf behind a slit, as near as possible to the flame of a candle, and then viewing the slit through a prism.

53. After this digression relating to the absorption of leaf-green, it is time to come to its internal dispersion. And first, when a cone of white light coming from the sun is admitted horizontally into the fluid, as close as possible to its upper surface, and the beautiful red beam of dispersed light is analysed by a prism, the spectrum is found to consist of a bright red band of a certain breadth, followed by a dark interval, and then a much broader green band not near so brilliant. There is usually but little false dispersion, and what there is may be almost entirely got rid of by analysing the beam by a Nicol's prism, so as to view it by light polarized in a plane perpendicular to the plane of dispersion. Now on raising the vessel without removing the prism from the eye, it was found that a dark band, which was in fact the absorption band No. 1, appeared almost exactly in the middle of the bright red band. On continuing to raise the vessel, so as to

make the dispersed rays pass through a still greater thickness of the medium before reaching the eye, the dark band increased in width, and when the red beam was almost absorbed, the part that was left consisted of two cones of red, one at each side of the dark band, which by this time had become broad. The whole appearance seemed to indicate that the bright red beam of dispersed light had a very intimate connexion with the intense absorption band No. 1.

54. Among coloured glasses, there is one combination which produces a very striking effect. When a deep blue glass is placed in the first position, the dispersed light, if the solution be at all strong, is confined to a very thin stratum adjacent to the surface, and is best seen by placing the vessel so that the surface of the fluid at which the light enters is situated at a little distance on either side of the focus of the lens, when there is seen a bright circle of a most beautiful crimson colour. It might be supposed that the red of which this circle mainly consists was nothing but the extreme red transmitted by the blue glass. But it is readily shown that such is not the case. For in the first place, the fluid transmits pretty freely the red transmitted by the blue glass, whereas the red light found in this circle is almost confined to the surface of the fluid. Again, it was found that a pale brown glass, which transmitted freely the extreme red, almost entirely cut off the bright circle, when placed in the first position without removing the blue glass, although it freely transmitted it when placed in the second position. It appears, therefore, that the bright circle is due, not to the red, but to the highly refrangible rays transmitted by the blue glass.

55. When a solution of leaf-green was examined by the third method, the appearance as seen from the outside was very singular. The fixed lines in all the more refrangible part of the spectrum were seen as interruptions in a bright red ground verging to crimson. The beauty and purity of the tint, and the strange contrast which it presented to the colours belonging to that part of the spectrum, were very striking. About H the tint began to verge towards brown, and the fixed lines beyond H were seen on a brownish red ground. That the ground on which the fixed lines of somewhat less refrangibility were seen was rather

crimson than red, arose, no doubt, from the mixture of a little blue or violet light due to false dispersion, and to the scattering which took place at the surface of the glass.

56. On looking down from above, the places of the more conspicuous bands of absorption were indicated by dark teeth, with their points turned towards the incident light, interrupting the dispersed light. It is to be understood that the light was transmitted as close as possible to the upper surface, so that the absorption by which these teeth were formed took place *before* dispersion. In this way the places of the absorption bands Nos. 1, 2 and 4, were perfectly evident. No. 3, it will be remembered, was by no means conspicuous. When the solution is of convenient strength, the absorption is so rapid beyond the bright band No. 5, that the dispersion is confined to a thin stratum close to the surface by which the light enters, and therefore no dark tooth would be seen corresponding to the dark band No. 5.

57. On following the active light through the spectrum, in the direction of increasing refrangibility, the dispersion was found to commence with a bright but narrow tail of pure red light, which shot right across the vessel. The light by which this tail was produced belonged to the more refrangible part of the extreme red band which is transmitted by a moderate thickness of the fluid. The activity of the incident light commenced almost abruptly: the same, it will be remembered, was the case with the absorbing power of the medium. After the tail of red light came the intense absorption band No. 1, where the dispersed light was confined to the immediate neighbourhood of the surface by which the active light entered. At this place a very bright band of dispersed light was visible on looking at the vessel from the outside. In this part of the spectrum the active and the dispersed light were both red; but that dispersion was accompanied by a change of refrangibility was shown by the effect of absorbing media. Thus the long red tail and the bright band adjacent to the surface were differently affected by a blue glass, according as it was held in the first or the second position; and the bright band, though much enfeebled, was still plainly visible through a considerable thickness of the fluid, although a stratum having a thickness of only a very small fraction of an inch was sufficient to

absorb the rays by which the band was produced. Although the dispersion continued throughout the whole of the visible spectrum and beyond, it was comparatively feeble in the brightest part of the spectrum. It became pretty copious again in the neighbourhood of the dark band No. 4, and remained copious throughout the blue and violet. In the green, the dispersed light was red, slightly verging towards orange, and in the blue and violet it was red verging a little towards brown.

58. It may seem superfluous, after what precedes, to bring forward any further proof of the reality of a change of refrangibility. Nevertheless the following experiment, which was in fact performed at an early stage of these researches, may not be deemed wholly unworthy of notice, as not involving the use either of absorbing media or of false dispersion.

A small narrow triangle of white paper was stuck on to the outside of the vessel containing the leaf-green, in such a manner that its axis was vertical, and its vertex, which was uppermost, was situated at the height of the middle of the spectrum. A narrow vertical slit was then placed at the distance of the image of the first slit, where the fixed lines were formed, and moved sideways till the light immediately beside the fixed line G passed through it. The vessel was then placed a few inches behind the slit, and moved sideways till the riband-shaped beam of homogeneous light, which passed through the second slit, was incident on the vertex of the triangle. On looking at the vessel from the front, as nearly as was convenient in the direction of the incident light, there appeared a bright vertical bar corresponding to a section of the incident beam. This bar was of two colours, namely, red in the upper half, where the light fell on the fluid, and indigo in the under half, where it fell on the paper. On refracting the whole system sideways, through a prism of moderate angle applied to the eye, the objects appeared in the following order as regards refrangibility. First came the upper half of the bright bar, which was only a very little widened by refraction, so that it consisted of red light which was approximately homogeneous. Next came the triangle, with its vertex a little rounded, and its edges tinged with prismatic colours. The vertex, which had formerly coincided with the bright bar, now lay a little to one side of its upper half. The triangle was of course seen by means of the diffused light of the room, which was not perfectly dark, and therefore its refrangibility must have corresponded to the brightest part of the spectrum, or nearly so. Lastly came the under half of the bright bar, which was much more refracted than the triangle, so as to be shifted almost completely off it. The paper triangle was far too close to the first surface of the fluid to allow of attributing the dislocation of the bright bar to any error depending upon parallax; but to prevent all possible doubts on this score, I took care to refract the system both right and left, and the result was the same in the two cases. The conclusion is therefore inevitable, that the indigo light which had changed its colour by dispersion from leaf-green had changed its refrangibility at the same time.

59. In viewing a solution of leaf-green in a pure spectrum, I noticed a phenomenon which further indicates the close connexion which seems to exist between the absorption and internal dispersion of this fluid. On holding the eye vertically over the fluid, and looking down at the dispersed light through a red glass, I observed five minima of illumination, having for the most part the shape of teeth with their bases situated at the surface by which the light entered, and their points turned inwards. These minima occupied positions intermediate between the bands of absorption, so far at least as the positions of the latter were indicated by dark teeth pointing in the contrary direction. The first minimum was situated a little beyond the intense absorption band No. 1, and corresponded in position to the bright band No. 2. The second was situated a little further on. The maximum intervening between this and the third was but slight, so that the second and third together formed pretty nearly one broad minimum. The third and fourth were situated one at each side of the dark band No. 4, so as to correspond in position to the bright bands Nos. 4 and 5. The fifth was situated a little way beyond the bright band No. 5. This last minimum was not tooth-shaped, inasmuch as it occurred at a part of the spectrum where the dispersed light was almost confined to the surface of the fluid. These minima are best seen when the solution is rather weak. They may be perceived without using a red glass, though not so easily as with its assistance. With a stronger solution it was observed that the

first minimum ran obliquely into the dark tooth corresponding to the absorption band No. 1.

60.   The reason of the occurrence of these minima appears to be simply this, that the more copiously dispersed light is produced, the more rapidly the incident light is used up in producing it, so that minima of activity correspond to points of the spectrum at which the incident light penetrates to comparatively great distances into the fluid before it is absorbed.   The oblique position observed in the first minimum is readily explained by considering that the illumination at any point of the field of view depends conjointly upon the activity of the incident light, which is a function of its refrangibility, and upon the fraction of the incident light left unabsorbed, which last is a function both of the refrangibility and of the distance from the first surface.

61.   It seems worthy of remark, that while the quantity of dispersed light is liable to fluctuations having an evident relation to the bands of absorption which occur throughout the spectrum, the quality of the light dispersed, as regards its refrangibility, appears rather to have reference to the intense absorption band No. 1.

Extract from blue leaves of the Mercurialis perennis.

62.   The juice of this plant has the property of turning blue by exposure to the air.   Some leaves and stalks which had turned blue were treated with alcohol, and a green fluid was thus obtained much resembling in colour the ordinary solutions of leaf-green, but I think of a rather bluer green than usual.   In its mode of absorption, too, it much resembled ordinary solutions of leaf-green, to which substance no doubt the greater part of its colour was due.   Its internal dispersion however was very peculiar, for it dispersed a copious orange in place of a blood red like the extracts from fresh green leaves in general, those of the *Mercurialis perennis* included.   On analysis the dispersed· beam was found to consist chiefly of a red band, similar to that which occurs in solutions of leaf-green, and of a yellow or orange and yellow band, a good deal brighter than the former, from which it was separated by an intervening dark band.   When the fluid was

examined by the second method, it was found that the yellow dispersion was produced principally by the brightest part of the spectrum. After a considerable time the fluid lost its fine green colour, as is very often the case with solutions of leaf-green, and became yellowish brown, but the red and yellow dispersions still continued.

When the fluid was examined by the fourth method, it was found that the red rays dispersed a red, just as in a solution of leaf-green. The additional dispersion which was so conspicuous in this fluid began almost abruptly about the fixed line D. When it was first observed, the refrangibility of the orange dispersed light could hardly, if at all, be separated from that of the active light. As the lens moved on, the orange beam rapidly grew brighter, and yellow entered into it; and now it was easy to see that the beam of falsely dispersed light lay at its more refrangible limit. The orange and yellow dispersed beam was brightest at about $D\frac{2}{3}E$; but though it decreased in intensity it could be traced far beyond that point, in fact, throughout the spectrum.

63. I have generally found that when a copious dispersion commences almost abruptly at a certain point of the spectrum, it is followed by a band of absorption in the transmitted light. This law did not seem applicable to the orange dispersion exhibited by the solution just mentioned; but then it is to be remembered that the solution contained a quantity of chlorophyll, which produces absorption bands with such energy that it would naturally mask the bands which might be due to another colouring principle with which it was mixed. To try whether the law would be obeyed if the chlorophyll were got rid of, I boiled in water some portions of the root and young shoots which had turned blue, chlorophyll being insoluble in water. The solution thus obtained was red, in small thicknesses pink, and dispersed copiously a yellow or rather orange light. On subjecting the fluid to prismatic analysis, a band of absorption was seen at the place expected. Since aqueous solutions of this nature are liable to decomposition, frequently decomposing before sunlight can be obtained by which to examine them, the red solution was concentrated by evaporation and purified by alcohol, in which the orange-dispersing principle is soluble, as had already

appeared from the properties of the alcoholic solution. The alcoholic solution thus obtained remained unchanged, at least for a long time, and had the further advantage over the aqueous solution of presenting the sensitive principle more nearly in a state of isolation, though it was still contaminated by some principle which dispersed a whitish light under the influence of rays of high refrangibility.

64. The blue colouring matter may be readily extracted by cold water, but is decomposed by boiling. The blue solution dispersed an orange light like the other, but the dispersed light could not be nearly so well seen, just as would be the case were the red orange-dispersing fluid mixed with an insensible blue fluid of a much deeper colour, so that the mixture of the two would be blue. And in fact when the blue fluid was changed to red by boiling the colour became far less intense.

Archil and Litmus.

65. It is stated by Sir David Brewster that a very remarkable example of internal dispersion, which had been pointed out to him by Mr Schunck, is exhibited in an alkaline or in an alcoholic solution of a resinous powder produced from orcine by contact with the oxygen of the air. Not being able readily to procure a specimen of orcine, I tried archil, and obtained from it and litmus some very remarkable solutions.

In the fluid state in which archil is sold, the colour is much too deep for convenient optical examination. When a small quantity of archil is diluted with a great deal of water, the diluted fluid is very sensitive. It is red by transmission, or in small thicknesses purple, but exhibits by dispersive reflexion a pretty copious but rather sombre green.

66. When the fluid was examined by different methods, it was found to disperse a little red, some orange, and a great deal of green. The red dispersion was so slight, that in observing by the third method it appeared doubtful whether there was any except false dispersion. It commenced in the red, when the active and dispersed lights had the same refrangibility, or nearly so. The orange dispersion commenced about the fixed line D,

the dispersed light being at first nearly homogeneous, and of the same refrangibility as the active light. On proceeding onwards in the spectrum, in observing by the fourth method, the orange beam became brighter, and yellow entered into it, but no colour beyond that, so that the orange and yellow beam was left behind by the beam of falsely dispersed light, from which it was separated by a perfectly dark interval. The green dispersion began about *b*, or a little beyond, coming on almost abruptly. The manner of its commencement was best observed by the fourth method, by holding a prism to the eye while the lens was moved through the spectrum. In this way it was found that on arriving at the point of the spectrum above mentioned, a gleam of green light shot across the dark space which before separated the beam of falsely dispersed light from the orange beam of truly dispersed light. As the lens moved on, the green dispersed light grew brighter, but its more refrangible limit did not seem to pass, or at least much to pass, the refrangibility it had at first; so that the green beam of truly dispersed light was almost immediately left behind by the beam of falsely dispersed light. The former, on being left behind, soon died away.

67. We might suppose either that the red, orange and green dispersions are due to the same sensitive principle, or that they are produced by three distinct sensitive principles mixed together in the solution. The latter would appear the more probable supposition, to judge by the apparent want of connexion between the three dispersions. This view is strongly confirmed by the following results. Some ether was poured on archil in the fluid state, and after being gently moved about and allowed to stand, a little was withdrawn without agitation. A purplish rose-coloured fluid was thus obtained, which was highly sensitive, exhibiting the orange and green dispersions but not the red. The orange dispersion was far more copious, in proportion to the whole quantity of dispersed light, than had been the case with archil diluted with water.

Some archil was violently agitated with ether, and after subsidence the ether was withdrawn. This ethereal solution was much deeper in colour than the former, and exhibited the red dispersion in addition to the orange and green. On adding a small quantity of water, and agitating, a separation, or at least partial separation, of the sensitive principles took place; for the upper fluid exhibited the orange dispersion abundantly, but none of the red, and little or none of the green, while the under fluid exhibited the green and red dispersions with little, if any, of the orange. The upper fluid exhibited a pretty copious dispersive reflexion of reddish orange, and the under fluid a remarkably copious reflexion of a fine green. A similar separation, more or less perfect, took place in other cases, the dispersion of orange bearing to that of green a greater ratio in the ether than in the water. Some of the green-dispersing fluids thus obtained were most remarkable on account of the extraordinary copiousness of the reflected green, and the strange contrast which it presented to the transmitted tint, which was a purplish red.

The red dispersion in the second ethereal solution, though decided, was by no means copious. In the case of archil merely diluted with water, it had been so slight that its existence might have been considered doubtful. It might be supposed that the first solution was not sufficiently concentrated to exhibit the red dispersion, in which case the red and green dispersions might have been due to the same sensitive principle. But an ethereal extract from dried archil, which was plainly concentrated enough, did not exhibit the red dispersion, although it did exhibit the orange and green dispersions. None of the sensitive principles appear to constitute the chief part of the colouring matter of this dye-stuff.

68.   When some of these ethereal solutions were examined by the third method, with a lens of shorter focus than usual, the appearance was very singular. At the less refrangible end of the spectrum the incident light was quite inactive; and then, on reaching a certain point, a copious dispersion of orange commenced abruptly. This continued with no particular change for some distance further on, when it passed abruptly into green. The fourth method showed however that the former dispersion continued, and was only masked, in the third method of observation, by a new and more powerful dispersion of green which then commenced. And in fact when the green-dispersing principle was separated, or partially separated, by water, the orange dispersion was seen to continue where before it appeared to have been exchanged for green.

69. I ought here to mention that a similar separation did not take place on the addition of water only to an ethereal extract from archil previously dried. The condition which determined the separation in the first case appeared to be the presence of a small quantity of ammonia, which would evaporate on drying the archil. And in fact when a small quantity of ammonia was added to the extract from dried archil, a partial separation was effected. I do not here enter into the question whether one of the sensitive principles may be obtained from the other, whether, for example, a chemical combination of the orange-dispersing principle with ammonia might disperse a green, or a green with a little orange. A solution containing a mixture of the same substance in two different states of chemical combination, both compounds being sensitive, is not the less justly regarded as containing two distinct sensitive principles.

70. The preceding results are mentioned, not for their own sake, but merely for the sake of the method of examination employed. The results indeed are so imperfect as to be worthless on their own account. A complete optico-chemical examination of archil and litmus would itself alone furnish a subject for research of no small extent; but it belongs rather to chemistry than to general physics. It is quite possible that internal dispersion may turn out of importance as a chemical test. The dispersing such a tint, and the having the dispersed light produced by light of such a refrangibility, form together a double character of so peculiar a nature that it enables us, so to speak, *to see* a sensitive principle in a solution containing many substances, some of them, perhaps, coloured, so that the colour of the solution may be very different from what it would be if the sensitive principle were present alone.

71. The law mentioned at the beginning of Art. 63 did not seem very applicable to archil when the fluid was merely diluted with water. But when the orange-dispersing and green-dispersing principles were obtained, as it would appear, more nearly in a state of isolation, by means of ether and water, the law was found to be obeyed. Thus, when the ethereal solution which exhibited the orange dispersion and little else was examined by the third method, the dispersion was found to commence with a tail of light followed by a dark tooth, indicating the position of a band of

absorption. When the light transmitted by a certain thickness of this fluid was subjected to prismatic examination, it was found to consist of red followed by some orange, when the spectrum was cut off with unusual abruptness. After a broad dark interval came the most refrangible colours faintly appearing. Those solutions which exhibited a copious dispersion of green gave, in addition to a band obliterating the yellow, a very distinct band separating the green from the blue. A similar band, but by no means distinct, might be seen in archil merely diluted ; and it is particularly to be observed that this band, which occurred a little above the point of the spectrum where the green dispersion commenced, became more conspicuous when the green-dispersing principle was present more nearly in a state of isolation.

72. Two portions of litmus were treated, one with ether and the other with alcohol, which were allowed to remain in contact with the solid. Both extracts, but especially the latter, were highly sensitive, exhibiting dispersions of orange and green similar to archil, and due apparently to the same sensitive principles. The ethereal extract dispersed chiefly orange, while the alcoholic extract dispersed orange and green in nearly equal quantities. The latter extract exhibited a remarkably copious dispersive reflection of a colour nearly that of mud, and was altogether one of the strangest looking fluids that I have met with. On viewing it in such a manner that no transmitted light entered the eye, one might almost have supposed that it was muddy water taken from a pool on a road. But when the bottle containing it was held between the eye and a window the fluid was found to be perfectly clear, and of a beautiful purple colour.

Canary Glass.

73. Among media which possess the property of internal dispersion in a high degree, Sir David Brewster mentions a yellow Bohemian glass, which dispersed a brilliant green light. This led me to seek for such a glass, and it proved to be pretty common in ornamental bottles and other articles. The colour of the glass by transmitted light is a pale yellow. Its ornamental character depends in a great measure upon the internal dispersion, which occasions a beautiful and unusual appearance in the articles made

of it. The commercial name of the glass is canary glass. The following observations were made with a small bottle of English manufacture.

74. When the sun's light was admitted without decomposition the dispersed beam was yellowish green. The dispersion was so copious that when a large lens was used the dispersed beam approached to dazzling. The prismatic composition of this beam was extremely remarkable. The beam was found on analysis to consist of five bright bands, which were equal in breadth and equidistant, or at least very nearly so, and were separated by narrow dark bands. The first bright band was red, the second reddish orange, the third yellowish green, the fourth and fifth green. I have very frequently observed dark bands, or at least minima, in the spectrum resulting from the prismatic analysis of dispersed beams, but I have not met with any example so remarkable as this, except in a class of compounds which the properties of canary glass led me to examine.

75. On analysing a beam of sunlight transmitted through a certain thickness of the glass, there was found to be a dusky absorption band a little below F, another less distinct at $F\frac{1}{2}G$, and the spectrum was cut off a little below G.

76. When the glass was examined by the third method, the dispersion was found to commence abruptly about the fixed line b. It remained remarkably copious throughout the whole of the visible spectrum and far beyond, with the exception of a band beginning a little above F, and having its centre at about $F\frac{1}{3}G$, where there was a remarkable minimum of activity. This band, it will be observed, was situated between the bands of absorption already mentioned. The tint of the dispersed light appeared to be uniform throughout, except perhaps where the dispersion was just commencing. This was the best medium I have met with for showing the fixed lines of extreme refrangibility, though some others were nearly as good.

77. On examining the glass by the fourth method, it was found that the dispersion commenced nearly where the dispersed light ended, that is, the lowest refrangibility of the rays capable of being dispersed was nearly the same as the highest refrangibility

of the rays constituting the dispersed beam exhibited by white light as a whole. The dispersion appeared indeed to commence a little earlier, at about the refrangibility of the fourth dark band in the spectrum of the entire dispersed beam. When the small prism was held to the eye with one hand, while the small lens in the board was gradually moved with the other, in a direction from the red to the violet, through the part of the spectrum where the dispersion commenced, it was found that the region of the first four bands was lighted up almost simultaneously, the whole field of view having been previously dark. When the lens was moved a very little further on the dispersed beam with its five bands was formed complete. Indeed the whole five appeared almost simultaneously. Speaking approximately, and in fact with almost perfect accuracy, we may say that if white light be conceived to be decomposed into two portions, the first containing rays of all refrangibilities up to that of the fixed line b, or thereabouts, and the second containing rays of all greater refrangibilities, the dispersed light produced by white light as a whole belongs exclusively to the first portion; and yet, were the bottle illuminated by the first portion alone, no dispersion whatsoever would be produced, whereas were it illuminated by the second portion alone, which contains not a ray having the same refrangibility as any one of the dispersed rays, the dispersion would be exhibited in full perfection.

Common Colourless Glasses.

78. Sir David Brewster states that he has met with many specimens, both of colourless plate and colourless flint glasses, which disperse a beautiful green light. All the colourless glasses which I have examined dispersed light internally to a greater or less extent, with the exception of some few specimens belonging to Dr Faraday's experiments. A beautiful green seems to be the commonest tint of the dispersed beam, and this I have found in wine glasses, decanters, apothecaries' bottles, pieces of unannealed glass, &c.; also in many specimens of plate and crown glass. The green was generally of a finer tint than that dispersed by the canary glass, but was not near so copious. On analysis it was found to consist usually of red and green separated by a dark band, or rather a minimum of brightness. Those specimens

which were examined by the third and fourth methods were found to exhibit a little false dispersion, produced chiefly in the brightest part of the spectrum, but the greater part was true dispersion. This dispersion was produced chiefly by a rather narrow band, comprising the fixed line G, where there appeared to be a remarkable maximum of sensibility. The line G lay a little above the lower limit of the band. Below the band dispersion also took place, though not near so copiously, and there appeared to be another maximum of sensibility some way further down in the spectrum; but above the band dispersion almost entirely ceased of a sudden; a very unusual circumstance when the active and the dispersed light are well separated in refrangibility. The position of the band in the spectrum, and the distribution of the illumination in it, which are very peculiar, were the same in all the specimens which were sufficiently sensitive to admit of being examined by the third method, but the tint of the dispersed light was not quite the same.

79. Orange-coloured glasses are frequently met with which reflect from one side, or rather scatter in all directions, a copious light of a bluish-green colour, quite different from the transmitted tint. In such cases the body of the glass is colourless, and the colouring matter is contained in a very thin layer on one face of the plate. The bluish-green tint is seen when the colourless face is next the eye. As this phenomenon was supposed by Sir John Herschel to offer some analogy with the reflected tints of fluor-spar and a solution of sulphate of quinine, I was the more desirous of determining the nature of the dispersion. It proved on examination to be nothing but false dispersion, so that the appearance might be conceived to be produced by an excessively fine bluish-green powder contained in a clear orange stratum, or in the colourless part of the glass immediately contiguous to the coloured stratum. The phenomenon has therefore no relation to the tints of fluor-spar or sulphate of quinine. It is true that the very same glass which displayed a superficial reflexion of bluish green, when examined by condensed sunlight exhibited also, in its colourless part, a little true dispersion, just as another colourless glass would do. But this has plainly nothing to do with the peculiar reflexion which attracts notice in such a glass.

Observations on the preceding results.

80. There is one law relating to internal dispersion which appears to be universal, namely, that when the refrangibility of light is changed by dispersion it is *always lowered*. I have examined a great many media besides those which have been mentioned, and I have not met with a single exception to this rule. Once or twice, in observing by the fourth method, there appeared at first sight to be some dispersed light produced when the small lens was placed beyond the extreme red. But on further examination I satisfied myself that this was due merely to the light scattered at the surfaces of the large prisms and lens, which thus acted the part of a self-luminous body, emitting a light of sufficient intensity to affect a very sensitive medium.

81. Consider light of given refrangibility incident on a given medium. Let some numerical quantity be taken for a measure of the refrangibility, suppose the refractive index in some standard substance. Let the refrangibilities of the incident and dispersed light be laid down along a straight line AX (fig. 2) taken for the axis of abscissæ; let AM represent the refrangibility of the incident light, and draw a curve of which the ordinates shall represent the intensities of the component parts of the truly dispersed beam. According to the law above stated, no part of the curve is ever found to the right of the point M; but in other respects its form admits of great latitude. Sometimes the curve progresses with tolerable uniformity, sometimes it presents several maxima and minima, or even appears to consist of distinct portions. Sometimes it is well separated from M, as in fig. 2; sometimes it approaches so near to M that the most refrangible portion of the truly dispersed beam is confounded with the beam due to false dispersion.

82. Let $f(x)$ be the ordinate of the curve corresponding to the abscissa x, a the abscissa of the point M. Since $f(x)$ is equal to zero when x exceeds a, the curve must reach the axis at the point M at latest, unless we suppose the function capable of altering abruptly, as is represented in fig. 3. I do not think that such an abrupt alteration, properly understood, is necessarily in contra-

diction with the law of continuity. For the sake of illustration, let us consider the phenomenon of total internal reflexion. Let P be a point in air situated at the distance z from an infinite plane separating air from glass. Conceive light having an intensity equal to unity, and coming from an infinitely distant point, to be incident internally on this plane at an angle $\gamma + \theta$, where γ is the angle of total internal reflexion. The intensity at P is commonly, and for most purposes correctly, considered as altering abruptly with θ, having, so long as θ is negative, a finite value which does not vanish with θ, but being equal to zero when θ is positive. The mode in which the law of continuity is in this case obeyed is worthy of notice. In the analytical expression for the vibration, when θ passes from negative to positive, the coordinate z passes from under a circular function into an exponential with a negative index, containing in its denominator λ, the length of a wave of light. As θ increases through zero, the expression for the vibration alters continuously; but if z be large compared with λ it decreases with extreme rapidity when θ becomes positive. On account of the excessive smallness of λ, it is sufficient for most purposes to consider the intensity as a function of θ which vanishes abruptly; and indeed it would be hardly correct to consider it otherwise. For the use of the term *intensity* implies that we are considering light as usual, whereas those phenomena which require us to take into account the disturbance in the second medium which exists when the angle of incidence exceeds that of total internal reflexion, lead us to consider the nature as well as the magnitude of that disturbance, which no longer consists of a series of plane waves constituting light as usual. It is in some similar sense that I mean to say that we may suppose the function $f(x)$, which expresses the intensity of the truly dispersed light, to alter abruptly, without thereby implying any violation in the law of continuity. In observing by the fourth method, the portion of the spectrum operated on, though it may be small, is necessarily finite, and in some cases no separation could be made out between the beams of truly and falsely dispersed light. Hence I cannot undertake to say from observation, whether the variation of $f(x)$ be always continuous, though sometimes very rapid, or be in some cases actually abrupt. I think, however, that observation rather favours the former supposition, a supposition which, independently of observation, seems by far the more likely.

83. Although the law mentioned in Art. 80 is the only one which I have been able to discover, relating to the connexion between the intensity and the refrangibility of the component parts of the dispersed beam, which appears to be always obeyed, and which admits of mathematical expression, there are some other circumstances usually attending the phenomenon which deserve notice.

When dispersion commences almost abruptly on arriving at a certain point of the spectrum, the dispersed beam is very frequently almost homogeneous at first, and of the same refrangibility as the active light. If the dispersed beam, when first perceived, be decidedly heterogeneous, its refrangibility extends almost, if not quite, to that of the active light, so that it is difficult, if not impossible, to separate the beams of truly and falsely dispersed light. On the other hand, when dispersion comes on gradually, it is generally found that the refrangibility of even the most refrangible part of the dispersed beam does not come up to that of the active light.

Thus in the cases of the red dispersion exhibited by a solution of leaf-green, and of the orange dispersions exhibited by solutions obtained from archil and from the *Mercurialis perennis*, the dispersed light was at first nearly homogeneous, and of the same refrangibility as the active light. In the case of the green dispersions shown by a solution obtained from archil, and by canary glass, the dispersed light was heterogeneous from the first; but still, when it first commenced, a portion of it had nearly the same refrangibility as the active light. In a solution of sulphate of quinine the dispersion came on gradually, being perceptible when the active light belonged to the middle of the spectrum; and in this case the dispersed light consisted of colours of low refrangibility. The bright part of the dispersion however came on pretty rapidly, when the active light approached the extreme limit of the visible spectrum, and accordingly the dispersed beam consisted in that case chiefly of light of high refrangibility.

84. The mode of absorption of any medium may very conveniently be represented by a curve, as has been done by Sir John Herschel. To represent geometrically in a similar manner the mode of internal dispersion, would require a curved surface. Let

the refrangibility of light be measured as before, and suppose for simplicity's sake the intensity of the incident light to be independent of the refrangibility, so that dy may be taken to represent the quantity of incident light of which the refrangibility lies between y and $y + dy$. Considering the effect of this portion of the incident light by itself, let x be the refrangibility of any portion of the dispersed light, and $z\,dx\,dy$ the quantity of dispersed light of which the refrangibility lies between x and $x + dx$. Then the curved surface, of which the coordinates are x, y, z, will represent the nature of the internal dispersion of the medium. We must suppose the intensity of the incident light referred to some standard independent of the eye, since the illuminating power of the rays beyond the violet, and even of the extreme violet, is utterly disproportionate to the effect which in these phenomena they produce.

From the nature of the case, the ordinate z of the surface can never be negative. The law mentioned in Art. 80 may be expressed by saying, that if we draw through the axis of z a plane bisecting the angle between the axes of x and y, at all points on the side of this plane towards x positive, the curved surface confounds itself with the plane of xy.

85. Let us consider the form of this surface in two or three instances of internal dispersion. For facility of explanation, suppose the plane of xy horizontal, let x be measured to the right, y forwards, and z upwards. Let a line drawn in the plane of xy through the origin, and bisecting the angle between the axes of x and y, be called for shortness the line L. In all cases the surface rises above the plane of xy only to the left of the line L.

In the case of a solution of leaf-green, the surface consists as it were of two mountain ranges running in a direction parallel to the axis of y, or nearly so. The first range, if prolonged, would meet the axis of x at a point corresponding to the place of the dark band No. 1 in the red, or nearly so. The second would meet it somewhere in the place corresponding to the green. The green range is much broader than the red, but very much lower, and is comparatively insignificant. The ridge of the red range is by no means uniform, but presents a succession of maxima and minima. The range commences at the end nearest to the axis of x with a very high peak, by far the highest in the whole surface. In following the ridge forwards, five minima or passes may be observed, with hills intervening. The ordinates y of the first four of these minima correspond to the refrangibilities of the bright bands Nos. 2, 3, 4 and 5. The last minimum lies a little further on. Whether similar minima exist in the green range is not decided by observation, on account of the faintness of the green dispersed light.

In the case of canary glass, the surface consists of five portions like mountain ranges running parallel to the axis of y, and having abscissæ belonging to the red, reddish orange, yellowish green, green, and more refrangible green, respectively. These ranges do not all start from the immediate neighbourhood of the line L, but on the side towards the axis of x end almost in cliffs, at points at which the ordinate y is nearly equal to the abscissa of the fifth range, perhaps a little less. Thus the first three ranges are well separated from the line L. The ranges are intersected by a sort of valley running parallel to the axis of x, and having for its ordinate y the refrangibility of $F\frac{1}{2}G$. With the exception of the minima which occur where the ranges are intersected by this valley, the ridges run on very uniformly, and it is only very gradually that the ranges die away.

The form of the surface which expresses the internal dispersion of a solution of sulphate of quinine, may be gathered from the description of that medium. In this case the surface resembles a rising country, not intersected by any remarkable mountain ranges or valleys.

Fig. 4 is a rude representation of the internal dispersion in a solution of leaf-green. The curves represented in the figure must be supposed to be turned through 90° about the lines on which they stand, and will then represent sections of the surface already described, made by vertical planes parallel to the axis of x. OL is the straight line bisecting the angle xOy. The figure is merely intended to assist the reader in forming a clear conception of the general nature of the phenomena, and must not be trusted for details. No attempt is made to represent the several maxima and minima in the intensity of the red beam of dispersed light. In any such figure, if we suppose homogeneous light to be incident on the medium, and wish to lay down the place of the falsely dispersed beam, we have only to draw a straight line parallel to the axis of x, through the point in the axis of y which corresponds to the refrangibility of the incident light, and find where this line cuts the straight line OL which bisects the angle xOy.

On the cause of the clearness of fluids, notwithstanding a copious internal dispersion which they may exhibit.

86. It has been already remarked, that though water holding a water colour in suspension makes an admirable imitation of a highly sensitive fluid, when the latter is viewed by dispersive reflexion alone, the two fluids have a totally different appearance when viewed by transmitted light. The cause of this difference appears to be plain enough. The light due to internal dispersion emanates from each portion of the fluid which is under the influence of the active light, and emanates apparently in all directions alike. I have not attempted to determine experimentally whether the intensity is strictly the same in all directions. The experiment would be very difficult, especially for directions nearly coinciding with that of the active light, because in that case the light which was really due to internal dispersion would be mixed up with the glare which is always found in the neighbourhood of light of dazzling brightness. However, I have seen nothing which led me to suppose that the intensity was different in different directions. We may express the results of observation extremely well, by saying that the fluid or solid medium is self-luminous so long as it is under the influence of the active light.

Accordingly, when a bright object, such as the sky, or the flame of a candle, is viewed through a highly sensitive fluid, the regularly transmitted light is accompanied by some side light due to internal dispersion. The latter, however, emanating in all directions alike from the influenced particles, is too faint, when contrasted with the regularly transmitted light, to make any sensible impression on the eye. But when a fluid, itself insensible, holds in suspension a great number of solid particles of finite size, the light reflected from such particles is reinforced, in directions nearly coinciding with that of the incident light, by a great quantity of diffracted light, so that a bright object viewed through such a fluid is surrounded by a sort of nebulous haze, giving the fluid a milky appearance.

Washed Papers.

87. In a paper "On the Action of the Rays of the Solar Spectrum on Vegetable Colours," Sir John Herschel mentions a peculiarity which he had observed in paper washed with tincture of turmeric, which consists in its being illuminated, when a pure spectrum is thrown on it, to a much greater distance at the violet end than is the case with mere white paper*. This phenomenon was attributed by Sir John to a peculiarity in its reflecting power, and was considered as a proof of the visibility of the ultra-violet rays. The colour of the prolongation of the spectrum was yellowish green. Sir John appears to have been in doubt whether the greenish yellow colour was to be attributed to the mixture of the true colour of the ultra-violet rays with the yellow of the paper due to diffused light, or to the real colour of the ultra-violet rays themselves, which on that supposition would have been incorrectly termed "lavender."

88. The fact of the change of refrangibility of light having been established, there could be little doubt that the true cause of the extraordinary prolongation of the spectrum on paper washed with tincture of turmeric was very different from what Sir John Herschel had supposed, and that it was due to a change of refrangibility in the incident light, which was produced by the medium in a solid state. Tincture of turmeric has already been mentioned as a medium which possesses in a high degree the property of internal dispersion. It was the observation of Sir John Herschel's already mentioned, which led me to try this medium. But it is by no means essential that a sensitive substance should be in solution, or in the state of a transparent solid, in order that the change of refrangibility which it produces should admit of being established by direct experiment, although of course the mode of observation must be changed.

89. A piece of paper was prepared by pouring some tincture of turmeric on it, and allowing it to dry. In this way the part which was deeply coloured by turmeric was in juxtaposition with the part which remained white, which was convenient in contrasting the effects of the two portions. The sun's light being reflected horizontally into a darkened room through a vertical

* *Philosophical Transactions* for 1842, p. 194.

slit, the paper was placed in a pure spectrum formed in the usual manner. On the coloured part the fixed lines were seen with the utmost facility far beyond the line H, on a yellowish ground. The colours too of all the more highly refrangible part of the spectrum were totally changed. From the red end, as far as the line F, or thereabouts, there was no material change of colour; but a little further on a very perceptible reddish tinge came on, which was quite decided at $F\frac{1}{2}G$, where it was mixed with the proper colour of that part of the spectrum. About $G\frac{4}{}H$ the colour became yellowish. The reality of a change of refrangibility was easily proved by refracting the spectrum on the screen by a prism applied to the eye. When the refraction took place in a plane parallel to the fixed lines, they were seen distinctly throughout the spectrum; but when it took place in a plane perpendicular to the former, the fixed lines in the less refrangible part of the spectrum, and as far as F, were distinctly seen; but in the rest of the spectrum they were more or less confused, or even wholly obliterated, according to their original strength, the refracting angle and dispersive power of the prism, and its distance from the paper. With a prism of small angle the edges of the broad bands H were seen tinged with prismatic colours.

90.   The change of refrangibility was further shown by the following observation. The paper was placed in the pure spectrum in such a manner that the line of junction of the coloured and uncoloured parts ran lengthways through the spectrum, so that the same fixed line was seen partly on the coloured and partly on the uncoloured portion. On viewing the whole through a prism of moderate angle applied to the eye, and so held as to refract the system in a direction perpendicular to the fixed lines, the line F was seen uninterrupted, but G was dislocated, the portion formed on the yellow part of the paper being a good deal less refracted than that formed on the white. The latter was indeed faintly prolonged into the yellow part of the paper, so that on this part G was seen double; but the image which was by far the more intense of the two was less refracted than that formed on the white paper. The whole appearance was such as to create a strong suspicion of some illusion, as if some other group of fixed lines formed on the yellow part of the paper

had been mistaken for G, though certainly no reason appears why such a group should not have had its counterpart on the white part. However, to remove all doubts, I refracted the system in the direction of the fixed lines, and then turned the prism round the axis of the eye through 90°, when the plane of refraction was situated as before. At first the two portions of the line G were of course seen in the same straight line; and the perfect continuity with which, as the prism turned round, the appearance changed into what had been first seen, left not the shadow of a doubt as to the reality of the dislocation.

91. The cause of the whole appearance is plain enough. The light coming from the illuminated part of the yellow paper consisted, in the neighbourhood of G, of two portions; the first, indigo light, which had been scattered in the ordinary way; the second and larger portion, heterogeneous light having a mean refrangibility a good deal less than that of G, which had arisen from homogeneous light of higher refrangibility. The absence of the first occasioned the faint prolongation of the more refracted part of the line G; the absence of the second gave rise to the less refracted part.

92. The broad bands H were seen faintly but quite distinctly on the white paper. On refracting them sideways by a prism of moderate angle held to the eye, they became confused, and tinged with prismatic colours. The confused images of these bands, seen in the white and coloured parts, were nearly continuous. It thus appears that the visibility of the bands H on the white paper was due to a change of refrangibility which that substance had produced in violet light of extreme refrangibility.

93. Effects similar to those produced by paper coloured by tincture of turmeric are also produced by turmeric powder, or even by the root merely broken across. Notwithstanding the roughness of the latter, the bands H and fixed lines far beyond are seen with the utmost facility.

94. These phenomena are much better observed by covering the slit with a deep blue glass, which absorbs all the bright part of the spectrum, while it freely transmits the violet and invisible rays, which are mainly efficient in this class of phenomena. In

this way fixed lines may be seen on common white paper far beyond *H*. These lines may be seen without the use of the blue glass, by allowing the bright colours to pass by the edge of the paper, and receiving on it only the extreme violet and invisible rays.

95.   Paper coloured by turmeric having exhibited so well the sensibility of that substance, I was induced to try various other washed papers, in fact, papers washed with most of the fluids with which I had made experiments.   I found almost always that sensitive solutions gave rise to sensitive papers, exhibiting a change of refrangibility of the same character as that shown by the solution.   Besides the turmeric paper, the two most remarkable were paper washed with a pretty strong solution of sulphate of quinine, and paper washed with the extract from the seeds of the *Datura stramonium*.   I should here observe, that it was not till long after the time when these experiments were made that I was acquainted with the high sensibility of a decoction of the bark of the horse-chestnut.   The former of the papers just mentioned exhibited the fixed lines of the invisible rays on a blue, and the latter on a green ground.   The dispersion produced by the quinine paper was not exhibited so early in the spectrum as in the case of turmeric, nor was it so copious in the extreme violet rays, and for some distance further on, but the quinine paper seemed superior to the other for showing the fixed lines of extreme refrangibility.   With the turmeric paper the group *n* was plain enough, but with the quinine paper I have seen some fixed lines of the group *p*.   The stramonium paper was, on the whole, I think, superior to the quinine paper in point of the copiousness of the dispersed light, but seemed hardly equal to it for showing the fixed lines of extreme refrangibility.   However, it is likely that paper washed with a solution of the sensitive principle in a state of purity would have been quite equal to the quinine paper in this respect.

96.   A washed paper is a little more convenient for use than a solution, but, as might be expected, it does not show the fixed lines with quite as much delicacy, nor is it quite so good for tracing the spectrum to the utmost limits to which it can be traced with the substance employed.

97. The sensibility of fresh leaf-green could not be made out on a washed paper by this mode of observation, but the sensibility of the substance extracted by alcohol from black tea, from which the brown colouring matter had been removed by hot water, was plainly exhibited by the redness which it produced in the highly refrangible part of the spectrum.

98. Paper washed with a solution of guaiacum seemed an exception to the general rule; but this is not to be wondered at, since a paper prepared in this manner is turned green when exposed to the light, and it is difficult to prevent some degree of discoloration. That the fluid state is not essential to the exhibition of the sensibility of this substance, was however plainly shown by the high degree of sensibility of the solid resin from which the solution was made. In this case the bands H were seen on a greenish ground. The dispersion of a fine blue light under the influence of rays of still higher refrangibility was hardly, or not at all, exhibited by the solid resin.

99. Shell-lac, common resin, glue, are all highly sensitive. The ground on which the fixed lines in the neighbourhood of H are seen is brown in the case of shell-lac, and greenish in the case of resin and glue. The sensibility of glue is evidently not due to gelatine, for isinglass is almost, if not quite, insensible. These are merely a few instances of sensibility: I shall defer further mention of the subject till I have described a better mode of observation. I will merely observe for the present, that several washed papers proved not greatly inferior to turmeric paper for showing the fixed lines about and beyond H.

Effect of refracting a Narrow Spectrum in a Vertical Plane.

100. In the arrangement last described, when a short slit is used, the spectrum received on the washed paper or other substance is of course narrow, so that the fixed lines formed on the paper are but short, and may roughly be regarded as mere points. If, now, the whole be viewed through a prism, so as to be refracted in a vertical plane, the effect is very striking. For facility of explanation suppose the red to be to the left, and the rays to be refracted upwards, so that to the observer the image

is thrown downwards. The original spectrum on the screen is decomposed by the prism held to the eye into two spectra, which diverge from each other. The first of these runs obliquely downwards from left to right, and contains the natural colours of the spectrum from red to violet. It consists of light which has been scattered in the ordinary way by the substance on which the primary spectrum is received, and the cause of its obliquity is evident. The second spectrum is horizontal, that is to say, it approximates to the form of a long rectangle having its longer sides horizontal. Of course it would be theoretically possible to render the vertical sides the longer, but when the whole arrangement of the apparatus is such as to be convenient for observation, the horizontal sides are much longer than the others. In this second spectrum the colours run *horizontally*, that is to say, the lines of equal colour are horizontal. The interruptions of the primary spectrum corresponding to fixed lines, almost reduced to points, are now elongated, so that in this strangely formed spectrum the principal fixed lines of the solar spectrum are seen running *across* the colours.

101. It will be convenient to have a name for the second of the two spectra above mentioned. As the term *secondary spectrum* is already appropriated to something altogether different, I shall call it the *derived spectrum*. The first of the diverging spectra may be called the *primitive spectrum*, while the original spectrum, considered as not yet decomposed by the prism held to the eye, may be called, for distinction, as in fact it has been already called, *primary*.

102. In accordance with the law enunciated in Art. 80, it is found that the derived spectrum appears *always on one and the same side* of the primitive, being *less refracted*.

103. The brilliancy of the derived spectrum, its extent, both vertically and horizontally, the colours of which it mainly consists, the distribution of its illumination in a horizontal direction, all depend upon the nature of the substance upon which the primary spectrum is received. As a general rule, it may be stated that it starts from the neighbourhood of the brightest part of the primitive spectrum, and extends from thence onwards to a good distance beyond the extreme violet; and that with a given substance its

colour is pretty uniform, that is, does not much change in passing from one vertical section to another. Sometimes the derived spectrum remains very bright up to its junction with the primitive, or at least till it gets so near that the superior brilliancy of the primitive spectrum prevents all observation on the derived; sometimes it remains dull to a considerable distance from the primitive spectrum, and then, opposite a highly refrangible part of the primitive spectrum, a strong illumination comes on in the derived, lasts for some distance, and afterwards gradually dies away. Many of the results mentioned in this paragraph are better observed by a somewhat different method, which will shortly be described.

104. It has been already stated that the bands H were distinctly seen on common white paper, the substance usually employed as a screen in experiments on the spectrum, but that this was due to a change of refrangibility produced in the extreme violet rays. These same bands have been seen on paper in the experiments of others, though of course their visibility was not attributed to its true cause. By the method of observation described in Art. 100, or still better, by a method not yet explained, it may be seen that the change of refrangibility produced by white paper is by no means confined to the extreme violet rays, and those still more refrangible, but extends from about the middle of the spectrum to a good distance beyond the extreme violet. The distance to which the illumination can be traced by means of light merely scattered in the ordinary way, may be seen by examining the primitive spectrum. In the primitive spectrum formed on white paper and other white substances, I have not been able to trace the illumination beyond the edge of the broad band H, which accords very well with the illuminating power of the extreme violet when received directly into the eye.

Illuminating Power of the Rays of High Refrangibility.

105. The prolongation of the spectrum seen on turmeric paper was brought forward by Sir John Herschel as a proof of the visibility of the ultra-violet rays, or rather as a confirmation of other experiments which had led him to the same conclusion. Of course, the experiment with turmeric must now be regarded as

having no bearing on the question; but from the way in which Sir John speaks of it, it would appear that he thought the other experiments not so conclusive as to be independent of the confirmation which they received from this. The experiment with the distorted spectrum, indeed, must now be put out of account, because in this experiment, as I have been informed by Sir John Herschel, the light was only thrown on a screen. Accordingly, the question of the visibility of these rays may be regarded as open to further investigation.

While engaged in some of the experiments described in Art. 89, I had occasion to form a pure spectrum in air in a well-darkened room, the slit itself by which the sun's rays entered being covered by a deep blue glass, so that no great quantity of light entered even at this quarter. Now, if ever, it would appear that the ultra-violet rays ought to be seen by receiving them directly into the eye; for the blue glass was so transparent with regard to these rays that the fixed lines far beyond H were seen with facility, even on substances, such as white paper, which stand low in the scale of sensibility; and the length of the spectrum from B to H was about an inch and a quarter, so that when the extreme violet rays entered the pupil, supposed to be held near the pure spectrum, not only the extreme red rays transmitted by the blue glass, but even the brighter part of the transmitted blue and violet rays, fell altogether outside it. However, on holding the eye a few inches in front of the pure spectrum, so as to see the fixed lines distinctly, the bands H were indeed seen with great facility; but I was not able to make out fixed lines beyond the end of the group l, that is, about the end of Fraunhofer's map. However, the eyes of different individuals may differ much in their power of being affected by the highly refrangible rays. It must be confessed that, on looking in the direction of the prisms, a good deal of blue light was seen, consisting of light which had been scattered at the surfaces of the prisms and lens. This light, though far from dazzling, was sufficient to prevent the eye from seeing excessively faint objects, even though they might be well defined. For want of a heliostat, I did not attempt an experiment I was meditating for securing a more perfect isolation of the ultra-violet rays*.

* See note B.

However, it seems to me to be a point of small importance, so far as regards its bearing on other physical questions, whether the illuminating power of these rays is absolutely null or only excessively feeble. It is quite certain that, if not absolutely null, their illuminating power is at least utterly disproportionate to the effect which they produce in the phenomena to which the present paper relates, and indeed that is true even of the violet rays. By *illuminating power*, I mean of course power of producing the sensation of light when received directly into the eye; for by giving rise to light of lower refrangibility, they are able to illuminate strongly an object on which they fall.

Mode of Observation specially applicable to Opaque Bodies.

106. In some of the experiments already described, the change of refrangibility was exhibited, which was produced by washed papers and solid bodies. There exists, however, a mode of observation far preferable to those which have already been explained as applicable to such cases, and which may even in some instances be employed with advantage in the examination of transparent bodies. In the experiment described in Art. 100, the primitive spectrum is pure, but the derived spectrum impure, on account of the finite length of the slit. Were the slit reduced to a point, it is true that the derived spectrum would become pure like the primitive, but then the quantity of light would be so small that the primary spectrum would hardly bear prismatic analysis. It is well, once for all, to examine a few sensitive opaque substances in a very pure spectrum, because then the exhibition of fixed lines running across the colours in the derived spectrum removes even the shadow of a doubt as to the reality of the change of refrangibility of the incident light. Besides this, the only theoretical advantage in having the primitive spectrum very pure is, that it might be expected to enable us to detect any very rapid fluctuations in the colour or intensity of the dispersed light. Of course, I am now speaking only with reference to experiments in which the observer is employing the spectrum to examine some substance, not employing the substance to examine the spectrum. But practically, I have not found any advantage on this account; for abrupt, or almost abrupt, changes in the colour or intensity of the dispersed light hardly ever, if ever, occur, except when the

active and the dispersed light have very nearly the same refrangibility. But such changes could not be observed even with a pure primitive spectrum, because in the place where they occur the primitive and derived spectra overlap; and independently of this, the brilliancy of the primitive spectrum would prevent all exact observation of the derived. It is true that, in the case of chlorophyll, or some of its modifications, changes of intensity having apparently somewhat the same nature were observed when the active and the dispersed light were widely separated in refrangibility. But the sensibility of this substance is difficult, if not impossible, to observe in the case of a washed paper or a green leaf, except by one of the methods not yet described, so that it is not to be expected that such fluctuations could be made out. Besides, it is to be remembered that the fluctuations observed in the case of solutions of chlorophyll were fluctuations in the rate at which dispersed light was produced, not fluctuations in the sum total of the dispersed light produced by the time the active light was exhausted. Fluctuations of the former kind by no means imply fluctuations of the latter; and indeed the circumstance, that maxima of activity in the solution correspond to minima of transparency, would seem to show that the total quantity of light dispersed, considered as a function of the refrangibility of the active light, is not subject to these fluctuations, or at least not to anything like the same extent. Now the total quantity of red light dispersed by a green leaf, or by a paper washed with a solution of chlorophyll, must depend upon the sensibility of this substance and upon its transparency conjointly, and therefore it is likely enough that such maxima and minima would not be observed, even were the dispersed light much stronger than it is.

107. Suppose now the slit by which the light enters to be placed in a horizontal instead of a vertical position, so as to lie in the plane of refraction. Corresponding to light of any given refrangibility, the image of the slit formed after refraction through the prisms and lens will now be a narrow parallelogram, which may be regarded as a horizontal line. The series of these lines, succeeding one another in a horizontal direction, and consequently overlapping, forms the spectrum incident on the body examined. This spectrum is now no longer pure, but only approximately so, a point, however, which, as we have seen, is not of much con-

sequence. But by this trifling sacrifice two very great advantages are gained. The first is increase of illumination. When the slit is vertical, the spectrum received on the body occupies a rectangle having for breadth the length of the image of the slit; but when it is horizontal, the same, or very nearly the same, quantity of light is concentrated into a rectangle having the same length as before (the length of the image of the slit being disregarded compared with that of the spectrum), but having for its breadth only the length of the image of a line drawn across the slit. Hence the intensity of the incident light is increased in the ratio of the breadth to the length of the slit. The second advantage is purity in the derived spectrum, a point of much consequence, because sometimes the composition of this spectrum presents very remarkable peculiarities. If the slit be not too long, the spectrum formed in air is still sufficiently pure to allow us to make out in a general way what are the refrangibilities of those portions of the incident light which are most efficient in producing dispersed light; and this is nearly all that can be done even when the spectrum is very pure.

108. The method of observation which has just been described is that which latterly I have almost exclusively employed in examining opaque substances. As it will be convenient to have a name for it, I shall speak of examining a substance in *a linear spectrum*. In examining substances which are only slightly sensitive, it is often highly advantageous to cover the slit with a blue glass.

109. Fig. 5 is intended to represent the usual appearance of the primary linear spectrum, and of the primitive and derived spectra. XY is the primary spectrum, as seen by the naked eye, RV, ST are the primitive and derived spectra into which it is separated by the prism held to the eye. The direction of the shading in RV is intended to represent the composition of this spectrum, which may be regarded as consisting of an infinite number of images of the slit arranged obliquely in the order of their refrangibility. The direction of the shading in ST is that of the lines of the same colour and same refrangibility. Of course the figure does not represent the amount of vertical displacement of the primary spectrum when viewed through the prism held to the eye.

110.   There is another mode of observation which I have occasionally found convenient when the object was to determine whether a substance exhibited so much as a low degree of sensibility.   In this method the sun's light was reflected horizontally through a large lens, and then transmitted through a small lens placed in the condensed beam.   The small lens was covered by a small vessel with parallel sides of glass, containing a blue ammoniacal solution of copper, or else by a deep blue glass combined with a weak solution of nitrate or sulphate of copper.   The object of the latter solution was to absorb the extreme red which is transmitted by a blue glass.   The light coming through the lens was then analysed by a prism, being received directly into the eye, or else allowed to fall on a white object which had been previously ascertained not to change the refrangibility of the light incident upon it.   I found clean white earthenware to serve very well for such an object, but each observer ought to test for himself the substance he employs. When a test object, such as white earthenware, is used, it is placed at the focus of the lens, and the spot of blue light formed upon it is analysed by a prism to see if the absorption is sufficient.   When the visible rays are considered to have been sufficiently absorbed, the object to be observed is placed at the focus of the lens, and the spot of light formed upon it is viewed through a prism.   The spectrum then seen is compared with that given by the test object.   This method of observation is rather easier than that of a linear spectrum, and is at least as delicate if the object be merely to determine whether a substance is sensitive or not, but on the whole it is not near so useful.   It may sometimes be used with advantage in the case of translucent bodies.

111.   An extremely pale solution of nitrate or sulphate of copper is sufficient to absorb the extreme red transmitted by a deep blue glass.   This is not the case with the ammoniacal solution, which does not absorb the extreme red till it is of a pretty deep blue.   Its absorbing power is greatest, not at the extreme red, but about the orange, as may be seen by using candle-light, which is richer in red rays than daylight.

112.   Another method of observation which is sometimes useful, consists in employing a large lens and absorbing medium,

as described in Art. 110, but leaving out the additional small lens. The substance to be examined is placed in the condensed beam, and viewed through an absorbing medium which is approximately complementary to the former. This method is chiefly useful in examining a confused mass of various substances. The most minute fragments of sensitive substances show themselves in this manner.

Results obtained with a Linear Spectrum.

113. When this method is applied to the examination of common objects, it is found that the property of producing a change of refrangibility in the incident light is extremely common. Thus, wood of various kinds, cork, horn, bone, ivory, white shells, leather, quills, white feathers, white bristles, the skin of the hand, the nails, are all more or less sensitive. To make a list of sensitive substances would be endless work; for it is very rare to meet with a white or light-coloured organic substance which is not more or less sensitive. I am not now speaking of organic substances obtained in a state of chemical isolation, of which some are sensitive and others insensible. That substances of a dark colour should frequently prove insensible is only what might have been expected, because the dispersed light is not reflected from the surface, but emanates from all points of a stratum of finite thickness; and in order that dispersed light should be forthcoming, it is necessary that the active light entering, and the dispersed light of a different refrangibility returning, should both escape absorption on the part of the colouring matter. Such substances usually consist of a mixture of various chemical ingredients, of which one or more may very likely be sensitive, in which case the substance may be compared to a solution of sulphate of quinine mixed with ink. Frequently however the colouring matter is itself sensitive.

114. Among sensitive substances I have mentioned the skin of the hand, which stands rather low in the scale. I have found the back of the hand a convenient test object. When the sunlight is not strong enough to show with ease the derived spectrum in the case of the hand, there is little use in attempting to observe.

115. It is needless to say that papers washed with tincture of turmeric, or with a solution of sulphate of quinine, display their sensibility in a remarkable manner when examined in a linear spectrum. The sensibility of turmeric paper is rather impaired by exposing the paper to the light, but on the other hand is materially increased by washing it with a solution of tartaric acid.

116. Paper washed with an ethereal solution from dried archil exhibited very well the sensibility of that substance. The derived spectrum consisted chiefly of two distinct portions, one containing orange and a little red, the other consisting chiefly of green, just as in the beam of dispersed light, produced by white light taken as a whole, which the solution itself exhibited. Indeed, I have found that the prismatic composition of dispersed light could be determined even more conveniently by means of a linear spectrum than by means of the beam dispersed by a solution.

117. The inside of the capsules of the *Datura stramonium* is nearly white, and apparently uniform. But when the capsules are examined in a linear spectrum, certain patches shine out like bright clouds in the invisible rays. The whole of the inside is sensitive, as such substances almost always are, but these patches, which are probably spots against which the seeds have pressed, are remarkably so. The capsules were examined after they had begun to burst.

118. By means of a linear spectrum the sensibility of chlorophyll may be detected in a green leaf. It is exhibited by the appearance in the derived spectrum of a narrow pure red band of remarkably low refrangibility. The refrangibility is so low that I have always found this band separated from the derived spectrum due to other sensitive substances with which chlorophyll or one of its modifications might have been mixed.

119. The petals of flowers, so far as I have examined, are as a class rather remarkable for their insensibility, some appearing quite insensible, and others only slightly sensitive. The bright yellow chaffy involucre of a species of everlasting, proved, however, highly sensitive, and its sensibility was also displayed in

an alcoholic solution. This medium was sensitive enough to exhibit a pretty copious dispersive reflexion of a pale greenish yellow light. Its sensibility was more confined than usual to the rays of very high refrangibility.

120. Among petals, the most remarkable which I have observed are those of the purple groundsel (*Senecio elegans*). These petals disperse a red light, more copious than is usual among petals. If a petal be placed behind a slit, and the transmitted light be analysed, it is found to exhibit three re- markable bands of absorption, much resembling those of blue glass, but closer together, and beginning later in the spectrum, the first appearing about the place of the orange. These bands are still better seen in a solution of the colouring matter in weak alcohol. On examining this medium by the third method, with a lens of shorter focus than usual, and looking down from above, the places of the absorption bands were indicated by tooth-shaped interruptions in the beam of light reflected from motes. The points of these teeth were occupied by red dispersed light, which did not appear in the intervening beams of light reflected from motes, from whence it appears that there is the same sort of connexion between the absorption and dispersion of this medium as was noticed in Art. 59, in the case of solutions of chlorophyll and its modifications.

121. A collection of sea-weeds appeared all more or less sensitive, most of them highly so. All, or almost all, except the white ones, exhibited in the derived spectrum the peculiar red band indicative of chlorophyll and its modifications. The trans- mitted light also exhibited more or less the absorption bands due to this substance, which was likewise, in the specimens tried, extracted by alcohol. But the most remarkable example of sensibility found in sea-weeds occurs in the case of the red colouring matter contained in orangy red, red, pink, and purple sea-weeds. To judge by its optical properties, this colouring matter appears to be the same in all cases, but to be mixed in different proportions with chlorophyll, or some modification of it, and probably other colouring matters, thus giving rise to the various tints seen in such sea-weeds. The derived spectrum exhibited by sea-weeds of this kind consists mainly of a band of unusual brightness, containing some red, followed by orange

and yellow. This band fades away gradually at its less refrangible limit, where it is separated by a dark interval from the narrow well-defined red band of still lower refrangibility due to chlorophyll. At its more refrangible limit, however, it breaks off with unusual abruptness.

122. When the light transmitted through such a sea-weed is subjected to prismatic analysis, in addition to one at least of the absorption bands due to chlorophyll, there is seen a band obliterating the yellow, another dividing the green from the blue, and a third, far less conspicuous, dividing the green into two. The whole of the green is absorbed more rapidly than the blue beyond, and not merely than the red, which last is the final tint.

123. The red colouring matter is easily extracted by cold water from certain kinds of red sea-weed, if fresh gathered; but when once the plant has been dried, the colouring matter cannot be extracted in any way that I know of. It is apparently insoluble in alcohol and ether, and is decomposed by boiling. Cold water extracts only a trace of it after a long time.

124. A piece of recently gathered red sea-weed, on being mashed with cold water, readily gave out its red colouring matter. When the residue was treated with alcohol, the fluid was almost immediately coloured green by chlorophyll, whereas this substance is only very slowly and sparingly extracted by alcohol from dried sea-weeds. A dried sea-weed may apparently be assimilated to an intimate mixture of gum and resin, which it would be very difficult to dissolve, whether it were attacked by water or alcohol.

125. The solution of the red colouring matter was highly sensitive, exhibiting a copious dispersive reflexion of a yellowish orange light. The transmitted light was pink or red, according to the thickness through which the light passed. When this light was analysed, the same three absorption bands which have been already mentioned were perceived. The analysis of the light transmitted by the fronds of various red sea-weeds had rendered it extremely probable that the faint division in the green did belong to the red colouring matter; but till I had obtained this matter in solution I did not feel certain that it

might not have been due to chlorophyll, the spectrum of which
exhibits a division in the green.

126.   When this fluid was examined in Sir David Brewster's
manner, and the dispersed beam was analysed, the spectrum was
found to consist of a broad band like that which has been already
described as seen in the derived spectrum given by a frond of red
sea-weed.   When the solution, which happened to be very weak,
was examined by the third method, the dispersion was found to
be produced chiefly by a portion of the incident spectrum, having
a breadth about equal to that of the interval between the two
principal bands of absorption.   To each of these bands corresponded a maximum of activity.   The tint of the dispersed light
was nearly uniform; but by the fourth method of observation
some faint dispersed red could be made out, which appeared
before the main part of the dispersion had come on.   This
medium affords a very good example of an intimate connexion
between absorption and internal dispersion.

127.   The colouring matters of bird's feathers appeared to be
insensible, white feathers being most sensitive, pale ones next,
and dark ones not at all: however, I have not examined a large
collection.

128.   Of coloured fruits, such as currants, &c., the colouring
matter appeared, in the very few cases which I have examined, to
be quite insensible.

129.   A set of water colours were by no means remarkable for
sensibility, but rather the contrary.   The inorganic colours
appeared quite insensible, except white lead, the sensibility of
which was perhaps due to size*, and offered nothing striking,
either as to its character or as to its amount.   Some lakes and
other organic colours proved moderately sensitive.   But I found
one water colour, called Indian yellow, which stands pretty high
among sensitive substances.   In its mode of dispersion it much
resembles turmeric, but it does not come up to that substance in
the amount of sensibility.   It is said to be composed of urate of

[* Meaning, of course, not magnitude, but the substance used to make the
powder stick together in a cake.]

lime*, but I do not know how far it may be regarded as chemically pure.

130. Many of the substances used in dyeing, and dyed articles in common use, furnish very remarkable examples of sensibility. Archil, litmus and turmeric have been already mentioned; and I have been recently informed by a friend that the *Mercurialis perennis*, in which a striking instance of sensibility was observed, was formerly employed in dyeing. A piece of scarlet cloth, examined in a linear spectrum, gave a copious derived spectrum which was very narrow, consisting chiefly of the more refrangible red. With a vertical slit the bands H and fixed lines beyond were seen on a red ground. Paper washed with a solution of cochineal and afterwards with a solution of alum, when examined in a linear spectrum, displayed a pretty high degree of sensibility, the derived spectrum consisting in this case of a red band. If tartaric acid be used instead of alum, the dispersion is a good deal more copious.

Common red tape is another example in which the derived spectrum is very copious, consisting mainly of a red band. Some red wool, dyed I suppose with madder, proved extremely sensitive. The derived spectrum in this case was pretty broad, but red was the predominant colour. Green wool, dyed I do not know with what, was also very sensitive, giving a pretty broad derived spectrum, in which green was the predominant colour. These examples may suffice, but the reader must not suppose that they form the only instances in which dispersion was observed among dyed substances. On the contrary, it is extremely common in this class.

131. Brazil wood, safflower, red sandal wood, fustic and madder, all gave rise to solutions having a pretty high degree of sensibility. The solutions here referred to were such as were obtained directly by water, &c., in which the colours which these substances are capable of producing were not brought out. The beautiful red colouring matters of logwood and camwood appear to be insensible; for a fresh-made solution of logwood in water exhibited no perceptible sensibility, and the slight sensibility

[* It was so stated in a book on artists' colours, but Dr Stenhouse told me it was a lake of some kind.]

exhibited by a similar solution of camwood seemed to have no relation to the red colouring matter.

132. Paper washed with a solution of madder in alcohol was sensitive in a pretty high degree, but the sensibility was greatly increased by afterwards washing with a solution of alum. Accordingly I found that a decoction of madder in a solution of alum exhibited a very high degree of sensibility, displaying a copious dispersive reflexion of a yellow light.   In this medium the dispersion commenced about the fixed line *D*, and continued from thence onwards far beyond the extreme violet, so that the group of fixed lines *n* was seen with great ease.

133. Safflower red, examined in the shape in which it is sold on what is called a *pink saucer*, proved highly sensitive, giving a bright and narrow derived spectrum, which consisted chiefly of the more refrangible red. This substance possesses some other remarkable optical properties, which however do not belong to the immediate subject of this paper.

134. Metals proved totally insensible. I have examined gold, platinum, silver, mercury, copper, iron, lead, zinc and tin. Brass is like simple metals in this respect; but if the surface be lackered the lacker displays its own sensibility.

135. The non-metallic elements, carbon, sulphur, iodine and bromine, are insensible.

136. Among common stones I have found dark flint, lime-stone, chalk and some others which were sensitive, though only in a low degree compared with organic substances. To guard against any impurity of the surface, the stones were broken across, and the fresh surface examined. In the cases mentioned, the sensibility observed is not to be attributed to the chief ingredient of the stone, for quartz, chalcedony, Iceland spar and Carrara marble were insensible.

Compounds of Uranium.

137. Towards the end of last autumn, when the lateness of the season afforded but few opportunities for observation, I learned from different sources that the kind of yellow glass which has been already mentioned as possessing in so high a degree the

property of internal dispersion was coloured with oxide of uranium. This rendered it interesting to examine other compounds of uranium; and I accordingly procured some crystallized nitrate of the peroxide, which, with a few other compounds formed from it, and some of the natural minerals which contain uranium, were examined by methods which have been already explained.

138. The crystals of the nitrate were not sufficiently large and perfect to admit of observation by the methods applicable to fluids and clear solids, but they could be readily observed by means of a linear spectrum. They proved to be sensitive in a very high degree, dispersing a green light which had the same very remarkable composition that has been already described in the case of the yellow glass. On placing a crystal in the continuation of the same linear spectrum with the glass, and viewing the whole through a prism, the five bright bands of which the derived spectrum given by each of the two media usually consisted, appeared to correspond to one another as regards their position in the spectrum. With great concentration of light I have seen an additional band of greater refrangibility in the spectrum of the crystals.

139. Some crystals of nitrate of uranium were gently heated so as to expel a good part at least of the water of crystallization. The residue after some time became opaque and nearly white. In this state it was still more sensitive than the crystals. The dispersed light was not exactly of the same tint, but more nearly white; and the derived spectrum was found on being analysed to contain, in addition to the bright bands usually seen in the derived spectrum of the crystals, another blue band still more refrangible. The fused mass gradually attracted moisture from the air, its colour changed to that of the crystals, and the most refrangible of the bright bands disappeared from the derived spectrum. Although when the incident light was very much concentrated I have seen this band even in the crystals, it was faint compared with the preceding bands, whereas in the case of the whitish mass its intensity was not very different from that of the others. It appears therefore that the quality as well as the quantity of the dispersed light was altered by depriving the crystals of a part of their water.

140. A solution of nitrate of uranium in water is decidedly sensitive, though not sufficiently so to exhibit much dispersive reflexion. When the dispersed beam is analysed it is resolved into bright bands. When the solution is examined in a pure spectrum, the mode of dispersion is found to agree with that of canary glass. The dispersion commences abruptly at the same part of the spectrum as in the case of the glass, and after a rather narrow band in which light is copiously dispersed, there follows a remarkable minimum of sensibility, just as in the glass (see Art. 76), where the dispersed light is almost imperceptible. After this the dispersion is resumed, and offers nothing remarkable. The minimum of sensibility occurs at the very same place in the spectrum, whether the sensitive medium be a solution of nitrate of uranium or glass coloured yellow by uranium.

141. *Yellow Uranite.*—This mineral, when examined in a linear spectrum, proved to be sensitive in an extremely high degree. The derived spectrum consisted, as in the case of the glass, of bright bands arranged at regular intervals, but in this case six were seen, a band being visible in the faint red at the extremity of the spectrum which could not be made out in the case of the glass.

142. *Green Uranite, or Chalcolite.*—According to M. Peligot the formula of the yellow uranite of Autun is PhO^5, CaO, $2(U^2O^2O)$, $8HO$, and the green uranite differs from the yellow only in having the lime replaced by oxide of copper*. Yet a specimen of green uranite on being examined in a linear spectrum proved totally insensible. The primitive spectrum showed however a very remarkable system of dark bands depending on the absorption of light by the mineral. In examining these bands, the previous prismatic decomposition of the light, so far from being necessary, is decidedly inconvenient. It is better to dispense with the prisms altogether, using only the lens, and placing the mineral so that the image of the slit is formed upon it. The bright line thus formed is viewed from a convenient distance through a prism, the eye being held out of the direction of regular reflexion. The position of any bands which may appear in the spectrum can then be determined by means of the fixed lines, which are seen at

* *Annales de Chimie*, Tom. v. (1842), p. 46.

the same time; or, if it be desired to see the latter more distinctly, it will be sufficient to attach a fragment of paper to the
mineral or other substance, placing it so that the image of the slit
is formed partly on the paper and partly on the substance to be
examined. I have frequently found this mode of observation
convenient in examining the absorption of light by opaque
substances. The manner in which the absorption of the medium
comes into play in this case will be considered in greater detail
further on (see Art. 176).

143. When green uranite was examined in this manner, it
showed a very remarkable system of dark bands of absorption.
These bands were seven in number, or at any rate six, and were
arranged with all the regularity of bands of interference. The first
was situated at about $b\frac{4}{4}F$, the second at F; the middle of the
sixth fell a very little short of G; the third, fourth and fifth were
arranged at regular intervals between the second and sixth; the
seventh was situated about as far beyond the sixth as the sixth
beyond the fifth. The spectrum was so faint in the region of the
seventh band as to leave some slight doubts respecting its existence. There would not have been light enough to see bands
further on.

144. Uranite is highly lamellar in its structure, from whence
it is otherwise called uran-mica. The reader may perhaps suppose
that the dark bands described in the last paragraph were bands of
interference, which I had mistaken for bands of absorption, and
that they were really of the nature of Newton's rings, or more
exactly of the bands seen in an experiment due to the Baron von
Wrede. There may, it will perhaps be said, have been a fissure
parallel to the first surface, so as to separate a thin plate; and the
interference of the two streams of light reflected respectively on
the upper and under surface of this plate may have produced the
bands observed. But various phenomena attending these bands
are irreconcilable with such a supposition. Towards the edges of
the crystal, where flaws did in fact exist, bands of the same nature
as Von Wrede's were actually observed. But these had an
appearance totally different from that of the others. The dark
bands of the interference system were more intensely black and
better defined than those of the other system, and were very

variable, depending as they did upon the thickness of the plate
by which they were formed, whereas the bands belonging to the
first system were always the same. Besides, were these bands due
to interference, there is no reason why they should be confined to
one region of the spectrum, and that by no means the brightest.
However, to take away all possible doubts respecting the nature
of the bands, I detached a small scale from the crystal, and having
placed it behind a slit in a beam of sunlight condensed by a lens,
I analysed the transmitted light by a prism. Were the bands
really due to absorption, they ought to be more distinct in the
transmitted light, whereas, were they of the nature of Von Wrede's
bands, they ought to be faint, and almost imperceptible. The
spectrum of the transmitted light contained however four dark
bands, which were well defined and intensely black. The whole
of the spectrum beyond the place of the next band was absorbed,
which is the reason why four bands only were visible.

145. The absorption bands of green uranite, though they
showed great regularity with respect to their positions, did not
appear very regular with regard to their intensities. The second,
fifth and sixth seemed to me to be more conspicuous than the
first, third and fourth. I cannot say for certain whether this
ought to be attributed to fluctuations in the absorbing power of
the medium, or fluctuations in the original intensity of the solar
spectrum, but I am strongly inclined to prefer the former view.

146. The intervals between the absorption bands of green
uranite were nearly equal to the intervals between the bright
bands of which the derived spectrum consisted in the case of
yellow uranite. After having seen both systems, I could not fail
to be impressed with the conviction of a most intimate connexion
between the causes of the two phenomena, unconnected as at first
sight they might appear. The more I examined the compounds
of uranium, the more this conviction was strengthened in my
mind.

147. Yellow uranite exhibits a system of absorption bands
similar to those of green uranite. Nitrate of uranium also shows a
similar system. In a solution I have observed seven of these bands
arranged at regular intervals. The first absorption band coincided
with F, the fifth with G nearly. The absorption bands may also

be seen by analysing the light transmitted through the crystals. The following arrangement exhibited at one view the absorption bands and those due to the light which had changed its refrangibility.

148. The sun's light was reflected horizontally by a mirror, and condensed by passing through a large lens. It was then transmitted through a vessel with parallel sides containing a moderately strong ammoniacal solution of a salt of copper. The strength of the solution, and the length of the path of the light within it, were such as to allow of the transmission of a little green besides the blue and violet. A crystal of nitrate of uranium was then attached to a narrow slit, and placed in the blue beam which had been transmitted through the solution, the crystal being turned towards the incident light. The light coming from the crystal through the slit was then viewed from behind, and analysed by a prism. A most remarkable spectrum was thus exhibited, consisting from end to end of nothing but bands arranged at regular intervals. The interval between consecutive bands appeared to increase gradually from the red to the violet, just as is the case with bands of interference. Although this interval appeared to alter continuously from one end of the spectrum to the other, the entire system of bands was made up of two distinct systems, different in appearance, and very different in nature. The less refrangible part of the spectrum, where only for the crystal there would have been nothing but darkness, was filled with narrow bright bands, due to the light which had changed its refrangibility. These bands were much narrower than the dark intervals between them, but they were not mere lines containing light of definite refrangibility. The more refrangible part of the spectrum was occupied by the system of bands of absorption. The interval between the most refrangible bright band and the least refrangible dark band of absorption appeared to be a very little greater than one band-interval, so that had there been one band more of either kind the least refrangible absorption band would have been situated immediately above the most refrangible bright band. With strong light I think I have seen an additional band of this nature.

149. *Pitchblende.*—This mineral proved to be quite insensible, and exhibited nothing remarkable.

150. *Hydrate of Peroxide of Uranium.*—Some crystallized nitrate of uranium was exposed to a heat a good deal short of redness, whereby most of the acid was expelled. The residue was of a deep brick-red colour, and consisted no doubt chiefly of anhydrous peroxide. It was quite insensible. In order to remove any undecomposed nitrate, it was boiled with water, whereby the undecomposed nitrate was dissolved, and the peroxide converted into a hydrate. This hydrate, after having been washed and dried at the temperature of the air, was of an extremely beautiful yellow colour, and was I suppose the hydrate $U^2O^3 + 2HO$ described in chemical treatises. It was tolerably sensitive, in fact for an inorganic substance extremely so, though the sensibility was much less than that of nitrate of uranium, yellow uranite, or canary glass. The derived spectrum consisted as before of separate bright bands. A small portion of the powder was attached by water to blotting-paper, and dried before a fire. The powder thus obtained on paper was duller than before, and inclined a little more to orange, though the colour was not much deeper than that of the former hydrate. From its colour and the circumstances of its formation, it was probably the other hydrate $U^2O^3 + HO$. It proved on examination to be totally insensible.

151. *Acetate of Peroxide of Uranium,* prepared by dissolving the yellow hydrate of the peroxide in acetic acid, and evaporating to crystallize.—This salt is extremely sensitive, about as much so as the nitrate. The derived spectrum consisted of six bright bands arranged at regular intervals. It seemed to me that the last five of these were respectively a little more refrangible than the five bands given by the nitrate, and then a sixth band was visible in the faint red in the case of the acetate which was not ordinarily seen in the nitrate. However, this observation has need to be repeated under more favourable circumstances.

152. Nitrate and acetate of peroxide of uranium, yellow uranite, and canary glass, are all so highly sensitive as to allow the primary spectrum to be examined with a prism at some distance. In the first three media the bright bands are narrow, much narrower than the dark intervals between; in the glass they appear much broader than in the other media.

153. *Oxalate of Peroxide of Uranium,* prepared in the manner mentioned by M. Peligot, namely, by adding a saturated solution

of oxalic acid to a solution of nitrate of uranium, washing and drying the precipitate.—This salt was sensitive, but only in a low degree. However, the derived spectrum bore prismatic examination sufficiently to show three or four bright bands. The absorption of the medium was examined by spreading some of the powder on glass along with water and allowing it to dry. The layer was then examined by different methods. The salt exhibits three very intense absorption bands in the highly refrangible part of the spectrum. The positions of these bands, by measurement, were $F\,0\text{·}31\,G$, $F\,0\text{·}58\,G$, $F\,0\text{·}85\,G$.

154. *Phosphate of Peroxide of Uranium*, prepared by precipitation from a solution of nitrate of uranium by adding a solution of common phosphate of soda.—This salt was sensitive, though not in a high degree. It was a good deal more sensitive than the oxalate, but I think not so much so as the hydrate of the peroxide. The derived spectrum consisted of bright bands as usual*.

155. *Uranate of Potassa*, prepared by dropping a solution of nitrate of uranium into a solution of caustic potash, stopping long before the alkali was neutralized.—This salt was found to be insensible, both in its original state as a gelatinous hydrate, and in various stages of drying.

156. *Uranate of Lime*, prepared in a similar manner with lime-water.—This salt, which after drying is of a fine orange colour, was like the preceding found to be insensible. It seemed interesting to examine these two salts, because the former contains two elements (not counting oxygen) in common with canary glass, and the latter two elements in common with yellow uranite. Yet the salts are insensible while the two other media are so remarkably sensitive.

157. *Solutions by means of Alkaline Carbonates.*—It is known to chemists that alkaline carbonates, added in solution to a solution of nitrate of uranium, give yellow precipitates which are redissolved in an excess of the precipitant. The solutions thus obtained with the carbonates of potassa and soda, which were of a greenish yellow colour, were found to be totally insensible.

* See note C.

They exhibited however four of those singular absorption bands so characteristic of salts of peroxide of uranium. Of these the third fell a little short of G, its more refrangible edge nearly coinciding with that fixed line; the first and second were situated between F and G, the distance of the first beyond F being somewhat greater than the interval between two consecutive bands. The fourth, which was situated beyond G, was fainter than the others. The second and third were the most conspicuous of the set.

158.   The absorption bands due to peroxide of uranium afford an easy mode of detecting that substance in solution.   For this purpose the solutions mentioned in the preceding paragraph are much preferable to the nitrate, for they produce much stronger bands when only a small quantity of uranium is present.   The absorption bands of nitrate of uranium are visible, as might have been expected, in presence of a large quantity of nitrate of copper*.

Optical Tests of Uranium in Blow-pipe Experiments.

159.   When a bead of microcosmic salt is fused with oxide of uranium, and brought to its highest state of oxidation, it is yellow by transmitted light.   Such a bead is sensitive in a very high degree, quite as much so as canary glass.   When the light falls sideways on it, and it is held against black cloth or a dark object, it exhibits plainly the green colour due to internal dispersion. When properly examined by means of sunlight its sensibility is evident at once, and when the dispersed light is viewed through a prism it is resolved into bright bands.   One of the most convenient modes of examining such minute objects consists in reflecting the sun's light horizontally through a large lens, intercepting by means of absorbing media all the rays except those of very high refrangibility, placing the object to be examined in the condensed beam, and viewing it through a prism.   So delicate is this test when applied to uranium, that on one occasion, when engaged in examining a bead coloured green by chromium, which had been fused in the exterior flame, I observed the appearance given by uranium.   This turned out to be actually due to uranium, of

* See note D.

which a mere trace was accidentally present without my knowledge.

160. The green communicated to microcosmic salt by uranium after exposure to the reducing flame has a very peculiar composition, by means of which the presence of uranium may be instantly detected. For this purpose it is sufficient to view through a prism the inverted image of the flame of a candle formed by the bead, the latter being so held as to be seen projected on a dark object. The observation is perfectly simple, and occupies only a few seconds. The spectrum exhibits an isolated band at the red extremity, followed by a very intense dark band of absorption. A similar dark band, but not quite so intense, occurs in the green : beyond the green there is usually but little light seen. As the absorption progresses the first dark band invades all the space from the red to the green, and the spectrum consists of an isolated red band and a green band divided into two. In its mode of absorption, the medium has a strong general resemblance to chlorophyll. The green due to copper or to chromium shows nothing remarkable when viewed through a prism, and could not possibly be confounded with the green due to protoxide of uranium. The absorption bands due to this oxide are not completely brought out till the bead is cold.

161. Uranium produces the same effects with borax as with microcosmic salt, but they are less distinct, or at least less easily produced.

162. When the uranium contained in a bead of microcosmic salt is thoroughly oxidized, and the bead is gently heated, so as just to be self-luminous, the light which it gives out is not red, like that of most substances at a low heat, but green, or rather greenish white.

163. Solutions of protoxide of uranium have a very remarkable effect on the spectrum, resembling more or less that of a bead of microcosmic salt coloured green by uranium. Of course the absorption can be observed much better by means of a solution than by a mere bead. I have observed several bands of absorption in such solutions, but the cases which I have hitherto

examined are too few to justify me in entering into detail. Besides, the absorption bands due to protoxide of uranium do not belong properly to my subject, the compounds of this oxide, so far as I have examined, being insensible.

Appearance of highly Sensitive Media in a Beam from which the Visible Rays are nearly excluded.

164. When a large beam of sunlight is reflected horizontally into a darkened room, and transmitted through an absorbing medium, placed in the window, of such a nature as to let pass only the feebly illuminating rays of high refrangibility and the invisible rays beyond, various sensitive media have a very strange and unnatural appearance when placed in the beam, on account of the peculiar softness of the dispersed light with which the media appear as it were self-luminous, and the almost entire absence of strong light reflected from convexities. Among substances eminently proper for this experiment, may be mentioned a solution of the bark of the horse-chestnut *, or of sulphate of quinine, or of stramonium seeds, a decoction of madder in a solution of alum, and above all, ornamental articles of canary glass. The appearance of a specimen of yellow uranite was curiously altered by this mode of examination. By daylight the mineral appeared much of the same colour as the stone in which it was imbedded, but when placed in a beam such as that above mentioned the uranite was strongly luminous, while the stone remained dark.

Natural Crystals.

165. Of natural crystals I have hitherto examined only a small number. For a long time I was occupied almost exclusively with vegetable products, the mineral kingdom not appearing promising. However, I have found internal dispersion in certain specimens of apatite, aragonite, chrysoberyl, cyanite, and topaz. In all these cases the dispersion appeared due, as in the case of

[* A solution which answers admirably, and is very easily prepared, is obtained by adding to a decoction when cold a suitable quantity of alum or a ferric salt, precipitating by ammonia, and filtering. The powerful fluorescence of the solution is due to mixed aesculin and fraxin.]

fluor-spar, to some substance accidentally present in small quantity; so that yellow uranite is at present the only natural crystal
to the essential constituents of which the property of internal
dispersion has been found to belong.

166. Among the minerals just mentioned apatite was the
most sensitive, though it fell very far short of yellow uranite.
That the sensibility was not due to phosphate of lime, was plain
from the circumstances that a colourless specimen was insensible,
and that the amount of sensibility was found to be different in
different parts of the same sensitive specimen. With the exception of the colourless crystal already mentioned, all the specimens
of apatite examined were of a greenish colour, and all were sensitive. The dispersed light was something of an orange colour,
but was not homogeneous orange. In one specimen it consisted
of three distinct bright bands at regular intervals. The mode in
which the sensibility of this crystal was connected with the
refrangibility of the incident rays was very peculiar. In aragonite dispersion was found in the transparent specimens examined;
the translucent specimens were found to be insensible. The dispersed light was of a brownish white colour. In the same crystal
some parts were insensible and others more or less sensitive. The
portions of equal sensibility were arranged in plane strata, just
as in the case of fluor-spar, as has been noticed by Sir David
Brewster. In a specimen which had been cut for showing conical
refraction, the strata were in some places perpendicular to the
plane of the optic axes, and in other parts parallel to the line
bisecting the axes, and inclined to their plane at such an angle
that the two directions of the strata must have been parallel to
two of the commonest lateral faces. Another specimen showed
strata parallel to an oblique terminal face. The strata are plainly
due, as Sir David Brewster has remarked with reference to fluor-
spar, to some substance taken up during crystallization. Accordingly, they preserve a sort of history of the growth of the crystal.
In a twin crystal of fluor-spar, the direction of the strata in that
part of the mass which was common to the geometrical forms of
both crystals, showed to which crystal it really belonged. In
fluor-spar the strata are parallel to the faces of the cube, at least
in the specimens which I have examined, and the same has been
observed by Sir David Brewster.

In chrysoberyl, cyanite and topaz, the dispersed light was red or reddish, and was too variable to allow of its being attributed to the essential constituents of the crystals. In these cases the sensibility was but slight; indeed in cyanite there was only a trace of dispersion when the crystal was examined under great concentration of light.

Coloured Glasses.

167. Besides canary glass, I have examined the common coloured glasses, including that coloured by gold, but with one exception have not met with any example in which the sensibility observed appeared to have any connexion with the colouring matter. The paler glasses exhibited a little internal dispersion, because the colour was not sufficiently intense to mask the dispersion which a common colourless glass would exhibit.

168. The exception occurred in the case of the pale brown glass, which has been already mentioned in connexion with my first experiment. This glass dispersed a red light under the influence of the highly refrangible rays. The colour of the light was not pure prismatic red, but red was predominant. A similar dispersion, due apparently to the same cause, was observed in the case of one of the common reddish brown German wine bottles. The sensibility of these glasses appears to be due to an alkaline sulphuret. [?] A bead purposely coloured in this manner was in fact found to disperse a red light like the glasses. Moreover, in the confused masses obtained by fusing sulphate of soda and sulphate of potash on charcoal before the blowpipe, certain portions were found which dispersed a red light, and that pretty copiously for an inorganic substance. A similar dispersion was observed among the products obtained by fusing together sulphur and carbonate of potash, while other parts of the confused mass exhibited dispersion of a different kind. It seems plain that among the combinations of sulphur with the alkalies sensitive compounds exist, but what they are I have not examined.

Cautions with respect to the discrimination between true and false internal dispersion.

169. In the early part of this paper certain tests were given for distinguishing between true and false internal dispersion in a fluid. But it requires some experience in observations of this kind to be able readily to decide, and a too rigid adherence to one of the tests to the exclusion of the others might lead to error.

The first test relates to the continuous appearance of a truly dispersed beam. But sometimes solid particles exist in mechanical suspension, which are so fine and so numerous, that this test alone might lead the observer to mistake a falsely for a truly dispersed beam. On the other hand, if a fluid which itself alone exhibits no internal dispersion, true or false, hold solid particles in what is obviously mere mechanical suspension, we must not immediately conclude that the medium, taken as a whole, is incapable of changing the refrangibility of any portion of the light incident upon it. For we have seen that the fluid state is not in the least degree essential to the exhibition of sensibility, and of course a fluid will serve as well as anything else for the mere mechanical support of a sensitive substance.

170. Thus lycopodium is very sensitive, as appears by examining the powder in a linear spectrum. Accordingly, I found that when a little lycopodium was mixed with water, and the whole medium was examined by the fourth method, it displayed its sensibility, although the beam of light which had changed its refrangibility was plainly discontinuous. When Indian yellow was used instead of lycopodium, the whole medium exhibited its sensibility when it was examined by the fourth method. In this case the suspended particles were so fine that the beam of light which had changed its refrangibility appeared to be continuous, though of course it was not really so. In observing with muddy fluids like these, it is almost necessary to employ absorbing media, since otherwise the effect of the light scattered at the surfaces of the prisms and large lens might lead the observer to conclusions altogether erroneous.

171. The next test relates to the polarization of a falsely dispersed beam. Being engaged on one occasion in examining

the effects of acids and alkalies on a weak solution of a sensitive substance, employing sunlight which had been merely reflected through a small lens, I met with a beam which had every appearance of having been only falsely dispersed, but on viewing it from above through a doubly refracting prism I was surprised at first by finding it unpolarized. It soon occurred to me that the beam must have been due, not to solid motes, but to excessively small bubbles of carbonic acid gas, the existence of which was thus revealed, though they were too small to be seen directly. The light being incident on these bubbles at an angle of about 45°, which is very little less than the angle of total reflexion, the reflected light would be almost perfectly unpolarized*.

172. Water which had been merely boiled in a test tube gave a similar result. The unpolarized beam of falsely dispersed light was of course due in this case to the air which had been held in solution. This shows why long-continued boiling should be necessary, in order to free water from air. It is not that the affinity of water for air is so great as to be only gradually overcome, but that the air, immediately expelled from solution when the temperature rises sufficiently, is still retained in a state of mechanical mixture, forming excessively minute bubbles, the terminal velocity of which is insensible. Accordingly it is not till larger bubbles are formed, by the casual meeting of a number of these small bubbles, that the air rises to the surface and escapes.

173. With respect to the test of true dispersion depending on the change of refrangibility, it has been already remarked that in some cases the change is so slight, that if this test alone were applied, the observer might mistake true dispersion for false. However, it is only in rare cases that there is any danger of being deceived in this manner in the application of the test; but on the other hand, in observing a muddy fluid or a translucent solid by the fourth method, the observer, if not on his guard, might easily be deceived by the effect of scattered light, and be led to mistake false dispersion for true. Thus suppose the medium to be water holding in suspension particles of an insensible water colour, and the small lens to be placed a little beyond the commencement of the violet. Two beams of light would enter the lens, namely, a

* See note E.

regularly refracted beam of violet, and a scattered beam of white
light. Of these the latter would be insignificant compared with
the former, were it not that the illuminating power of the colours
belonging to the middle of the spectrum is so very much greater
than that of the violet. When the dispersed beam was analysed
by a prism, it would be decomposed into a violet beam of definite
refrangibility, followed by a dark interval, and then a broad band
containing the colours of the brighter part of the spectrum in
their natural order. This is what is constantly seen in cases of
true dispersion; but the polarization of the beam, and its beha-
viour under the action of absorbing media, would reveal the coun-
terfeit character of the dispersion.

On the Colours of Natural Bodies.

174. By this expression I mean to include only the colours to
which it is usually applied, namely, those of leaves, flowers, paints,
dyed articles, &c., which form the great mass of the colours that
fall under our observation. I do not refer to colours due to refraction, such as those of the rainbow, or to diffraction, such as those
of the coronæ seen about the sun and moon, or to interference, such
as those seen in the clear wings of small flies, or to the colours
which accompany specular reflexion, which last are usually but
slight, though sometimes pretty intense.

In some few instances, as for example in the case of fluor-spar,
various salts of peroxide of uranium, acid solutions of disulphate of
quinine, &c., colours are observed, sufficiently strong to arrest attention, which have a remarkable and hitherto unsuspected origin.
But I am not now speaking of colours arising from a change of
refrangibility in the incident light. In the vast majority of cases
these colours are far too feeble to form any sensible portion of
the whole colour observed. The colours which dyed articles give
out under the influence of the highly refrangible rays usually
agree more or less nearly with those of which such substances
commonly appear, and it is possible that the colour arising from a
change of refrangibility may contribute in some slight degree to
the brilliancy of the tint observed. If, however, the effect be
sensible I am persuaded that it is but slight; and very brilliant
colours may be produced without a change of refrangibility, as for

example in the case of biniodide of mercury.   For the present I shall neglect the light which may have changed its refrangibility.

175.   Few, I suppose, now attach much importance to the bold speculations in which Newton attributed the colours of natural bodies to the reflexion of light from thin plates.   Sir David Brewster has shown how extremely different the prismatic composition of the green of the vegetable world is, from what it ought to be, according to Newton's theory, and what Newton supposed that it was.   It is now admitted that the various colours of natural bodies are merely particular instances of one general phenomenon, namely, that of absorption.   Absorption is most conveniently studied in a clear fluid or solid, but it does not the less exist in a body of irregular structure, such as a dyed cloth or a coloured powder.

The green colouring matter of leaves affords an excellent example of the identity of the effect produced on light by natural bodies and of ordinary absorption; for the same very peculiar system of absorption bands which are displayed by a clear solution of the colouring matter may be observed directly in the leaf itself. However, it is needless to bring forward arguments to support a theory now I suppose universally admitted; my present object is merely to point out the mode in which the colours which bodies reflect, or more properly scatter externally, depends upon the absorbing power of the colouring matter, so as to justify the conclusions deduced in Art. 142, from observations made in the manner there described.

176.   Let white light be incident on a body having an irregular internal structure, such as a coloured powder.   A portion will be reflected at the first irregular surface, but the larger portion will partly enter the particles, partly pass between them, and so proceed.   In its progress the light is continually reflected in an irregular manner at the surfaces of the particles, and a portion of it is continually absorbed in its passage through them.   For simplicity's sake, suppose the light incident in a direction perpendicular to the general surface, and neglect all light which is more than once reflected.   Let t be the thickness of a stratum which the light has penetrated, I the intensity of the light at that depth, or rather the intensity of a given kind of light, so that the whole

intensity may be represented by $\int I d\mu$, μ being the refractive index in some standard substance. In passing across the stratum whose thickness is dt, suppose the fraction $q dt$ of the light to be absorbed, and the fraction $r dt$ to be reflected and scattered in all directions, then

$$dI = - (q + r)\, I dt.$$

Integrating this equation, and supposing I_0 to be the initial value of I, when $t = 0$, we have

$$I = I_0 e^{-(q+r)t} \dots\dots\dots\dots\dots\dots\dots(a).$$

For the sake of simplicity, suppose the body viewed in a direction nearly perpendicular to the general surface; and of the light reflected and scattered in passing across the stratum whose thickness is dt, suppose that the fraction n would enter the eye if none were lost by absorption, &c. Then the intensity of the light coming from that stratum would be $nrIdt$. But in getting back across the stratum whose thickness is t, the intensity is diminished in the ratio of I_0 to I. Hence if I' be the intensity of the light actually entering the eye,

$$dI' = nrI_0^{-1}I^2 dt = nrI_0 e^{-2(q+r)t} dt.$$

If we suppose the thickness of the body sufficient to develope all the colour which the body is capable of giving, the superior limit of t will be ∞, and we shall have

$$I' = \frac{nr}{2(q+r)}\, I_0 \dots\dots\dots\dots\dots\dots(b).$$

177. The colour which accompanies ordinary reflexion being usually but slight, I shall neglect the chromatic variations of r. It is q which is subject to extensive and apparently capricious variations, depending upon the refrangibility of the light. Imagine two curves drawn whose abscissæ are proportional to μ, and ordinates proportional to the ratio of I to I_0 for the first, and the ratio of I' to I_0 for the second. These curves will serve to represent to the mind the composition of the light transmitted through a stratum of the body having a thickness t, and of that reflected from the body when seen in mass. It is plain that the maximum and minimum ordinates in the two curves will correspond to the same abscissæ; but unless t be very small, so small as to be insufficient to bring out the colour of the medium seen by trans-

mission, the maxima and minima will be much more developed in the first curve, whose ordinates vary as e^{-qt}, than in the second, whose ordinates vary as $(q + r)^{-1}$. If, then, the absorbing power be subject to fluctuations depending on the refrangibility of the light, the bands of absorption may be observed either in the reflected or in the transmitted light, but they admit of being better brought out in the latter.

178. If the nature of the substance be given, q will be given. If now the body be of a loose nature, as for example blue glass reduced to a fine powder, r will be considerable. Hence, in accordance with the expression (b), the quantity of light scattered externally will be considerable, but the tint will be but slight. If the powder be now wetted, the reflexions at the surfaces of the particles will be diminished, r will be diminished, and, as appears from (b), the quantity of light scattered externally will be diminished, but at the same time the tint will be deepened, since the chromatic variations of I' are increased. If the body be compact and nearly homogeneous, r will be small, and therefore very little light will be returned, except what is regularly reflected at the first surface. The tint of the small quantity of light which is reflected otherwise than regularly, will be somewhat purer than before, inasmuch as the chromatic variations of I' tend to become the same as those of q^{-1}.

On the nature of False Dispersion, and on some applications of it.

179. It has been already stated that a beam of falsely dispersed light seen in a fluid has generally more or less of a sparkling appearance, indicating that it owes its origin merely to motes held in mechanical suspension. Sometimes, however, no defect of continuity is apparent. This is especially the case when two fluids are mixed together, of which one contains in solution a very small quantity of a substance which we might expect to be precipitated by the addition of the other, or when a slightly viscous fluid has remained quiet for a long time. If some part at least of a falsely dispersed beam be plainly due to motes, that does not of course prove for certain that there is no part which may have a different origin, and may be essentially connected with

true dispersion; nor do the theoretical views which I entertain of the cause of the latter lead me to regard it as at all impossible that a beam polarized in the plane of reflexion, and having the same refrangibility as the incident light, may be a necessary accompaniment of true dispersion. However, observation, I think, points in a contrary direction; for although more or less of false dispersion is almost always exhibited along with true dispersion, the quantity of the former seems to have no relation to the quantity of the latter, but does seem to have relation to the greater or less degree of clearness which we should be disposed to attribute to the fluid.

180. The phenomenon of false internal dispersion seems to admit of being applied as a chemical test to determine whether or not precipitation takes place. Thus, if a little tincture of turmeric be greatly diluted with alcohol, and then water be added, a yellow fluid is obtained which appears to be perfectly clear, exhibiting no sensible opalescence; but the occurrence of a copious false dispersion when the fluid is examined by sunlight, reveals at once the existence of suspended particles, though they are too minute to be seen individually, or even to give a discontinuous appearance to the falsely dispersed beam. Although such a precipitation could not, I suppose, be used as a means of mechanical separation, it might still be useful as pointing out the possibility of an actual separation under different circumstances as to strength of solution, &c.

181. One of the best instances of false dispersion that I have met with, best, that is, in forming a most excellent imitation of true dispersion, occurred in the case of a specimen of plate-glass which was made, as I was informed, with a quantity of alkali barely sufficient. This glass, which was very slightly yellowish brown, when viewed edgeways by transmitted light, had a bluish appearance when viewed properly, strongly resembling that of a decoction of the bark of the horse-chestnut, diluted with water till the dispersed light is no longer concentrated in the neighbourhood of the surface. But when the glass was examined by sunlight, the polarization of the dispersed beam, and the identity of its refrangibility with that of the incident light, showed that this was merely an instance of false dispersion. Another very good example of

false dispersion is afforded by chloride of tin dissolved in a very large quantity of common water.

182. When a horizontal beam of falsely dispersed light is viewed from above, in a vertical direction, and analysed, it is found to consist chiefly of light polarized in the plane of reflexion. It has often struck me, while engaged in these observations, that when the beam had a continuous appearance, the polarization was more nearly perfect than when it was sparkling, so as to force on the mind the conviction that it arose merely from motes. Indeed, in the former case, the polarization has often appeared perfect, or all but perfect. It is possible that this may in some measure have been due to the circumstance, that when a given quantity of light is diminished in a given ratio, the illumination is perceived with more difficulty when the light is uniformly diffused than when it is spread over the same space, but collected into specks. Be this as it may, there was at least no tendency observed towards polarization in a plane perpendicular to the plane of reflexion, when the suspended particles became finer, and therefore the beam more nearly continuous.

183. Now this result appears to me to have no remote bearing on the question of the direction of the vibrations in polarized light. So long as the suspended particles are large compared with the waves of light, reflexion takes place as it would from a portion of the surface of a large solid immersed in the fluid, and no conclusion can be drawn either way. But if the diameters of the particles be small compared with the length of a wave of light, it seems plain that the vibrations in a reflected ray cannot be perpendicular to the vibrations in the incident ray*. Let us suppose for the present,

[* The way in which I at the time regarded the problem was as follows. Suppose polarized light to be passing through a medium which holds in suspension a vast number of excessively fine particles of some substance different from the medium itself, the dimensions of the particles being for simplicity supposed extremely small compared with the length of a wave. The ether in the medium will be vibrating to and fro in a direction perpendicular to the direction of propagation, and either in or perpendicular to the plane of polarization. The inertia of the particles being presumably very great compared with that of a corresponding volume of the ether alone, the ponderable particles may be supposed to remain at rest, and they will therefore disturb the motion of the ether, and cause vibrations to spread out from them in the ether. Now the repose of the particles may be regarded as the resultant of two equal and opposite motions, one that in the case of the beams actually observed, the suspended particles were small compared with the length of a wave of light. Observation showed that the reflected ray was polarized. Now all the appearances presented by a plane-polarized ray are symmetrical with respect to the plane of polarization. Hence we have two directions to choose between for the direction of the vibrations in the reflected ray, namely, that of the incident ray, and a direction perpendicular to both the incident and the reflected rays. The former would be necessarily perpendicular to the directions of vibration in the incident ray, and therefore we are obliged to choose the latter, and consequently to suppose that the vibrations of plane-polarized light are perpendicular to the plane of polarization, since experiment shows that the plane of polarization of the

the same as that of the ether itself, the other a to and fro motion along the same line as the former but in the opposite direction; and we may superpose these motions as regards their effect on the ether. In the former the particles would be moving with the ether, and therefore would not disturb it; as regards the latter we may think of the particles as moving to and fro in otherwise still ether, and producing therefore an ethereal disturbance emanating in all directions from the particle. This disturbance having to be transversal will evidently be *nil* in a polar direction and a maximum in an equatorial direction, varying in fact in amplitude as the sine of the polar distance, the polar line being a line through the middle of the particle drawn in the direction of the incident vibrations. The direction of propagation of the incident light and that of the line of sight being as in the text, if the incident light be common light we may replace it by two independent streams, of equal intensity, polarized the one in a vertical and the other in a horizontal plane; and of these the one for which the plane of vibration is vertical will not give rise to any diffracted light entering the eye, while the other will give rise to a stream for which the direction of vibration is horizontal, and which is therefore polarized in such a manner that the plane of vibration passes through the line of sight and is perpendicular to the direction of propagation of the incident light, and which therefore may be extinguished by an analyser suitably turned; and in a similar way, as stated in the text, the light entering the eye may be quenched by polarizing the light before incidence on the particles instead of analysing it after diffraction. The conclusions of theory, which are enunciated with reference to the plane of vibration, exactly agree with the results of experiment, which are described with reference to the plane of polarization; and to make the two fit we must suppose the direction of vibration in polarized light to be perpendicular, not parallel, to the plane of polarization.

There can be little doubt that in several cases which fell under my notice, especially in that of the glass mentioned in Art. 181, the particles were sufficiently fine to render the above reasoning applicable. Still more must that have been the case in the beautiful experiments of Tyndall on the decomposition of gases and vapours by rays of high refrangibility, who was led independently to the same conclusions as those stated in the text regarding the phenomena of polarization exhibited by fine particles in suspension.]

reflected ray is the plane of reflexion. According to this theory, if we resolve the vibrations in the incident ray horizontally and vertically, the resolved parts will correspond to the two rays, polarized respectively in and perpendicularly to the plane of reflexion, into which the incident ray may be conceived to be divided, and of these the former alone is capable of furnishing a reflected ray, that is of course a ray reflected vertically upwards. And in fact observation shows that, in order to quench the dispersed beam, it is sufficient, instead of analysing the reflected light, to polarize the incident light in a plane perpendicular to the plane of reflexion.

Now in the case of several of the beams actually observed, it is probable that many of the particles were really small compared with the length of a wave of light. At any rate they can hardly fail to have been small enough to produce a tendency in the polarization towards what it would become in the limit. But no tendency whatsoever was observed towards polarization in a plane perpendicular to the plane of reflexion. On the contrary, there did appear to be a tendency towards a more complete polarization in the plane of reflexion.

M. Babinet has been led by the same reasoning to an opposite conclusion respecting the direction of the vibrations in polarized light, resting on an experiment of M. Arago's, in which it appeared that when light was incident perpendicularly on the surface of white paper, and the reflected or rather scattered light was viewed in a direction almost grazing the surface, it was found to be partially polarized in the plane of the sheet of paper*. But the actions which take place when light is incident on a broad irregular surface, like that of paper, bounding too a body which is so translucent that a great part of the light must enter it and come out again, appear to me to be too complex to allow us to deduce any conclusion from the result respecting the direction of vibration. Besides, the result itself admits of easy explanation, by attributing it to the light which has entered the substance of the paper and come out again, which might be expected to be polarized by refraction.

* *Comptes Rendus*, Tom. xxix. p. 514.

Effect of Heat on the Sensibility of Glass, &c.

184.   The sensibility of glass is temporarily destroyed by heat. The glass may be heated by holding it in the flame of a spirit-lamp, as a heat much short of redness is sufficient.   This takes place even with glass coloured by oxide of uranium, which is in general so highly sensitive.   The sensibility returns again as the glass cools.   A bead of microcosmic salt, containing uranium in its highest state of oxidation, is very sensitive when cold, but insensible when hot.   The sensibility gradually comes on as the bead cools.   A solution of nitrate of uranium in water on being heated has its sensibility impaired, very much so by the time the temperature reaches the boiling-point.   The sensitive compounds, whatever may have been their precise nature, obtained by fusing the sulphates of soda and potassa on charcoal before the blowpipe, were insensible while hot.   The few vegetable solutions which I have examined with this object did not seem to have their sensibility affected by being heated.

Effect of Concentration and Dilution.

185.   In investigating the change of refrangibility produced by a sensitive substance in solution, it is almost always convenient to have the solution weak.   This however is by no means merely a matter of convenience, for the quantity of light which the medium is capable of giving back with a changed refrangibility is often materially diminished by increasing the concentration of the solution.   Thus a solution which, when in a concentrated state, exhibits no sensible dispersive reflexion, will often exhibit when much diluted a very copious appearance of that nature. On the other hand, the dilution may of course be carried too far, so as to render imperceptible the peculiar properties of the substance dissolved.   Yet it is wonderful what a degree of dilution a highly sensitive solution will bear before its sensibility ceases to be perceptible.

That the sensibility will be diminished, and will at last become imperceptible, if only the dilution be carried far enough, is nothing more than might have been predicted with the utmost confidence.   In such a case the light passes completely through the fluid long before it has produced all the effect which it is capable of producing. But that concentration should be an obstacle to the exhibition of the phenomenon is not perhaps what we should have expected, and deserves an attentive consideration.

186.   Imagine a given sensitive substance to be held in solution, in a vessel of which the face towards the eye is plane, and the breadth in the direction of vision as great as we please; and suppose the solvent, or at least the fluid used for diluting the solution, to be itself colourless and insensible. Suppose the fluid to be illuminated by light of given intensity and given refrangibility entering at the face next the eye, and let the eye E from a given position look in the direction of a given point P in the nearer surface of the vessel. In short, let everything be given except the strength of the solution. For the sake of simplicity regard the eye as a point, and make E the vertex of an indefinitely thin conical surface surrounding the line EP. Call this conical surface C, and let c be the surface within the fluid generated by right lines coinciding with the refracted rays which would be produced by incident rays coinciding with the generating lines of the surface C. This latter surface we may if we please regard as cylindrical, since we shall only be concerned with so much of the fluid contained within it as lies at a distance from P less than that at which the light entering the eye in consequence of internal dispersion ceases to be sensible; and in the cases to which the present investigation is meant to apply this distance is but small compared with PE. Let the fluid within c be divided into elementary portions by planes parallel to the surface of the fluid at P, and at distances from P proportional to the strength of the solution. It is evident that an element of a given rank, reckoned from P, will contain a constant number of sensitive molecules, and the incident light in reaching this element has to pass through a thickness of the medium such that a plate of the same thickness, and having a given area, contains a given number of sensitive or absorbing molecules. The same is true of the dispersed light which proceeds from the element and enters the eye. Now it seems natural to suppose that if the strength of a solution be doubled, trebled, &c., or reduced to one-half, one-third, &c., the quantity of light absorbed will be the same provided the length of the path of the light be

reduced to one-half, one-third, &c., or doubled, trebled, &c.   This comes to the same thing as supposing that each absorbing molecule stops the same fractional part of the light passing it, whether the solution be more or less dilute. We should similarly be inclined to suppose that each sensitive molecule would give out the same quantity of light, when influenced by light of given intensity, whether it belonged to a stronger or a weaker solution. If we admit these suppositions, it is plain that the quantity of dispersed light which reaches the eye from the element under consideration will be independent of the strength of the solution. This being true for each element in particular will be true for the aggregate effect of them all, and therefore the quantity of light exhibited by dispersive reflexion will be independent of the strength of the solution. It may be readily seen that the result will be the same if we take into account the finite size of the pupil.

187. Now this is by no means true in experiment. On examining in a pure spectrum a highly concentrated solution of sulphate of quinine, a copious dispersion was observed to commence a little below the fixed line G. It remained very strong as far as H, and beyond. In the weak solution first mentioned in this paper, it will be remembered that the dispersion seemed to come on about $G\frac{1}{2}H$. The reason of this, or at least one reason, is evident, and was very prettily shown by the form of the space to which the dispersed light was confined. On looking down from above, so that this space was seen in projection, it appeared in the case of the weak solution to have approximately the form of the space contained between one branch of a rectangular hyperbola, one asymptote, and a line parallel to the other, the first asymptote being the projection of the anterior surface, and the line parallel to the other being the course of the least refrangible of the active rays which were capable of producing a sensible quantity of dispersed light. The breadth of the illuminated space, which among the most highly refrangible rays was almost insensible, continually increased, until the space ended in a blue beam which went quite across the vessel. But in the case of the strong solution the illuminated space had throughout an almost insensible breadth, except just close to its lower limit, that is, the limit corresponding to the

least refrangible of the active rays, where it ended in a sort of tail or plano-concave wedge, which penetrated to a moderate distance into the fluid. Hence one reason, though perhaps not the only reason, why the strong solution showed a copious dispersion from G to $G\frac{1}{2}H$, where the weak solution showed hardly any, is plain enough. But in the region of the invisible rays beyond the violet, the dispersion was plainly more copious with the weak than with the strong solution. It appears then that in such a case the sensitive molecules do not act independently of each other, but the quantity of light emitted by a given number of molecules is less, in proportion to the light (visible or invisible) consumed, than when a solution is more dilute. We should expect à priori that when a solution is tolerably dilute further dilution would make no more difference in this respect. This seems to agree very well with experiment. For when a pretty dilute solution and one much more dilute are compared with respect to the quantity of dispersed light given out in a given portion of the incident spectrum, they appear to be alike. I suppose the comparison to be made with respect to such a portion of the incident spectrum, or in the case of solutions of such strength, that the dispersed light is confined to a space extending to no great distance into the fluid in either solution. Under these circumstances the comparison may be made easily enough.

188. In the actual experiment, the elementary portions of light coming from the elementary strata of fluid situated at different distances from the anterior surface enter the eye together. Let us however trace the consequences of the very natural supposition, that in passing across a given stratum of fluid the quantity of light absorbed, as well as the quantity given out by dispersion, is proportional, cœteris paribus, to the intensity of the incident light. The incident light is here supposed to be homogeneous, and to belong indifferently to the visible or invisible part of the spectrum. In crossing the elementary stratum having a thickness dt, let the fraction qdt of the incident light be absorbed, and the fraction rdt dispersed in such a direction as to reach the eye; and of the latter portion let the fraction sdt be absorbed in crossing a stratum having a thickness dt, s being different from q on account of the change

of refrangibility. Then by a very simple calculation similar to that of Art. 176, we find for the intensity I' of the dispersed light which enters the eye

$$I' = \frac{r}{q+s} I_0,$$

I_0 being the intensity of the incident light. Since a sensitive fluid is in general coloured, and the dispersed light is in general heterogeneous, s will in general be different for the different portions into which the dispersed light would be decomposed by a prism. However, if the fluid be colourless, or all but colourless, as is the case with a solution of sulphate of quinine, s will be insensible, so that I' will be proportional simply to rq^{-1}. Hence from the observed variations in I', arising from variations in the strength of the solution, we may infer the corresponding variations in rq^{-1}.

If, then, we represent by the ordinate of a curve the ratio of the quantity of light given out to the quantity of light absorbed by a given number of active molecules, the abscissa being the ratio of the quantity of diluting fluid to the quantity of the sensitive substance in solution, it appears that the curve will be concave towards the axis of the abscissæ, and will have an asymptote parallel to that axis.

On the Choice of a Screen.

189.   We have seen that white paper, the substance commonly employed as a screen on which to receive the spectrum, gives back with a changed refrangibility a portion of the light incident upon it. This might in some cases lead an observer not aware of the circumstance to erroneous conclusions. Since the colour of dispersed light depends upon its refrangibility, which is different from that of the active light, the colours of a spectrum received on white paper must be somewhat modified. In truth the intensity of the light dispersed is so small compared with the intensity of the light scattered, that the modification is quite insensible except in the extreme violet. But beyond the extreme violet the spectrum seems to be prolonged with a sort of greenish gray tint, which belongs neither to that nor to any other part of the true spectrum. In experiments on absorption, if instead of

receiving the light directly into the eye it be found convenient to form a pure spectrum on a screen of white paper, then, if the absorbing medium be placed in the path of the incident light, the scattered light forming any part of the spectrum cannot be cut off or weakened without at the same time cutting off or weakening the dispersed light coming from the same part of the screen. But if the absorbing medium be held in front of the eye, its effect on the spectrum will sometimes be very sensibly different from what it would be were the screen to send back none but scattered light.

It is true that the quantity of light dispersed by white paper is so small that this substance may very well continue to be used as a screen, without any danger of the observer's being deceived, if only he be aware of the fact of dispersion, so as to be on his guard. Still, it is not unreasonable to seek for a substitute for paper, which may be free from the same objection.

190. A porcelain tablet appeared to be unexceptionable in this respect, for it exhibited no perceptible sensibility, even when examined by a linear spectrum.   However, the translucency of the substance gave the spectrum a blurred appearance, and the fixed lines were not shown so well as on paper.

Chalk scraped smooth is well adapted, from its fineness, its whiteness and its opacity, for showing the most delicate objects. The finest fixed lines are beautifully seen on it, decidedly better than on paper. Its sensibility too, though not absolutely null, is much less than that of most kinds of white paper. Indeed, it would be an unnecessary refinement to seek for anything better, were it not that a piece of sufficient size might not always be at hand. From what I have seen, I believe that the best kind of screen will be obtained by the use of some white inorganic chemical precipitate, but my experiments in this department have not yet been sufficiently extended to authorize me in recommending any particular process.

191. The object of the observer may however be altogether different, and he may wish to extend the spectrum as far as possible, for the purpose of viewing the fixed lines belonging to the invisible part beyond the extreme violet, or making experiments on the invisible rays. For this purpose it would be proper to

employ a clear and highly sensitive solid or fluid. A weak solution of sulphate or phosphate of quinine would do very well, or a weak decoction of the bark of the horse-chestnut (no doubt a solution of pure esculine would be better), or an alcoholic solution of the seeds of the *Datura stramonium*. But perhaps the most convenient thing of all would be a slab of glass coloured yellow by oxide of uranium. This would be always ready, and in point of sensibility the glass does not seem to yield to any of the solutions above mentioned, at least so far as relates to those rays which are capable of passing through glass *.

192. In making experiments on the invisible rays, it is well to get rid, as far as possible, of the glare arising from the bright part of the spectrum, and therefore a clear solid or solution is preferable to an opaque screen. If it be desired to show the fixed lines in the visible and invisible parts of the spectrum at the same time, a screen may be employed consisting of paper washed with a moderately strong solution of sulphate of quinine, or an alcoholic solution of stramonium seeds. Turmeric paper is not, I think, quite so good for showing the fixed lines of very high refrangibility, but is at least equally good for the extreme violet and for the rays a good distance further on, especially if it has been washed with a solution of tartaric acid. It is likely that many other acids would do as well. Very excellent screens might probably be prepared by washing paper with a solution of esculine, or even of the bark of the horse-chestnut†, or by covering pasteboard with yellow uranite reduced to fine powder, and made to adhere by a weak solution of pure gum Arabic; but these I have not tried.

Application of internal dispersion to demonstrating the course of rays.

193. Solutions of quinine have already been employed for this purpose, and a weak decoction of the bark of the horse-chestnut appears to be decidedly better. But the effect is immensely improved by using absorbing media to cut off all the rays belonging to the bright part of the visible spectrum. A deep blue glass will answer very well for this purpose if its faces be even, so as

* See note F.   † See note G.

not to disturb the regularity of the refraction. The appearance of the general pencil refracted through a rather large lens, with its caustic surface, its geometrical focus, &c., is singularly beautiful when exhibited in this way, on account of the perfect continuity of the light, and the delicacy with which the different degrees of illumination belonging to different parts of the pencil are represented by the different degrees of brightness of the dispersed light. The solution should be contained in a vessel with plane sides of glass, and ought to be very weak, or else only the part of the pencil which lies near the surface by which the light enters will be properly represented.

Application of internal dispersion to the determination of the absorbing power of media with respect to the invisible rays beyond the violet, and the reflecting power of surfaces with respect to those rays.

194. Hitherto no method has been known by which the absorbing power of a medium with respect to these rays could be determined for each degree of refrangibility in particular, except that which consists in taking a photographic impression of a pure spectrum, the light forming the spectrum having been transmitted through the substance to be examined. It is needless to remark how troublesome such a process is when contrasted with the mode of determining the absorption which media exercise on the visible rays. But the phenomenon of internal dispersion furnishes the philosopher, so to speak, with *eyes to see the invisible rays,* so that the absorbing power of the medium with respect to these rays may be instantly observed. For this purpose it is sufficient to form a pure spectrum, using instead of a screen a highly sensitive fluid or solid, such as one of those mentioned in Art. 191, and to hold before it the medium to be examined, or else to place the medium over the whole or a part of the slit.

195. In this way the transparency of glass coloured yellow by oxide of silver with respect to the violet rays and some of those still more refrangible, which has been remarked by Sir John Herschel*, may be at once observed. A set of green glasses were found to be very variable in the mode in which they absorbed the

---

* *Philosophical Transactions for* 1840, p. 89.

invisible rays, some absorbing the more refrangible of the rays capable of affecting a dilute solution of sulphate of quinine and transmitting the less refrangible, others absorbing the less and transmitting the more refrangible, and others again absorbing them all. These rays were absorbed by solutions of chromate and bichromate of potash so weak as to be almost colourless. A thickness of about a quarter of an inch of sulphuret of carbon was sufficient to absorb all the rays beyond $Hk1$, so that a hollow prism filled with this fluid would be useless in experiments on these rays. It should be remarked that the sulphuret of carbon employed was not yellow from dissolved sulphur, but apparently as colourless as water.

196. To determine qualitatively the reflecting power of a polished surface with respect to the invisible rays of each particular degree of refrangibility, it would be sufficient to form a pure spectrum as usual, reflect the rays sideways before they come to the focus of the larger lens, place a sensitive medium to receive them, and compare the effect with that produced on the same medium when the rays are allowed to fall directly upon it.

Effect of different Flames.

197. Want of sunlight proved to be such an impediment to the pursuit of these researches that I was induced to try some bright flames, with the view of obtaining some convenient substitute. Candle-light is very ill adapted to these experiments. The flame of a camphene-lamp proved no better, perhaps rather worse, for it abounds so much in rays belonging to the bright part of the spectrum that the glare of the light prevents all observation of faint objects; and the flame does not appear to be rich in invisible rays in anything like the proportion in which it is rich in visible ones. The flame of nitre burning on wood or charcoal produced a very good effect, exhibiting, when the combustion was most vivid, a copious dispersive reflexion in a weak solution of sulphate of quinine contained in a bottle held near it. The tint of the dispersed light appeared to be not quite the same as that given by daylight, but to verge a little towards violet. However, I do not place very strong reliance on the judgment of the eye under such circumstances. A still stronger dispersive reflexion was produced

by a flash of gunpowder. The tint in this case appeared to be the same as that seen by daylight.

198.   While engaged in some of these experiments on bright flames, I was surprised by discovering the strong effect produced by the flame of a spirit-lamp, the illuminating power of which is so feeble.   When this flame was held close to a bottle containing sulphate of quinine, a very distinct dispersive reflexion was exhibited.   The same was the case with several other sensitive solutions.   However, the full effect of the flame is not thus exhibited, because a considerable portion of the rays which it emits is stopped by glass.   It is best observed by pouring the solution into an open vessel, such as a wine-glass or tumbler, holding the flame immediately over it, and placing the eye in or very little below the plane of the surface.   In this way nothing is interposed between the flame and the fluid, except an inch or two of air, the absorption produced by which, it is presumed, is insensible ; and the plane strata, parallel to the surface, into which the illuminated portion of the fluid may be conceived to be divided, are all projected into lines, whereby the intensity of the blue light is materially increased.   It is to be observed further, that if the eye be held a little below the plane of the surface, there enters it, not only the light coming directly from the blue stratum itself, but also that coming from its image formed by total internal reflexion. This mode of observation has already been employed by Sir John Herschel in the case of sunlight.   As it is frequently useful in these researches it will be convenient to have a name for it, and I shall accordingly speak of it as the method of observing by *superficial projection*.

199.   The opacity of a solution of sulphate of quinine appears to increase regularly and rapidly with the refrangibility of the light.   Hence we may form an estimate of the refrangibility of any light by which the solution may be affected, by observing the degree in which the illumination is concentrated in the neighbourhood of the surface.   For this purpose it is essential to employ a weak solution, since otherwise streams of invisible light of various degrees of refrangibility produce each their full effect in strata so very narrow, that they cannot be distinguished by the breadth of the stratum.   Now to judge by the great concentration of the illumination produced by a spirit-lamp, even in

the case of an extremely weak solution, as well as by the considerable degree in which the active rays were intercepted by
glass, these rays, taken as a whole, must have been of very high
refrangibility, such as to place them among the most refrangible
of the fixed lines represented in the map, or perhaps even altogether beyond them. In making observations on the solar spectrum, it was plain that the prisms were by no means transparent
with respect to the rays belonging to the group p of fixed lines.
Yet these rays, before they produced their effect, had to pass
twice through the plate-glass belonging to the mirror (except
so far as regards the rays reflected at the first surface), then
through three prisms, though to be sure as close as possible to
the edges, then through a lens by no means very thin, and
lastly, through the side of the vessel containing the fluid. Such
a train of glass would be sufficient materially to weaken, if not
even wholly to cut off the active rays coming from the flame of a
spirit-lamp.

200.   The flame of naphtha* produces nearly the same effect
as that of alcohol. The flame of ether is not so good; but
whether this arises solely from its richness in visible rays, which
only produce a glare, or likewise from a comparative poverty in
highly refrangible invisible rays, it is not easy to say. The flame
of hydrogen produces a very strong effect. The invisible rays in
which it so much abounds, taken as a whole, appear to be even
more refrangible than those which come from the flame of a
spirit-lamp. In making some observations with the flame of
hydrogen, when the gas was nearly exhausted, so that the flame
was reduced to a roundish knob no larger than a sweet pea, and
giving hardly any light, it was found still to produce a very
marked effect when held over the surface of a solution of sulphate of quinine. The flame of sulphuret of carbon produces on
most objects a much stronger effect than that of alcohol. It
exhibits distinctly the blue light dispersed close to the surface
of a solution of guaiacum in alcohol, which the flame of alcohol
does not. It appears then that the flame of sulphuret of carbon
is rich in invisible rays of such a refrangibility as to place them
among the groups of fixed lines m, n, or a little beyond, since

[* By this was meant wood-spirit, commercially called naphtha, not the
hydrocarbon to which the name more properly belongs.]

when a solution of guaiacum is examined in the solar spectrum, it is found that that is the region in which the blue dispersed light is produced. The blue light dispersed by a solution of guaiacum may also be seen by using the blue flame of sulphur burning feebly. The poverty of the flame of a spirit-lamp, not only with respect to visible rays, but also with respect to invisible rays, except those of very high refrangibility, accounts for the circumstance that it does not exhibit, or at least hardly at all exhibits, the blue light dispersed by fluor-spar.

Mode of determining, by means of the light of a spirit-lamp, the transparency of bodies with respect to the invisible rays of high refrangibility.

201. If the body be a solid, and be bounded by parallel surfaces, its transparency with regard to these rays is easily tested. For this purpose it is sufficient to hold the flame of a spirit-lamp a little way above the surface of a weak solution of sulphate of quinine contained in an open vessel in a dark room, and then, placing the eye so as to see the dispersed light in projection, alternately to interpose and remove the plate to be examined.

202. On examining in this way various specimens of glass, I found none which did not show evident defects of transparency. The purest specimens of plate-glass appeared, I think, to be the least defective. I cannot say whether the observed defects of transparency were due to the essential ingredients of the glass, or to accidental impurities. It is possible that glass made with chemically pure materials might be transparent*. I believe that a mere trace of peroxide of iron, or of sulphuret of soda or potassa, would be sufficient to impair materially the transparency of glass with respect to these rays, and such impurities are very likely to be present. Quartz, however, appeared to be perfectly transparent, the active rays passing through the thickness of one

---

* Some specimens of glass belonging to Dr Faraday's experiments, which from the absence of colour and of internal dispersion seemed hopeful, could not be examined for transparency, on account of their irregular figure; and as they were only lent to me by a friend, I did not feel myself at liberty to get them cut and polished.

or two inches, whether parallel or perpendicular to the axis, without any perceptible loss. The contrast between quartz and mica was very striking, for a plate of mica no thicker than paper produced a very sensible diminution in the illumination.

203. For the purpose of observing fluids, I procured two vessels consisting of sections of a wide glass tube, about an inch long, closed at one end with a disc of quartz. I shall call these for brevity quartz vessels, though of course the bottom is the only part in which there is any occasion to use quartz. When a fluid is to be examined it is poured into a quartz vessel, and then the vessel with its fluid contents is examined in the manner of a solid plate, as described in Art. 201. On account of the perfect transparency of quartz, the fluid is as good as suspended in air. When a quartz vessel was partly filled with water, the addition of a very small quantity of nitrate of iron was sufficient to cause the absorption of the active rays. The solution was so weak as to be almost colourless when viewed through the thickness through which the rays would have to pass. A solution of perchloride of iron had a similar effect. These fluids I had specially examined by sunlight, and had not found in them the least trace of internal dispersion. When a fluid exhibits internal dispersion, it is almost always very opaque with regard to rays of high refrangibility, as is shown, without any special experiment, in the course of the observations by which the internal dispersion is exhibited; but it by no means follows conversely, that when a fluid is very opaque with regard to these rays, though nearly transparent with regard to the visible rays, it exhibits the phenomenon of internal dispersion.

204. I have little doubt that the solar spectrum would be prolonged, though to what extent I am unable to say, by using a complete optical train in every member of which glass was replaced by quartz. Such a train would be rather expensive, but would not involve any particular difficulty of execution. If solid prisms of quartz were used, half of the incident light would be lost, on account of the double refraction of the substance, unless the prisms were cut in a particular manner, which however would seem likely to involve some difficulties, both in the execution and in the observations. But hollow prisms holding fluids might be employed, having the two faces across which the light has to

pass made of quartz plates. For a reason already mentioned, sulphuret of carbon cannot be employed for filling the prisms, and the dispersive power of water is very low, but there appears to be no objection to the use of a solution of some colourless metallic salt. At least saturated solutions of sulphate of zinc and of acetate of lead, the only salts I have tried with this view, showed no defects of transparency when examined in quartz vessels by means of the flame of a spirit-lamp and a solution of sulphate of quinine*.

Effect of Hydrochloric Acid, &c. on Solutions of Quinine.   Optical evidences of combination in other instances.

205.   Sir John Herschel, in his interesting paper already so often referred to, observes that it is only acid solutions of quinine which exhibit the peculiar blue colour, and that among different acids the muriatic seems least efficacious (page 145).

For my own part I have tried solutions of quinine (not di-sulphate) in dilute sulphuric, phosphoric, nitric, acetic, citric, tartaric, oxalic, and hydrocyanic acids, and also in a solution of alum. In all these cases the blue colour of the dispersed light was plainly seen by ordinary daylight, especially when the fluid was examined by superficial projection. It was not easy to say which solution answered best, but I am inclined to think that in which phosphoric acid was used.

206.   But when quinine was dissolved in dilute hydrochloric acid the blue colour was not exhibited, not even when the fluid was held in the sunlight, and examined by superficial projection. Certain theoretical views led me to regard this as an evidence of a more intimate union between quinine and hydrochloric acid than between quinine and the acids first mentioned, and to try whether the addition of hydrochloric acid to the solutions mentioned in the preceding paragraph would not destroy the blue colour. On trial this proved to be actually the case, so that even sulphuric acid is incapable of developing the blue colour in a solution of quinine in hydrochloric acid.

* See note H.

207. That the quinine was not decomposed when the blue colour due to sulphate of quinine was destroyed by hydrochloric acid, but only differently combined, was shown by adding a solution of carbonate of soda, which produced a white precipitate; and when this was collected on a filter, washed, and redissolved in dilute sulphuric acid, it exhibited the blue colour as usual.

208. The addition of a solution of common salt, instead of hydrochloric acid, to the solutions mentioned in Art. 205, likewise destroyed the blue colour. In the case of sulphuric acid this is only what might have been confidently anticipated; but we should not perhaps have expected that quinine in combination with a weak acid, such as citric, would decompose hydrochlorate of soda, giving rise to citrate of soda and hydrochlorate of quinine; yet this appears to be the nature of the reaction.

209. It might perhaps be supposed that the sulphuric acid was only partially expelled from sulphate of quinine by hydrochloric acid, and that the salt in solution was really a sort of double salt, in which the same base, quinine, was combined with sulphuric and hydrochloric acids in atomic proportion. But if so, it is probable, though not certain, that the same salt would be formed on adding hydrochloric acid to a solution of disulphate of quinine, even though the quantity were not sufficient to combine with the whole of the disulphate. On this supposition, if hydrochloric acid were added by small quantities at a time to a solution of disulphate of quinine, the blue colour ought not to be developed; and when acid enough had been added it ought to be incapable of being developed by the addition of sulphuric acid; whereas, if the whole of the sulphuric acid be expelled by hydrochloric acid, the blue colour ought to be first developed, by the conversion of a portion of the disulphate of quinine into a sulphate, and then destroyed, on the addition of more acid, by the conversion of the sulphate into a hydrochlorate. On trying the experiment with a solution of disulphate of quinine in warm water, it was found that the blue colour was actually first developed and then destroyed.

210. A practical conclusion which seems to follow from these results is, that in the employment of quinine in medicine it is of

little consequence whether the sulphate, phosphate, acetate, or hydrochlorate be used, since the first three salts would be immediately converted by the common salt in the body into the hydrochlorate, and the small quantity of a neutral salt of soda resulting from the double decomposition could hardly, one would suppose, be worth considering. However, the common quinine is associated with cinchonine, the reactions of which may be different. According to Sir John Herschel, the latter alkaloid does not exhibit the blue colour, and therefore the optical tests do not apply to it. If it be desired to obtain a soluble salt of quinine which shall not be converted by common salt, by double decomposition, into a hydrochlorate, it must apparently be sought for among the combinations of quinine with very weak acids, the affinity of which for soda does not much help that of hydrochloric acid for quinine. It seems likely enough that such salts may exist; for though acetate or citrate of quinine decomposes hydrochlorate of soda, hydrochlorate of quinine is decomposed by carbonate of soda; and it is probable that many vegetable acids behave like the carbonic in this respect.

211. The blue dispersion of a solution of sulphate of quinine is destroyed by hydrobromic and hydriodic acids just as by hydrochloric. In the experiment, solutions of bromide and iodide of potassium were used; but as a considerable excess of sulphuric acid was purposely added to the solution of quinine, the potassa introduced would merely remain inert in the solution as a sulphate, without impeding the observation. The same experiment was tried with phosphate of quinine with the same result.

212. It is stated in Turner's *Chemistry*, that the play of colours observed in solutions of polychrome (*i.e.* esculine) is destroyed by acids, and heightened by alkalies. The destruction, or at least almost complete destruction, of the blue colour due to dispersed light in a decoction of the bark of the horse-chestnut, which is produced by acids, is readily observed; but I could not perceive that the addition of alkalies in the first instance to a fresh solution made any difference one way or other. If the blue colour had previously been destroyed by an acid, it was restored by the alkali. If the horse-chestnut had never been examined chemically, these observations alone would indicate that in all

probability the principle to which the blue colour was due was
capable of entering into firm combination with acids, but did not
combine with alkalies. It is, in fact, as we know, a vegetable
base*.

213. A solution of nitrate of uranium in ether is insensible,
as if some of the elements of the ether entered into firm combination with the oxide of uranium. In connexion with this circumstance, it is rather remarkable, that although the ether passes off
by evaporation when the solution is left to itself in an open
vessel, if heat be applied chemical action sets in, and the residue
consists chiefly of a salt which has all the appearance of oxalate
of uranium. This salt, when washed and examined in the moist
state, without very great concentration of light, was found to be
insensible†.

214. It is rare to meet with solutions so highly sensitive as
those of quinine and esculine, but similar observations may be
made on a great number of solutions, by employing suitable
methods. The most searching method consists in forming a
bright and tolerably pure spectrum, by transmitting the sun's
light through a very broad slit, or even leaving out the slit
altogether. It is desirable to use a lens of only moderate focal
length in connexion with the prisms. The solution having been
placed in the spectrum, the acid, or other agent whose reactions
it is desired to study, is to be added, and the effect, if any,
observed. It is usually advantageous to cover the slit with a blue
glass, or similar absorbing medium; but sometimes effects take
place in the bright part of the spectrum, which is intercepted by
such a medium. When false dispersion abounds, it is well to look
down on the fluid through a Nicol's prism, so as to stop all light
which is polarized in the plane of reflexion.

*Negative results with reference to a mutual action of the rays
incident on sensitive solutions.*

215. The antagonistic effects of the more and less refrangible
rays, which have been observed in certain phenomena, induced

* [It is not a base but a glucoside.]
† See note I.

me to try whether anything of the kind could be perceived in the case of internal dispersion. The following arrangement was adopted for putting this question to the test of experiment.

A tumbler was filled with a very dilute solution of sulphate of quinine, and placed in a pure spectrum. As usual, the illuminated portion of the fluid consisted of two distinct parts, one the blue beam of truly dispersed light, corresponding to the highly refrangible rays, the other the beam reflected from motes, exhibiting the usual prismatic colours, and corresponding to the brighter of the visible rays. The fluid was nearly free from motes, so that the first beam was by far the brighter of the two; and the second beam, without being bright enough at all to interfere with the observation, was useful as serving to point out where the red, yellow, &c. rays lay. A flat prism, having an angle of about 130°, was then held in front of the vessel, with its edge vertical, and situated in the more refrangible part of the visible rays. The rays forming the two beams were thus bent in opposite directions, and the beams made to cross each other within the fluid; and by turning the prism a little in both directions in azimuth, that is, round an axis parallel to the incident rays, it was easy to make sure that the beams did actually cross. But not the slightest perceptible difference in the blue beam was made by the passage of the red and other lowly refrangible rays across it.

216.   Certain theoretical views having led me to regard it as doubtful whether the intensity of light internally dispersed was proportional to the intensity of the incident rays, other circumstances being the same, I was induced to try the following experiment.

The sun's light was reflected horizontally through a large lens, which was covered by a screen containing two moderately large round holes, situated in the same horizontal plane, and a good distance apart. The beams coming through the two holes converged of course towards the focus of the lens, and at the same time contracted in width, and became brighter from the concentration of the light. For our present purpose, they may be regarded as cylindrical beams converging towards the focus of the lens. When they had approached each other sufficiently, they were transmitted through a blue ammoniacal solution of copper, contained in a vessel with parallel sides. The object of this was of course to absorb all the bright visible rays, which would not only be useless for exciting the solution which it was meant to try, but would materially hinder the observation by the glare which they would produce. The beams were then admitted into a vessel containing a decoction of the bark of the horse-chestnut, greatly diluted with water. In passing through the fluid they produced two blue beams of truly dispersed light, which converged towards a point a little way outside the vessel. A flat prism, with an angle of about 150°, was then held in front of the vessel, with its edge vertical, and situated between the incident beams. The blue beams of dispersed light were thus made to cross within the fluid; and by moving the prism in azimuth, it was easy to make one beam either fall above the other, cross it, or fall below it. Now on looking down from above with one eye only, and moving the prism backwards and forwards in azimuth, I could not perceive the slightest difference of illumination, according as the blue beams actually crossed each other, or were merely seen projected one on the other. In this experiment, then, it appeared that one beam of incident rays produced as much additional dispersed light in a portion of fluid already excited by the other beam, as it was capable of producing in a similar portion of fluid not otherwise excited.

Effect of an electric spark.   Nature of its phosphorogenic rays.

217.   For the use of the apparatus with which the following experiments were made, I am indebted to the kindness of Professor Cumming.

An electric spark produces an internal dispersion of light in a very striking manner in the case of an extremely dilute solution of sulphate of quinine. Having prepared a solution so weak, that when it was examined by superficial projection by the light of a spirit-lamp, nothing was seen but a pale gleam of light extending a good way into the fluid, and not only not confined to the surface, but not even showing any particular concentration in the neighbourhood of the surface, I placed it so as to be illuminated by the sparks from the prime conductor of an electrifying machine, which passed at no great distance over the surface.

A very marked internal dispersion was produced, but the nature
of the effect depended in a good measure on the character of the
spark. A feeble branched spark, giving but little light, and
making little noise, produced an illumination extending to a
considerable depth, and very much stronger than that occasioned
in the same solution by the flame of a spirit-lamp. The rays by
which this was produced passed in a great measure through a
plate of glass interposed between the spark and the surface of the
fluid. But a bright linear spark, making a sharp crack, produced
an illumination almost confined to an excessively thin stratum
adjacent to the surface of the fluid; and the rays by which this
was produced were cut off by glass, though transmitted through
quartz. The same was the case with the discharge from a Leyden
jar, which produced a bright light almost confined to the surface*.

218. The opacity of a solution of sulphate of quinine appears
to increase regularly and rapidly with the refrangibility of the
rays incident upon it. Hence we are led to the conclusion that a
strong electric spark is excessively rich in invisible rays of extremely high refrangibility. Glass is opaque with respect to these
rays, but quartz transparent.

219. It is known that the phosphorogenic rays of an electric
spark, at least those which affect Canton's phosphorus, pass very
freely through quartz, but are stopped by a very moderate thickness
of glass. This alone, after what has been already mentioned,
would lead us to suppose that the phosphorogenic rays coming
from such a spark are merely rays of very high refrangibility. If
so, they ought to be intercepted by a very small quantity of a
substance known to absorb such rays with energy.

After having made some experiments on the production of
phosphorescence in Canton's phosphorus by means of an electric
discharge, and observed how the influence of the discharge was
transmitted through quartz and stopped, or almost entirely stopped,
by glass, I felt confident that my own observations were comparable with those of others. A small portion of the phosphorus
was then placed on card, covered by an empty quartz vessel, and
had the discharge of a Leyden jar passed over it. The phos-

* See note J.

phorescence was powerfully excited, being visible in a room which was by no means quite dark; and when the card was carried into a dark place, the phosphorescent light remained plainly visible for a good while. The experiment was then repeated with a fresh portion of the same phosphorus, the vessel this time containing water. The phosphorescence was produced as before, though not I think so copiously. But on taking a fresh portion of the phosphorus, and substituting for water a very dilute solution of sulphate of quinine, the influence of the spark was arrested, and the phosphorus was not rendered luminous. It was found that a solution containing only about one part of quinine in 10,000, with a depth of half an inch, was sufficient to prevent the generation of phosphorescence.

220. This result, it seems to me, would be sufficient, were proof wanting, to show that no part of the effect is attributable *directly* to the electrical disturbance. The effect produced when the phosphorus is at the distance of an inch or so from the points of the discharger seems exactly the same as when it is nearer, being merely somewhat weaker, as would naturally be expected, whatever view were taken of the nature of the influence. But at the distance of an inch, the influence of the spark, though it passes freely through quartz and water, is cut off by adding to the water an excessively small quantity of sulphate of quinine. It cannot be supposed that the electrical relations of the medium, or its permeability to electrical attractions and repulsions, are utterly changed by such an addition; while, on the other hand, the result is in perfect conformity with what we know respecting the stoppage of radiations by absorbing media. However, the principal object of the experiment was not to confirm the view which makes the influence of the spark to consist in the rays which emanate from it, a view which I suppose is pretty generally adopted, but to investigate more fully the nature of these rays. Enough has, I think, been adduced to show that they are merely rays which there is no reason to suppose are physically different from those of light, but quite the contrary, and which are of very high refrangibility, and are therefore invisible, since they fall far beyond the limits of refrangibility within which the retina is affected. Indeed, it seems very likely that the highly refrangible rays never reach the retina, but are absorbed by the coats of the

eye*. Hence the phenomena relating to the phosphorescence produced by an electric discharge afford no countenance to the supposition that it is possible to divide rays of a given refrangibility into phosphorogenic, chemical, luminous, &c. Of course the most unexceptionable mode of determining the refrangibility of the phosphorogenic rays would be by actual prismatic decomposition, but this would require the employment of a quartz train.

Points of resemblance and contrast between internal dispersion and phosphorescence.

221. As the term *phosphorescence* has been applied to several different phenomena, I must here explain that I mean the spontaneous exhibition of a soft light, independently of chemical changes, which some substances exhibit for a time after having been exposed to the sun's rays, or to an electric discharge, or to light from some other sources.

In many respects the two phenomena have a strong resemblance. Thus, the general features of internal dispersion cannot be better conceived than by regarding the sensitive medium as self-luminous while under the excitement of the active rays. Again, it is well known that the rays of the solar spectrum by which the phosphorescence of Canton's phosphorus, sulphuret of barium, and other phosphori, is produced, are those of high refrangibility, as well as the invisible rays beyond; and these are precisely the rays which in the great majority of cases are most efficient in producing internal dispersion. I do not however know how far it may be true that when phosphorescence is excited by homogeneous light the refrangibility of the incident light is a superior limit to the refrangibilities of the component parts of the light emitted. Indeed, according to Professor Draper, when the phosphorescence of Canton's phosphorus is excited by the rays from incandescent lime, the active rays belong to the red extremity of the spectrum†. If this result be confirmed‡, it follows that the

---

* See note K.

† *Philosophical Magazine*, Vol. xxvii. (Dec. 1845) p. 436.

‡ [Early in 1853 I was engaged along with Faraday in preparing in the laboratory of the Royal Institution some experiments to show at an evening lecture on the subject of this paper. I expressed to him a wish to repeat Draper's experiment,

most striking law relating to internal dispersion is not obeyed in the case of phosphorescence.

In the same paper Professor Draper remarks, "Some time ago I determined the refrangibility of the rays of an electric spark which excite phosphorescence in sulphuret of lime; they are found at the violet extremity of the spectrum." In what way Professor Draper determined the refrangibility of rays with respect to which glass is so opaque, he does not give the least hint. Being perfectly in the dark as to the evidence on which the conclusion is based, I cannot accept it in contradiction to my own experiments. Perhaps, however, "at the violet extremity" may mean nothing more than somewhere in the highly refracted region beyond the visible rays. If so, Professor Draper's statement is in accordance with my own conclusions.

222. When one part of a phosphorus has been excited, the phosphorescence is found gradually to extend itself to the neighbouring parts*. In this respect a substance which exhibits internal dispersion presents a striking contrast. The finest fixed lines of the spectrum are seen sharply defined, whether in a solution, or in a clear solid, or on a washed paper.

223. Of course, theoretically, there ought, to a certain extent, to be a communication of illumination from one part of a sensitive fluid to another, on account of the light which is twice, three

as it had such an important bearing on the subject, and he immediately tried it. We obtained, however, only a negative result, as the Canton's phosphorus which had been acted on by bright light which had been passed through a solution of bichromate of potash did not give out in the dark any sensible light. Long afterwards, as I was engaged with some experiments on Balmain's luminous paint, it occurred to me that possibly Draper's result might have been due to a latent effect of a previous exposure to light; and I wrote to him to enquire whether this might have been possible. He replied that he generally heated his phosphori before proceeding to an experiment, in order to guard against the possible existence of a latent effect of previous exposure; that at that distance of time he could not be certain whether that had been done in the particular experiment referred to; that if it had not, the result he had mentioned might have been brought about in that way.]

* [M. Becquerel afterwards explained this apparent result by attributing it to the increased sensitiveness of the eye arising from continuance in the dark, which enabled the observer to see outlying portions of a phosphorescing patch which were not seen at first, thereby giving the impression that the phosphorescence was extending.]

times, &c., dispersed. This however must be excessively small; for the mean refrangibility of the dispersed light is usually much lower than the refrangibility of the active light, perhaps lower than that of any light capable of exciting the solution. However, generally some few of the dispersed rays would have a refrangibility sufficiently high to be dispersed again. But practically the intensity of the light twice dispersed in this manner would be so very small that it may safely be altogether disregarded.

224. But by far the most striking point of contrast between the two phenomena, consists in the apparently instantaneous commencement and cessation of the illumination, in the case of internal dispersion, when the active light is admitted and cut off. There is nothing to create the least suspicion of any appreciable duration in the effect. When internal dispersion is exhibited by means of an electric spark, it appears no less momentary than the illumination of a landscape by a flash of lightning. I have not attempted to determine whether any appreciable duration could be made out by means of a revolving mirror*.

225. There appears to be no relation between the substances which exhibit a change of refrangibility and those which phosphoresce, either spontaneously, or on the application of heat. Thus the sulphurets of calcium and barium, on being examined for internal dispersion, were found to be insensible, as was also Iceland spar. The last substance phosphoresced strongly on the application of heat. So far as was examined, the minerals which did exhibit a change of refrangibility showed no special disposition to phosphoresce. Sir David Brewster has remarked, that a specimen of fluor-spar which exhibited a blue light by internal disper-

---

* [The experiment I had in my mind was to view in a revolving mirror the substance to be examined while illuminated in a dark room by a succession of sparks from the prime conductor of an electrifying machine, taking one's chance for the casual appearance of images in the field of view. The experiment was afterwards tried with apparatus kindly lent me, but whether with sparks from a prime conductor or with an induction coil I am not now sure. Notwithstanding what M. Becquerel had in the mean time done with his beautiful phosphoroscope, the results obtained are not perhaps wholly without interest. Thus yellow uranite instead of its usual appearance showed a well-defined image of a very ordinary looking yellow stone and a long drawn out gleam of green light due of course to the fading phosphorescence.]

sion, exhibited when heated a blue phosphorescent light; but this appears to have been merely a casual coincidence*.

On the Cause of True Internal Dispersion, and of Absorption.

226. In considering the cause of internal dispersion, we may I think at once discard all supposition of reflexions and refractions of the vibrations of the luminiferous ether among the ultimate molecules of bodies. It seems to be quite contrary to dynamical principles to suppose that any such causes should be adequate to account for the production of vibrations of one period from vibrations of another.

All believers, I suppose, in the undulatory theory of light are agreed in regarding the production of light in the first instance as due to vibratory movements among the ultimate molecules of the self-luminous body. Now in the phenomenon of internal dispersion, the sensitive body, so long as it is under the influence of the active light, behaves as if it were self-luminous. Nothing then seems more natural than to suppose that the incident vibrations of the luminiferous ether produce vibratory movements among the ultimate molecules of sensitive substances, and that the molecules in turn, swinging on their own account, produce vibrations in the luminiferous ether, and thus cause the sensation of light. The periodic times of these vibrations depend upon the periods in which the molecules are disposed to swing, not upon the periodic time of the incident vibrations.

227. But in the very outset of this theory an objection will probably be urged, that it is quite as much contrary to dynamical principles to suppose the periodic time of the ethereal vibrations capable of being changed through the intervention of ponderable molecules as without any such machinery†. The answer to this objection is, that such a notion depends altogether on the applicability of a certain dynamical principle relating to indefinitely small motions, and that we have no right to regard the molecular vibra-

---

* *Report of the Meeting of the British Association at Newcastle in* 1839, p. 11.

† [The attempt here made to account for the lowering of refrangibility has long since been given up, and I have been led to adopt what seems to me a far more probable explanation. See Addition to this paper, p. 410.]

tions as indefinitely small. The excursions of the atoms may be, and doubtless are, excessively small compared with the length of a wave of light; but it by no means follows that they are excessively small compared with the linear dimensions of a complex molecule. It is well known that chemical changes take place under the influence of light, especially the more refrangible rays, which would not otherwise happen. In such cases it is plain that the molecular disturbances must not be regarded as indefinitely small. But vibrations may very well take place which do not go to the length of complete disruption, and yet which ought by no means to be regarded as indefinitely small. Furthermore, it is to be observed that if in the cases of indefinitely small molecular displacements the forces of restitution be not proportional to the displacements, the principle above alluded to will not be applicable however small the disturbance may be; and if in the expressions for the forces of restitution the terms depending on first powers of the displacements (supposed finite), though not absolutely null, be very small, the principle will not apply unless the molecular excursions be extremely small indeed. In consequence of the necessity of introducing forces not proportional to the displacements, it would be very difficult to calculate the motion, even were we acquainted with all the circumstances of the case, whereas we are quite in the dark respecting the actual data of the problem. But certainly we cannot affirm that in the disturbance communicated back again to the luminiferous ether none but periodic vibrations would be produced, having the same period as the incident vibrations. Rather, it seems evident that a sort of irregular motion must be produced in the molecules, periodic only in the sense that the molecules retain the same mean state; and that the disturbance which the molecules in turn communicate to the ether must be such as cannot be expressed by circular functions of a given period, namely, that of the incident vibrations.

228.   It is very remarkable with what pertinacity a particular mode of internal dispersion attaches itself to a particular chemical substance. Thus the singular dispersion of a red light exhibited by the green colouring matter of leaves is found in a green leaf, or in a solution of the green colouring matter in alcohol, ether, sulphuret of carbon, or muriatic acid. The dispersion exhibited by nitrate of uranium is found in a solution of the salt in water, as

well as in the crystals themselves, which are doubly refracting. In all probability therefore the molecular vibrations by which the dispersed light is produced are not vibrations in which the molecules move among one another, but vibrations among the constituent parts of the molecules themselves, performed by virtue of the internal forces which hold the parts of the molecules together. It is worthy of remark that it is chiefly among organic compounds, the ultimate molecules of which we are taught by chemistry to regard as having a complicated structure, that internal dispersion is found. It is true that peroxide of uranium furnishes many examples of internal dispersion ; but then the anhydrous peroxide is itself insensible, it is only some of the compounds into which it enters that are so remarkably sensitive ; and the chemical formulæ of these compounds, so far as they are known, are not by any means extremely simple, although it is true that they may not be more complicated than formulæ relating to other oxides. Why this particular oxide should be disposed to enter into tottering combinations I do not pretend even to conjecture ; but it seems not a little remarkable that peroxide of uranium, which is so peculiar with respect to its optical properties, should also present some singularities in its mode of chemical combination, which led M. Peligot to regard it as the protoxide of a compound radical.

229. We are, I conceive, at present far from an explanation of the phenomena of internal dispersion in all their details. They appear to be associated with the inmost structure of chemical molecules, to such a degree as to throw even the phenomena of polarization into the shade. In this respect, indeed, absorption seems superior to polarization, since most of the phenomena of polarization refer rather to the state of crystalline aggregation of the molecules than to their constitution ; but the phenomena of internal dispersion appear to be much more searching than those of absorption. There is one law however relating to internal dispersion so striking and so simple, that it seems not unreasonable to look for an explanation of it ; I allude to that according to which the refrangibility of light is always lowered in the process of dispersion. I have not hitherto been able altogether to satisfy myself respecting a dynamical explanation of this law, but the following conjectures will not perhaps be deemed altogether unworthy of being mentioned.

230. Reasons have already been brought forward for regarding the molecular vibrations as performed under the influence of forces not proportional to the displacements. For simplicity's sake, let us suppose for the present the parts of the forces of restitution depending upon first powers of the displacements to be absolutely null. Then, when a molecule is disturbed, its atoms will be acted on by forces depending upon the second and higher powers of the displacements. These forces must tend to restore the atoms to their mean positions; otherwise the equilibrium would be unstable, and the atoms would enter into new combinations, either with one another, or with the atoms of the surrounding medium; so that, in fact, such compounds could never be formed. The condition of stability would require the parts of the forces depending upon squares of the displacements to vanish, but this is a point which need not be attended to, all that is essential to bear in mind being, that we have forces of restitution varying in a higher ratio than the displacements. If the parts of the forces of restitution which depend upon first powers of the displacements, though not absolutely null, be very small, the remaining parts must still be such as to tend to restore the atoms to their positions of equilibrium; otherwise the stability of the molecule, though not mathematically null, would be so very slight, that such compounds would probably never form themselves, but others of more stability would be formed instead. Or, even were such unstable compounds formed, they would probably be decomposed on attempting to excite them in the manner in which sensitive substances are excited in observing the phenomena of internal dispersion; so that whether they exist or not, they may be set aside in considering these phenomena.

231. Now when vibrations are performed under the action of forces which vary in a higher ratio than the displacements, the periodic times are not constant, but depend upon the amplitudes of vibration, being greater or less according as the amplitudes are less or greater. Suppose the molecular and ethereal vibrations already going on, and imagine the amplitudes of the former kept constant by the application of external forces. According to the value of the epoch of the vibrations of a particular molecule, the ethereal vibrations will tend, in the mean of several successive undulations, to augment or to check the vibrations of the molecule.

For some time there will be a tendency one way, then for some time a tendency the other way, and so on, the opposite tendencies balancing each other in the long run. The lengths of the times during which the tendency lies in one direction, will depend upon the periodic times of the molecular and ethereal vibrations, being on the whole greater or less according as the two periodic times are more or less nearly equal. But since no external forces actually act to keep the amplitudes constant, when the ethereal vibrations are favourable to disturbance the molecule is further disturbed, and therefore its periodic time is diminished; and when they are favourable to quiescence the disturbance of the molecule is checked, and therefore its periodic time is increased. If, then, the ether be vibrating more rapidly than the molecule, when the action is favourable to disturbance the periodic time of the molecular vibrations is rendered more nearly equal to that of the ethereal vibrations, and therefore the time during which the action is favourable to disturbance is prolonged; but when the action is favourable to quiescence, the effect is just the reverse. Hence, on the whole, there is a balance outstanding in favour of disturbance. But if the ether be vibrating more slowly than the molecule, it appears from similar reasoning that there will be a balance the other way. Hence it is only when the periodic time of the ethereal vibrations is less than that of the molecular, that the latter vibrations can be kept going by the former.

232. But it will probably be objected to this explanation, that when a periodic disturbing force affects the mean motion of a planet, the mean motion is a maximum, not when the force tending to augment it is a maximum, but at a time later by a quarter of the period of the force, namely, when the force vanishes in changing sign; and that in a similar manner the change in the periodic time of the vibrations of a disturbed molecule will affect equally the duration of the time during which the action is favourable to increased disturbance, and that during which it is favourable to quiescence, or more exactly will not alter either, since the effects in the first and second halves of those times will neutralize each other. The answer to this objection is, that we must not treat a molecule as if it were isolated, like a heavenly body, since it is continually losing its motion by communication,

perhaps to neighbouring molecules, but at any rate to the luminiferous ether; for without a communication of the latter kind
there would be no dispersed light.   Hence we must consider the
immediate tendency of the disturbing forces rather than their
tendency in the long run.

233.   When a molecule itself vibrates in an irregularly
periodical manner, the vibrations which it imparts to the ether
are of course of a similar character.   The resolution of these
into vibrations corresponding to different degrees of refrangibility
involves some very delicate mathematical considerations, into
which I do not propose to enter.   But without this it is evident
that when the ether is agitated by the vibrations of an immense
number of molecules, in all possible states as regards amplitude,
and consequently periodic time of vibration, the disturbance of
the ether must consist of a mixture of periodic vibrations, having
their periods comprised between the greatest and least of those
belonging to the molecular vibrations; and corresponding to these
different periods there will be portions of light of different degrees
of refrangibility found in the dispersed beam.   These refrangibilities will range between two limits, an inferior limit equal to
the refrangibility corresponding to the periodic time of indefinitely
small vibrations, and a superior limit equal to the refrangibility of
the active light.

234.   This theory seems to accord very well with the general
character of dispersed beams, as regards the prismatic composition
of the light of which they consist.   When analysed by a prism,
these beams are sometimes found to break off abruptly at their
more refrangible border, but I do not recollect ever to have met
with an instance in which a beam broke off abruptly at the
opposite border, except when the whole beam was almost homogeneous.   This is just as it ought to be according to the above
theory, because the amplitude of vibration decreases indefinitely
in approaching the less refrangible limit.   In the case of a
solution of chlorophyll, we may suppose that the part of the
molecular forces of restitution depending on first powers of the
displacements is considerable, on which supposition, the effect
ought to approach to what would take place were there no other
part.   But were the forces of restitution strictly proportional to

the displacements, the vibrations would be isochronous, and could only be excited by ethereal vibrations having almost exactly the same period, but would be powerfully excited by such. Accordingly, in a solution of chlorophyll the dispersion comes on very suddenly; a large part of it is produced by active light of nearly the same refrangibility as the dispersed light; and the latter, by whatever active light produced, has nearly the same refrangibility that it had at first. This supposition, combined with the preceding theory, accounts also for the transparency of the fluid with respect to rays of less refrangibility than the first absorption band, for the great intensity of that band, for the rapidity with which opacity comes on at its less refrangible border, and the comparatively slow resumption of transparency on the other side. A difference of the same nature on opposite sides of a maximum of opacity seems to be a very common phenomenon in absorption. On the other hand, in those numerous cases in which the dispersion comes on gradually, in the manner described in Art. 44, we may suppose the part of the forces of restitution depending on first powers of the displacements to be but small.

235. It may appear at first sight to be a formidable objection to the theory here brought forward, that in the experiment mentioned in Art. 216, the intensity of the dispersed light did not appear to be more than doubled when the intensity of the incident disturbance was doubled; and that in the experiment described in Art. 215, the rays of low refrangibility did not appear to exercise any protecting influence. But the difficulty may, I think, be got over by a very reasonable supposition. It seems very natural to suppose that a given molecule remains for the greater part of the time at rest, or nearly so, and only now and then gets involved in vibrations. On this supposition, it is only a very small per-centage of the molecules that at a given instant are vibrating to an extent worth considering. Conceive now a stream of light consisting of the highly refrangible rays to be incident on a sensitive medium, and to cause 1 per cent. of the sensitive molecules to vibrate considerably, the rest vibrating so little that they may be regarded as at rest. Now imagine a second stream, similar in all respects to the first, to influence the medium which is already under the influence of the first stream.

Of the 1 per cent. of the molecules already vibrating, many are vibrating, we may suppose, nearly with their maximum amplitude, and consequently are not much affected. Besides, it is a great chance if the epoch of the ethereal vibrations belonging to the second stream is such as to produce any great tendency either towards quiescence or towards disturbance in a molecule just for the short time that it is vibrating strongly under the influence of the first stream. But of the 99 per cent. of quiescent molecules 1 per cent. are made to vibrate. Hence the effect of the two streams together is very nearly the same in kind as that of one alone, but double in intensity.

236. The apparent absence of a protecting influence in the less refrangible rays seems at first more difficult to account for, but perhaps the following reasoning may be thought satisfactory. We ought not to attribute more influence in the direction of protection to a second beam of rays of low refrangibility, than in the contrary direction to a second beam of rays of high refrangibility. Now if the effect of a beam of rays of high refrangibility be to throw 1 per cent. of the molecules into a state of vibration, it would be a commensurate effect in a beam of rays of low refrangibility to stop the vibrations of 1 per cent. of the molecules, if they were all vibrating. But since only 1 per cent. are actually vibrating, the real protecting effect amounts to no more than stopping the vibrations of one molecule in every 10,000, an effect which may be regarded as insensible.

237. The simple consideration that work cannot be done without the expenditure of power, shows that when light incident on a medium gives rise to dispersed light, a portion at least of the absorption which the medium is observed to exercise must be due to the production of the dispersed light. If the dispersed light really arises from molecular disturbances, and for my own part I think it almost beyond a question that it does, it follows that in these cases light is absorbed in consequence of its being used up in producing molecular disturbances. But since we must not needlessly multiply the causes of natural phenomena, we are led to attribute the absorption of light in all cases to the production or augmentation of molecular disturbances, unless reason be shown to the contrary. It might seem at first sight that the

production or non-production of dispersed light establishes at once a broad distinction between different kinds of absorption. I do not think that much stress can be laid on this distinction. In the first place it may be remarked, that we have no reason to suppose that vibrations which are of the same nature as those of light are confined to the range of refrangibility that the human eye can take in. If, therefore, no dispersed light be perceived, it does not follow that no invisible rays are dispersed. If the incident light belong to the visible part of the spectrum, the dispersed rays (if any), being of lower refrangibility than the incident light, can only be invisible by having a refrangibility less than that of red light, and would manifest themselves solely or mainly by their heating effect. However, though invisible rays of this nature are in all probability emitted by the body in consequence of the absorption of visible light, we are not bound to suppose that in their mode of emission they precisely resemble the visible rays observed in the phenomena of internal dispersion. In most cases, perhaps, they are more nearly analogous to the visible rays emitted by solar phosphori. It is possible to conceive, and it seems probable that there exist, various degrees of molecular connexion from mere casual juxtaposition to the closest chemical union. A compound molecule may vibrate as a whole, by virtue of its connexion with adjacent molecules, or it may vibrate by itself, in the manner of an isolated vibrating plate or rod, and between these extreme limits we may conceive various inter- mediate modes of vibration. Hence, without departing from the general supposition that the absorption of light is due to the production of molecular disturbances, we may conceive that the modes in which the ether communicates its vibrations to the molecules, and the molecules in turn communicate their dis- turbances to the ether, are very various.

I do not bring forward the idea that the absorption of light is due to the production of molecular disturbances as new, though possibly the communication of the ethereal vibrations to the molecules may hitherto have been supposed necessarily to imply the existence of synchronous vibrations among the molecules. The change in the periodic time of vibrations which takes place in the process of internal dispersion would hardly have been suspected, had it not been for the singular phenomenon which pointed it out.

238. The only theory of absorption, so far as I am aware, in which an attempt is made to deduce its laws from a physical cause is that of the Baron Von Wrede, who attributes absorption to interference*. The Baron's paper is in many respects very beautiful, but it has always appeared to me to be a fatal objection to his theory that it supposes vibrations to be annihilated. It is true that two streams of light may interfere and produce darkness, but then to make up for it more light is produced in other quarters. Light is not lost by interference, but only the illumination differently distributed. Were the disappearance of light in the direction of a pencil admitted into a medium merely a phenomenon of interference, the full quantity of light admitted ought to be forthcoming in side directions. Were a series of vibrations incident on a medium, without producing any progressive change in its state, or any disturbance issuing from it, it would follow that work was continually being annihilated. But we have reason to think that the annihilation of work is no less a physical impossibility than its creation, that is, than perpetual motion.

List of highly sensitive substances.

239. For the sake of any one who may wish to make experiments in this subject, I subjoin a list of the more remarkable of the substances which have fallen under my notice. It will be seen that most of these substances were suggested by the papers of Sir David Brewster and Sir John Herschel.

Glass coloured by peroxide of uranium: yellow uranite: nitrate or acetate of the peroxide. Probably various other salts of the peroxide would do as well. The absorption bands of the salts, whether sensitive or not, of peroxide of uranium ought to be studied in connexion with the change of refrangibility.

A solution of the green colouring matter of leaves in alcohol. To obtain a solution which will keep, it is well previously to steep the leaves in boiling water. The alcohol should not be left permanently in contact with the leaves, unless it be wished to observe the changes which in that case take place, but poured off

* Poggendorff's *Annalen*, B. xxxiii. S. 353 ; or Taylor's *Scientific Memoirs*, Vol. i. p. 477.

when the strength of the solution is thought sufficient. Also, the solution when out of use must be kept in the dark.

A weak solution of the bark of the horse-chestnut.

A weak solution of sulphate of quinine, *i.e.* a solution of the common disulphate in very weak sulphuric acid. Various other salts of quinine are nearly if not quite as good.

Fluor-spar (a certain green variety).

Red sea-weeds of various shades : a solution of the red colour-ing matter in cold water. If a solution be desired, a sea-weed must be used which has never been dried. Sometimes even a fresh sea-weed will not answer well.

A solution of the seeds of the *Datura stramonium* in not too strong alcohol.

Various solutions obtained from archil and litmus (see Arts. 65 to 72).

A decoction of madder in a solution of alum.

Paper washed with a pretty strong solution of sulphate of quinine, or with a solution of stramonium seeds, or with tincture of turmeric. The sensibility of the last paper is increased by washing it with a solution of tartaric acid. This paper ought to be kept in the dark.

A solution, not too strong, of guaiacum in alcohol.

Safflower-red, scarlet cloth, substances dyed red with madder, and various other dyed articles in common use.

Many of the solutions here mentioned are mixtures of various compounds. Of course if the sensitive substance can be obtained chemically pure it will be all the better.

Conclusion.

240.   The following are the principal results arrived at in the course of the researches detailed in this paper :—

(1)   In the phenomenon of true internal dispersion the refrangibility of light is changed, incident light of definite refrangibility giving rise to dispersed light of various refrangi-bilities.

(2)   The refrangibility of the incident light is a superior limit

to the refrangibility of the component parts of the dispersed light.

(3) The colour of light is in general changed by internal dispersion, the new colour always corresponding to the new refrangibility. It is a matter of perfect indifference whether the incident rays belong to the visible or invisible part of the spectrum.

(4) The nature and intensity of the light dispersed by a solution* appear to be strictly independent of the state of polarization of the incident rays. Moreover, whether the incident rays be polarized or unpolarized, the dispersed light offers no traces of polarization. It seems to emanate equally in all directions, as if the fluid were self-luminous.

(5) The phenomenon of a change of refrangibility proves to be extremely common, especially in the case of organic substances such as those ordinarily met with, in which it is almost always manifested to a greater or less degree.

(6) It affords peculiar facilities for the study of the invisible rays of the spectrum more refrangible than the violet, and of the absorbing action of media with respect to them.

(7) It furnishes a new chemical test, of a remarkably searching character, which seems likely to prove of great value in the separation of organic compounds. The test is specially remarkable for this, that it leads to the independent recognition of one or more sensitive substances in a mixture of various compounds, and shows to a great extent, before such substances have been isolated, in what menstrua they are soluble, and with what agents they enter into combination. Unfortunately, these observations for the most part require sunlight.

(8) The phenomena of internal dispersion oppose fresh difficulties to the supposition of a difference of nature in luminous, chemical, and phosphorogenic rays, but are perfectly conformable

* [The statement was designedly limited to "a solution" (which I thought might include glass, as a solution made at a high temperature) because it seemed à priori very likely that there might be a dependence in the case of a crystal; and I think that I had already seen reason from observation to at least suspect that such there was in the case of crystals of nitrate of uranium. Some time afterwards almost marked dependence was observed in the case of platinocyanides.]

400 ON THE CHANGE OF REFRANGIBILITY OF LIGHT.

to the supposition that the production of light, of chemical changes, and of phosphoric excitement, are merely different effects of the same cause. The phosphorogenic rays of an electric spark, which, as is already known, are intercepted by glass, appear to be nothing more than invisible rays of excessively high refrangibility, which there is no reason for supposing to be of a different nature from rays of light.

NOTES ADDED DURING PRINTING.

Note A. Art. 23.

Shortly after the preceding paper was forwarded to the Royal Society, I found M. Edmond Becquerel's map of the fixed lines of the chemical spectrum, which is published in the 40th volume of the *Bibliothèque Universelle de Genève* (July and August, 1842). I had seen in Moigno's *Repertoire d'Optique Moderne*, that the map had been presented to the French Academy, and naturally felt anxious to obtain it; but not finding any further notice of it either in that work or in the *Comptes Rendus*, I supposed that it had not yet been published. The principal lines in this map I recognized at a glance. M. Becquerel's broad band I is my l; his group of four lines M with the preceding band forms my group m; his group of four lines N forms the first four of my group n; his line O is my n. It is only in the last group that there can be any doubt as to the identification; but I feel almost certain that M. Becquerel's P is my o, and the next two lines, the last in his map, are the two between o and p. It is difficult at first to believe that the strong line p should have been left out, while the two faint lines between o and p are represented, but the difficulty is, I think, removed by considering the feeble photographic action in that part of the spectrum. M. Becquerel expressly states that lines were seen beyond the last he has represented, though they were hardly distinct; and on comparing together his map, Mr Kingsley's photographs, and my own map, I think hardly any doubt can remain as to the identification.

I take this opportunity of referring to another very interesting paper of M. Becquerel's, entitled "Des effets produits sur les corps par les rayons solaires," which is published in the *Annales de Chimie*, tom. IX. (1843), p. 257, and with which I was not acquainted till lately, or I should have referred to it before. This paper contains, among other things, an investigation of the effects of transparent and coloured screens on the luminous, chemical, and phosphorogenic rays, in which it is shown that, notwithstanding the great difference in the action of a given screen on the three classes of rays when we study the effect of the incident rays as a whole, its action is the very same when we confine our attention to rays of any one refrangibility. Among the media employed by M. Becquerel, are some whose absorbing effect I have mentioned in the present paper, as having been determined by methods depending upon the change of refrangibility. In such cases my own results, as might have been anticipated, are in perfect harmony with those of M. Becquerel. With respect to the results at which I have arrived regarding the nature of the phosphorogenic rays of an electric spark, which are mentioned towards the end of the paper, I have been in a good measure anticipated by M. Becquerel. Yet I do not think that even he was aware that so much of the effect of the spark was due to rays of such high refrangibility.

Note B. Art. 105.

I have since succeeded, by a particular arrangement, in seeing so far into the "lavender" rays as to make out the groups of fixed lines m, n, p by means of light received directly into the eye, and even to perceive light beyond that *.

As to the colour of these rays when they are well isolated, I think the corolla of the lavender gives as good an idea of it as could be expected from the circumstances. They seem to me to want the luminousness of the blue and the ruddiness of the violet.

---

* [The arrangement actually adopted was to form a pure spectrum with a quartz train in the usual way, to isolate and at the same time condense a small portion of the spectrum by a small quartz lens of short focus placed in or near the pure spectrum, and to view the spot of light so formed from some distance behind through a quartz prism applied to the eye. This prism was cut to show practically single refraction, and its office was of course to remove to one side the scattered light belonging to the ordinarily visible spectrum.]

No doubt much error and uncertainty has hitherto existed both as to the colour and as to the illuminating power of these rays, because the gray prolongation of a spectrum formed on paper by projection has been mistaken for the lavender rays.

Note C.   Art. 154.

On adding common phosphoric acid to a solution of nitrate of uranium no effect seemed to be produced, but on examining the vessel some days afterwards, a precipitate was found to have fallen. This precipitate proved to be sensitive in a very high degree.

Note D.   Art. 158.

I have since observed in a mineral solution a system of absorption bands so remarkable, and so closely resembling in many respects those found in the salts of peroxide of uranium, though they occur in a totally different part of the spectrum, that I think no apology is needed for mentioning the circumstance. The medium referred to is a solution of permanganate of potassa, in fact, red solution of mineral chameleon. In order to see the bands, it is essential to employ a dilute solution, or else to view it in small thickness, since otherwise the whole of the region in which the bands occur is absorbed. The bands are five in number, and are equidistant, or at least very nearly so. The first is situated at about three-fifths of a band-interval above D; the last coincides with F, or, if anything, falls a little short of it. The second and third are the most intense of the set. I have carefully examined the solution for change of refrangibility, and have not found the least trace. Ferrate of potassa shows nothing remarkable.

By means of the bands just mentioned, the colour of permanganate of potassa may be instantly and infallibly distinguished from that of certain other red solutions of manganese, the colour of which some chemists have been disposed to attribute to permanganic acid (see a paper by Mr Pearsall "On red Solutions of Manganese," *Journal of the Royal Institution*, New Series, No. IV. p. 49).

Note E.   Art. 171.

If we suppose the angle of incidence *exactly* equal to 45°, assume $\frac{4}{3}$ for the refractive index of the fluid, and apply Fresnel's formulæ to calculate the ratio of the intensity of light reflected at the exterior surface of a bubble, and polarized in a plane perpendicular to the plane of incidence, to that of light similarly reflected and polarized in that plane, we find 0·228 to 1, a ratio which certainly differs much from one of equality.   But in order to render the two intensities equal, it is sufficient to increase the angle of incidence by only 3° 35′; and in fact, as a matter of convenience, the position of the observer was usually such that the deviation of the light was somewhat greater than 90°, and therefore the angle of incidence somewhat greater than 45°.

Note F.   Art. 191.

I have since received a slab of glass of the kind here recommended, which has been executed for me by Mr Darker of Lambeth, and which answers its purpose admirably, the medium being eminently sensitive.   Besides its general use as a screen, this slab, from its size and form, has enabled me to trace further than I had hitherto done (Arts. 75, 76) the connexion between certain fluctuations of transparency which the medium exhibits and corresponding fluctuations of sensibility.

Note G.   Art. 192.

Paper washed with a mere infusion of the bark of the horse-chestnut is quickly discoloured; but a piece washed with a solution which had been purified by chemical means remained white, and proved exceedingly sensitive.

Note H.   Art. 204.

I have since ordered a complete train of quartz, of which a considerable portion, comprising among other things two very fine prisms, has been already executed for me by Mr Darker.   With these I have seen the fixed lines to a distance beyond H more than double that of p; so that the length of the spectrum, reckoned from H, was more than double the length of the part previously known from photographic impressions. The light was reflected by the metallic speculum of a Silbermann's heliostat, which I have received from M. Duboscq-Soleil. With the glass train the group p was faint, but with the quartz train there was abundance of light to see not only the group p, but the fixed lines as far as $Hp1$, or thereabouts. From the group n to about the middle of the new region, the lines are less bold and striking than in the region of the groups H, l, m, n, but the latter part of the new region contains many lines remarkable both for their strength and for their arrangement. I hope to make a careful drawing of these lines as shown by the complete train with a summer's sun.

I have some reasons for believing that the photographic action of these highly refrangible rays is feeble, perhaps almost absolutely null. In the second of the papers referred to in Note A (p. 300), M. Becquerel describes an experiment in which a prism of quartz was employed to form a spectrum; and yet the impressed spectrum formed by rays which had traversed the quartz alone was hardly longer than that formed by rays which, in addition to the quartz, had traversed a screen of pure flint-glass a centimetre in thickness. It is possible, I am inclined to think probable, that glass made with *perfectly* pure materials would be transparent like quartz, but all the specimens I have examined were decidedly defective in transparency. Besides, M. Becquerel, who may be allowed to be the best judge of his own experiments, considered the result just mentioned as a proof that the impressed spectrum formed by rays which had traversed quartz only did not extend, except a very trifling distance, beyond that formed by his train of glass; and yet his map, formed by means of the latter, does not take in the line p.

However, among the multitude of preparations capable of being acted on by light, it is probable that there may be some which are acted on mainly by rays of unusually high refrangibility, and which, on that very account, would not be suitable for the ordinary purposes of photography. With these it is possible that the new region of the solar spectrum might be taken photographically.

Note I. Art. 213.

I have since examined the salt, or product, whatever it may be, in the dry state, and under more favourable circumstances, and have found it sensitive, though not by any means in a high degree. It exhibits also the absorption bands which seem to run through the salts of peroxide of uranium.

In connexion with the insensibility of a solution of nitrate of uranium in ether, it seems interesting to mention a fact which I have since observed, namely, that the sensibility of a solution of nitrate of uranium in water is destroyed by the addition of a little alcohol.

Note J. Art. 217.

On repeating this experiment on a subsequent occasion, I could not satisfactorily make out the difference of character of a strong and of a weak spark from the prime conductor, perhaps because the machine was in less vigorous action; but the difference between the effects of a mere spark and of the discharge from a Leyden jar was plainly evident. I would here warn the reader, that in order to perform the experiment in such a manner as to obtain a striking and perfectly decisive result, it is essential to employ an excessively weak solution. The reason of this is evident.

A severe thunder-storm which visited Cambridge on the evening of July 16, 1852, afforded me a good opportunity of observing the effect of lightning on a solution of quinine, and other sensitive media. From the copiousness of the dispersed light, it was evident that the proportion of the active, and therefore highly refrangible, rays to the visible rays was very far greater in the radiation from lightning than in daylight. A difference of character was observed between the effects of a weak distant flash, and of a bright flash nearly overhead, similar to that which has been described with reference to the effects of a spark from a machine, and of the discharge from a Leyden jar. In artificial discharges, the stronger the spark the more the rays of excessively high refrangibility seem to abound, in proportion to the whole radiation. Now a flash of lightning is a discharge

incomparably stronger than that of a Leyden jar.   It might
have been expected, therefore, that the radiation from lightning
would be found to abound in invisible rays of excessively high
refrangibility.   Yet I could not make out in a satisfactory manner
the absorption of the rays by glass, even by common window-
glass.   I do not wish to speak positively regarding the result
of this observation, for of course observations with lightning
are more difficult than those made with a machine which is
under the control of the observer.   Yet it did seem as if the
spark from a Leyden jar was richer than lightning in rays of
so high a refrangibility as to be stopped by glass.   If this be
really true, it must be attributed to one of two things, either
the non-production of the rays in the first instance, in the
case of lightning, or their absorption by the air or clouds in
their passage from the place of the discharge.   If they were
not produced, that may be attributed to the rarity of the air
at the height of the discharge, that is, at the height of the
thunder-cloud.   No doubt the metallic points of the discharger
belonging to the electrical apparatus may have had an influence
on the nature of the spark; but I am inclined to think that this
influence, so far as it went, would have acted in the wrong
direction, that is, would have tended to produce rays of lower, at
the expense of those of higher refrangibility.

Note K.   Art. 220.

My attention has recently been called to a paper by M. Brücke
(Poggendorff's *Annalen*, B. v. (1845) S. 593), in which he describes
some experiments which show that the different parts of the eye,
and especially the crystalline lens, are far from transparent with
respect to the rays of high refrangibility.   The eyes employed
were those of oxen and some other animals; and the inquiry was
carried on by means of the effect which light that had passed
through the part of the eye to be examined produced on a film of
tincture of guaiacum that had been dried in the dark.   Of course
the phenomena described in the present paper afford peculiar
facilities for such an inquiry, and I had frequently thought of
entering upon it, but have not yet made any observations.
Independently of the facility of the observations, and the ad-

vantage of being able to examine readily light of each degree of refrangibility in particular, the results obtained by means of sensitive media seem to be more trustworthy on this account, that it would be possible to employ fresh eyes.   The experiments of M. Brücke necessarily occupied a considerable time, and it may be doubted whether the eye, especially after dissection, might not have changed in the interval, and whether the results so obtained are applicable to the eye as it exists in the living animal.

INDEX TO THE PRECEDING PAPER.

N.B. *The figures refer to the articles, the letters to the original notes,*
pp. 400—407.

ADDITION TO THE PRECEDING PAPER.

[For a long time, in fact from before 1867, I have taken a different view of the nature of the phenomenon from that indicated in Arts. 63 and a few following. I was led to it by a phenomenon I casually noticed in making observations on the fluorescence of various glasses. The sun's light was reflected horizontally into a darkened room, passed through a convex lens, and with or without previous filtration by passing through a blue glass placed at or near the focus, was admitted into the glass to be examined. In general the fluorescence when examined in this way showed no appreciable duration, but in one class of glasses when the glass was merely moved sideways by hand somewhat rapidly across the incident pencil, a luminous trail was seen extending from the focus at the moment into the part of the glass on which the focus had previously fallen. The colour of the fluorescent cone as a whole when the glass was at rest, and that of the root of the trail (or part nearest to the focus) when the glass was in motion, varied with the composition of the glass, but the distribution of the fluorescence in the spectrum when the glass was examined in an intense and fairly pure spectrum was the same or very nearly so, and was that characteristic distribution alluded to in § 78. I may mention that I have circumstantial evidence that the substance in the glass to which this effect was due was manganese in a lower state of oxidation than that which gives the purple colour. Now the colour of the trail changed from its root to its end, becoming reddish at last, in such a manner as to indicate a decreasing mean refrangibility, except in one case, that of a phosphatic glass prepared by the Rev. W. Vernon-Harcourt, where the colour was red to start with, so that there was no opportunity to change.

The change of colour with the lapse of time since excitement of the glass led me to regard the alteration in the molecular disturbance as brought about in the following manner. The incident vibrations of the ether agitate the sensitive molecule or complex molecule, and this agitation gradually spreads outwards by the molecular forces into the complex system formed of the surrounding molecules and molecular groups, which may be mostly themselves insensible, very much in the manner in which if a very minute portion of a metallic mass be supposed strongly heated to start with, the heat is rapidly diffused into the neighbourhood by conduction; and if the heated spot is at the surface, so as to permit of radiation outwards, the radiant heat is of a mean refrangibility which decreases as the time goes on. Indeed the phenomenon of fluorescence, which is a brief phosphorescence, seems to be closely akin to the very familiar one of the heating of a body by sunshine, and the consequent emission of heat-rays of low refrangibility.

Being desirous of seeking an analogy in some simple dynamical problem which could be actually worked out, I took the case of an infinite weightless flexible and inextensible stretched string, loaded at equal intervals with equal masses, and supposed one of the masses permanently acted on by a small transverse disturbing force expressed by $c \sin nt$, t being the time, and I demanded the permanent state of motion of the masses. It proved to be of a different character according as the frequency of the disturbing force was greater or less than that corresponding to the shortest of the periods, infinite in number, belonging to the natural oscillations of the loaded string when no disturbing force is acting. Take the mean position of the mass on which the disturbing force acts for the origin of abscissæ. The motion will of course be symmetrical on the two sides of the origin, and it will suffice to refer to that on the positive side. When the period of the force is greater than the critical period, the disturbance extends to infinity. It is of the nature of an undulation propagated outwards from the origin, and the various masses will at any moment lie on a curve of sines. But when the period of the force is less than the critical period, the disturbance instead of extending to infinity decreases on receding from the origin, and the masses at any moment lie in a curve derived from a fixed curve of sines by diminishing at a given moment the ordinates in geometric progression as we recede from the origin, while as the time changes all the ordinates vary as $\sin nt$.

Suppose now that the system is at rest and that the disturbing force begins to act. A motion will be produced gradually tending to become that permanent kind of motion above referred to. If the period of the force be greater than the critical period, no significant local disturbance can result, because the disturbance near the origin which the force tends to produce is carried away in both directions. But when the period of the force is less than the critical period the local disturbance mounts up continually, tending towards its permanent state.

Suppose now the disturbing force, after having acted long enough to bring the motion into its permanent state, were then to cease. In the first case there would indeed be a change of motion in the neighbourhood of the origin; but as the excursions of the masses would be but small, for the reason already mentioned, the effect of the change would be quite insignificant. But in the second case the disappearance of the force would leave the masses in the neighbourhood of the origin in a condition of displacement or motion which is not, as in the former case, insignificant, and which may be thought of as an initial disturbance in a system now left to itself. This disturbance would gradually widen out so as to involve a continually increasing group of masses, and the variation of the motion as at a given moment we proceed along the string would become less and less sharp as the time progresses, so that the disturbance if harmonically resolved would give a result in which the elements of the resolution corresponding to longer periods would acquire an increasing relative importance.

The general analogy of these dynamical results to the phenomena of fluorescence will I think be readily perceived. The periodic force acting on one of the masses of our loaded string is analogous to the action of the incident ethereal vibrations on a constituent part of a complex molecule or molecular group. The considerable but finite regularity of the disturbing force is analogous to what we must suppose to be going on as regards the ethereal vibrations in a small portion of a nearly pure spectrum. The agitation of the masses when the disturbing force ceases and the loaded string is left to itself is analogous to the agitation of the excited molecules when one series of regular incident ethereal vibrations comes to an end and is replaced by another such series, or a secular change takes place in the incident vibrations which is equivalent to a succession of such independent series.

We are far from being able to subject molecular disturbances to strict mathematical calculation; for in the first place we are ignorant of the molecular structure to which the calculations would have to be applied, and in the second place even if we did know the structure, the calculations would very probably prove to be too complicated to be satisfactorily carried out. Nevertheless the calculation in some assumed system of a tolerably simple character which possesses some features in common with that which is supposed to lie at the base of some observed phenomenon may not be without use, as indicating in what direction we are to look, or are not to look, for an explanation of the phenomenon. In the present case if we vary the problem by merely supposing that the equidistant masses are of two magnitudes, coming alternately, a result is obtained having some features strikingly resembling those of absorption.]

INDEX TO VOLUME III.

[*The references are to pages.*]

CAMBRIDGE: PRINTED BY J. AND C. F. CLAY, AT THE UNIVERSITY PRESS.

Printed in the United States
By Bookmasters